Cambridge Studies in Applied Ecology and Resource Management

The rationale underlying much recent ecological research has been the necessity to understand the dynamics of species and ecosystems in order to predict and minimise the possible consequences of human activities. As the social and economic pressures for development rise, such studies become increasingly relevant, and ecological considerations have come to play a more important role in the management of natural resources. The objective of this series is to demonstrate how ecological research should be applied in the formation of rational management programmes for natural resources, particularly where social, economic or conservation issues are involved. The subject matter will range from single species where conservation or commercial considerations are important to whole ecosystems where massive perturbations like hydro-electric schemes or changes in land-use are proposed. The prime criterion for inclusion will be the relevance of the ecological research to elucidate specific, clearly defined management problems, particularly where development programmes generate problems of incompatibility between conservation and commercial interests.

World soil erosion and conservation

WORLD SOIL EROSION AND CONSERVATION

Edited by
David Pimentel
Chairman, Study Group for World Soil Conservation, Commission on Ecology, IUCN

CAMBRIDGE UNIVERSITY PRESS
Cambridge, New York, Melbourne, Madrid, Cape Town, Singapore, São Paulo, Delhi

Cambridge University Press
The Edinburgh Building, Cambridge CB2 8RU, UK

Published in the United States of America by Cambridge University Press, New York

www.cambridge.org
Information on this title: www.cambridge.org/9780521104715

First published 1993
This digitally printed version 2009

A catalogue record for this publication is available from the British Library

Library of Congress Cataloguing in Publication data

World soil erosion and conservation / edited by David Pimentel.
    p.   cm. – (Cambridge studies in applied ecology and resource management)
    Includes bibliographical references and index.
    ISBN 0-521-41967-0
    1. Soil erosion.   2. Soil conservation.   I. Pimentel, David, 1925–   .   II. Series.
    S623.W67   1993
33.76'16—dc20   92-17945   CIP

ISBN 978-0-521-41967-3 hardback
ISBN 978-0-521-10471-5 paperback

CONTENTS

CONTRIBUTORS

Allen, J., College of Agriculture and Life Sciences, Cornell University, Ithaca, NY 14853, USA

Arden-Clarke, Charles, WWF-International, Avenue du Mt Blanc, 1196 Gland, Switzerland

Beers, A., College of Agriculture and Life Sciences, Cornell University, Ithaca, NY 14853, USA

Craswell, Eric T., Australian Centre for International Agricultural Research, PO Box 1571, Canberra, ACT 2601, Australia

Edwards, Keith, PO Box 348, Campbelltown, NSW 2560, Australia

El-Swaify, Samir A., Chair, Department of Agronomy and Soil Science, G. Donald Sherman Laboratory, University of Hawaii at Manoa, 1910 East–West Road, Honolulu, Hawaii 96822, USA

Evans, R., Geography Department, Cambridge University, Downing Place, Cambridge CB2 3EN, UK

Guinand, L., College of Agriculture and Life Sciences, Cornell University, Ithaca, NY 14853, USA

Hawkins, A, College of Agriculture and Life Sciences, Cornell University, Ithaca, NY 14853, USA

Hurni, Hans, Geography Institute, University of Berne, Hallerstrasse 12, CH-3012 Berne, Switzerland

Khoshoo, T. N., Tata Energy Research Institute, 7 Jor Bagh, New Delhi 110003, India

Lal, Rattan, Department of Agronomy, Ohio State University, Room 200, 2021 Coffey Road, Columbus, OH 43210, USA

Linder, R., College of Agriculture and Life Sciences, Cornell University, Ithaca, NY 14853, USA

McLaughlin, Lori, 301 Maple Avenue #B-5, Ithaca, NY 14850, USA

McLaughlin, P., College of Agriculture and Life Sciences, Cornell University, Ithaca, NY 14853, USA

Meer, B., College of Agriculture and Life Sciences, Cornell University, Ithaca, NY 14853, USA

Molina Buck, Jorge S., President, Friends of the Land of Argentina, Olaguer Feliu178 (1640), Martinez, Argentina

Musonda, D., College of Agriculture and Life Sciences, Cornell University, Ithaca, NY 14853, USA

Perdue, D., College of Agriculture and Life Sciences, Cornell University,
Ithaca, NY 14853, USA

Pimentel, David, Department of Entomology, Comstock Hall, Cornell
University, Ithaca, NY 14853, USA

Poisson, S., College of Agriculture and Life Sciences, Cornell University,
Ithaca, NY 14853, USA

Ryszkowski, Lech, Research Centre for Agricultural and Forest
Environment, Polish Academy of Sciences, Bukowska St 19, 60 809
Poznan, Poland

Salazar, R., College of Agriculture and Life Sciences, Cornell University,
Ithaca, NY 14853, USA

Scott, Christopher, A., Department of Agricultural and Biological
Engineering, Riley-Robb Hall, Cornell University, Ithaca, NY 14853,
USA

Siebert, S., College of Agriculture and Life Sciences, Cornell University,
Ithaca, NY 14853, USA

Sivamohan, M.V.K., Department of Entomology, Comstock Hall, Cornell
University, Ithaca, NY 14853, USA

Stoner, K., College of Agriculture and Life Sciences, Cornell University,
Ithaca, NY 14853, USA

Tejwani, K.G., Director, Land-Use Consultant, 25/31 Old Rajindra Nagar,
New Delhi-110-060, India

Walter, M.F., Department of Agricultural and Biological Engineering,
Riley-Robb Hall, Cornell University, Ithaca, NY 14853, USA

Wen Dazhong, Institute of Applied Ecology, Chinese Academy of
Sciences, PO Box 417, Shenyyang, China

FOREWORD

Soil is one of the most fundamental resources we have. For some years, reports have told of its alarming degradation by erosion and salinization. As long ago as 1984, one report argued that at current rates of loss, topsoil reserves would disappear in about 150 years, a few decades after the exhaustion of the planet's recoverable oil reserves. Yet these warnings have had little impact. The public has clamoured for action to plug the ozone hole over Antarctica, stop acid rain, halt the greenhouse effect and save the forests – yet little has been said about the soil loss that could destroy the hopes of a stable and fruitful earth a century from now.

Action, if it is to be sound, needs to be based on facts. There has been a shortage of them regarding the scale, regional pattern, rates and causes of soil loss – and possible remedial measures. This book is a contribution to filling that gap. Professor Pimentel, its editor, is a world-famous ecological scientist and he has assembled a team of distinguished collaborators. Chapter by chapter they review the situation around the world. At the outset, the editor reminds us that some 30–50% of the earth's land surface is affected by soil degeneration. In West Africa, low crop yields are due partly to severe past erosion and reduced fertility. Recent famines in Ethiopia are described not as natural disasters, but as cultural catastrophes following natural events, in a country which, given wise management and environmental restoration, has the capacity to be self-sufficient and sustainable within the next 50 years. In China, about one-third of the total cultivated surface is undergoing serious water and wind erosion. In India, fertile lands are being squeezed between the advance of the deserts and extending erosion in the hills. Over half the area of Australia used for agricultural and pastoral purposes needs treatment for at least one form of degradation. Water erosion affects over half of Argentina's land area, and wind erosion nearly as much. Even in Britain, soil erosion, largely caused by water, has come to be recognized as having an important agricultural impact. The same holds in Poland. In the tropics generally and in the humid tropics in particular, soil erosion is perhaps the most serious mechanism of land degradation.

These summary sentences, taken from the different chapters of this book, make disturbing reading. Yet this volume is not a chronicle of doom and pessimism. It is a practical book which sets about finding remedies. The problems are catalogued: their causes discussed and urgent remedial action proposed. These policies involve the management of agriculture, the proper use of water resources, the sound application of science, but, above all, the provision of advice, guidance and help to human communities. These policies, directly linked to the wise management of the earth, need, of course, to be related to others that this book does not address in any detail. Among these are the needs for policies that will lead to restraint on the current escalation in human numbers, which is one of the causes of the intolerable pressure being put on so many lands.

Such policies in turn interlink with measures to promote a better quality of life for the rural poor including a more equitable access to land property; to protect catchment forests and other areas where natural vegetation is the optimum form of land cover; to develop new trading relations between nations; and to curb the warming of the biosphere which may enhance the aridification process in many parts of the globe.

We hear much today about 'sustainable development'. It demands an approach that caters for all aspects of human society, and all components of the resource base on which that society depends. Few such resources are more fundamental than soil. If soil erosion can be stemmed, and the fertility of the land insured, that in itself will be a good indication that we are beginning to achieve that sustainability of development without which there can be no human future. This book is commended as a contribution to that vital process.

This volume is a product of IUCN's Commission on Ecology, one of six networks of expert volunteers, working together to analyse global environmental problems, and propose practical solutions. It is a good example of the work of such a commission at its best. As Director General of IUCN, I compliment the Commission and the Chairman and members of the task force on their achievement.

Martin W. Holdgate
Director General IUCN

Overview

D. PIMENTEL

A perspective

Soil erosion is one of the most serious environmental problems in the world today, because it seriously threatens agriculture and the natural environment. Adequate food supplies depend on productive land. Over one thousand million humans today are malnourished – more than ever before in the history of human society. Today 5.5 thousand million humans exist on earth, and an additional 250 000 humans are born each day and must be fed.

More than 97% of the world's food comes from the land rather than the oceans and other aquatic systems, with an ever-increasing portion being derived from steep lands in tropical climates. Therefore, the control of soil erosion for a sustainable agriculture and environment is essential for all programs of food security and environmental conservation.

At a time when agricultural efforts are focused on increasing food production, soil degradation worldwide is increasing. The dimensions of land degradation are alarming; it affects from 30 to 50% of the earth's land surface. In some areas the productivity of eroded soils cannot be restored, even with heavy applications of fertilizers and other fossil energy inputs. (Severely eroded soils can be restored in some cases with enormous investments including adding either 2000 tonnes/hectare [t/ha] of quality soil or 500 tonnes [dry]/ha of rotted cattle manure to the soil.) In addition, water resources and aquatic environments are degraded because of sediments and other pollutants washed from the land.

Land degradation

Soil erosion throughout the world is proceeding largely uncontrolled, with minimal conservation. This is in part because the amount of soil that erodes with each rain or windstorm is almost imperceptible. For instance, 15 tonnes of soil lost from a hectare of land during a single storm removes only about 1 mm of soil from the surface. Such soil losses are measurable by a number of standard methods, but quantitative monitoring is almost a non-existent feature of development programs. Under tropical and temperate agricultural conditions, at least 500 years are required for the formation of 2.5 cm of topsoil – the best estimate of a renewal rate is about 1 t/ha/yr.

Severe soil erosion is occurring in most of the world's agricultural regions, and the problem is growing as more marginal land is brought into production and less crop residues are returned to the soil for protection and improvement. Soil loss rates in Europe range between 10 and 20 t/ha/yr. In the United States, soil erosion on cropland averages 16 t/ha/yr. In Asia, Africa, and South America, soil erosion rates on cropland range between 20 and 40 t/ha/yr.

Worldwide degradation of agricultural land by erosion, salinization, and waterlogging is causing the irretrievable loss of an estimated 6 million hectares each year. More than half of the estimated 11.6 million hectares of forests cleared annually is to compensate for degraded agricultural lands. In fact, approximately 80% of the world's forests destroyed each year are for agriculture.

Reduced crop productivity

Erosion adversely affects crop productivity by reducing the availability of water, nutrients, and organic matter, and, as the topsoil thins, by restricting rooting depth. Reduction in the water available to plants is erosion's most harmful effect. During water erosion, most of the water is lost due to rapid runoff. In addition, both water and wind erosion reduce the water-holding capacity of soil by selectively removing organic matter and the finer soil particles. In soils degraded by erosion, water infiltration may be reduced as much as 90%. Thus, runoff may result in water shortages for the crops even in years with good rainfall.

After water, shortages of soil nutrients removed by erosion are an important limiting factor in crop productivity. One tonne of rich agricultural soil may contain a total of 4 kg of nitrogen, 1 kg of phosphorus, 20 kg of potassium, and 2 kg of calcium. Soil nutrients are abundant in the finer soil particles and organic matter that is selectively removed by erosion. In addition to directly providing energy and nutrients, organic matter is important to productivity through improved infiltration, water retention, soil structure, and cation exchange capacity. Also, organic matter and soil biota are interdependent in maintaining soil quality and recycling nutrients. In the tropics, where the 'typical' farmer cannot afford optimum replacement of lost soil qualities, erosion impacts are more severe than under temperate conditions.

Effects on natural environment

In addition to reducing the productivity of the land, erosion and water runoff cause serious offsite environmental effects. Sediments from erosion fill riverbeds, lakes, and reservoirs, significantly reducing their usefulness for navigation, irrigation, hydroelectric power, fisheries, and recreation. Sediments may interfere with fish spawning and general survival of the fishery. Eutrophication from the nutrient load added to lakes and other water bodies is also a major problem for fish productivity of aquatic systems.

Sediments deposited in lowland agricultural lands often reduce plant growth and have a negative impact on productivity. Rapid water runoff from agriculture and other cleared land results in flooded lowlands, destruction of crops, livestock, property, and sometimes people.

Conservation technologies

The principal method of controlling soil erosion and its accompanying rapid water runoff is maintenance of adequate vegetative cover. Plants and biomass intercept and dissipate the enormous energy in raindrops, enabling the water to reach the soil without damage. Furthermore, plant stems, roots, and organic matter help reduce runoff 90% or more and enhance water infiltration and percolation into the soil. Lessons from natural forest performance indicate the extreme importance of forest litter, and thus low-lying ground covers, in imparting effective protection against runoff and erosion.

A variety of conservation technologies are available and are effective in preventing soil erosion. The proven conservation technologies for different soils, slopes, crops, and rainfall and wind conditions include the following: crop rotations; strip cropping; contour planting; terraces; mulches; no-till planting; ridge planting; grass strips; tree/shrub hedges; rock hedges; pit-cultivation; agroforestry; shelter-belts; and various combinations of these conservation technologies. The matching of alternative practices with specific soils, climates, and topographies remains a high-priority subject for future quantitative research. It is encouraging to learn about the actions of individual farmers in all nations who have implemented integrated soil conservation programs.

Causes of erosion problems

Although many effective conservation technologies are available for farmers in all nations, serious soil erosion problems exist in all regions of the world. In a few cases, governments have instituted relatively sound conservation projects for some agricultural regions; however, several events occurring at the same time in these nations may offset any gains. This can be illustrated in the United States. For example, while the US Soil Conservation Service was successful in encouraging farmers to adopt contour planting and no-till technologies, the farmers at the same time were abandoning crop rotations and removing hedgerows and tree shelter-belts. Therefore, in the United States overall the total amount of soil erosion has remained about the same since 1935.

In other nations where programs of soil conservation were added in some regions, the growth in human population has forced farmers onto marginal land that is very susceptible to erosion. In addition, the rapidly growing use of crop residues for fuel and other purposes by the rural people is exposing agricultural land to intensified erosion.

Conservation economics

Implementing soil and water conservation technologies benefits both farmers and society as a whole. Conservation technologies help land retain water, nutrients, and soil organic matter and generally increase crop yields. For example, $1 invested in conservation has been found to return from $1.30 to $3.00 in increased crop yields. The return depends on the specific crop, location, season, and environmental situation.

Some controversy and confusion exists concerning the costs of soil erosion and benefits of conservation. The roots of the controversy are the fact that a full accounting of the onsite and offsite impacts of erosion is difficult to provide or quantify, and the problems that exist with the economic models and the data used in the models. Most such models concerned with soil erosion deal only with the loss of topsoil, available nutrients, and soil depth, while ignoring water runoff loss, reduced soil water-holding capacity, total nutrient loss, and organic matter loss. In these economic models in which approximately 1 mm of soil (about 15 tonnes) is lost together with its available nutrients, the impact on reduced yield may be only 0.1 to 0.5%. Using such a model for the United States, for instance, suggested that only $500 million of available fertilizer nutrients were lost annually. However, a knowledgeable agronomist assessing the total fertilizer nutrients lost (both available plus those nutrients that would be mineralized = total nutrients) reported an $18 thousand million loss annually. Note, this is a 36-fold discrepancy between the model and reality!

In addition, $1 invested in soil and water conservation on the farm watershed may save from $5 to $10 invested in dredging rivers, building levees, and restoring property damaged by floods near the mouths of rivers. Thus, clearly, soil and water conservation pays a significant return in benefits to farmers and society as a whole.

Recommendations

1. The public, scientists, and political leaders need to be better informed concerning the seriousness of soil erosion worldwide and the complex effects upon agriculture and the natural environment.
2. The benefits of soil and water conservation need to be demonstrated to farmers, and incentives provided in some cases to farmers to prevent soil erosion. Evidence for short as well as long-term benefits abounds.
3. Governments should implement soil conservation programs to complement their agricultural development and food security programs.
4. Soil and water conservation must be recognized as integral parts of all land and environmental management programs by governments.

1

Soil erosion and conservation in West Africa

R. LAL

In context of the present chapter, West Africa is defined as the region lying between 0° N, 18° E and 18° N, 30° E. The region has an annual rainfall ranging between 300 and 3000 mm received in one or two distinct seasons. Alfisols are the predominant soils of the West African Savanna region with a rainfall of 800 to 1500 mm. The West African Alfisols are derived from crystalline acid rocks and basement complex and are usually characterized by a gravelly horizon at shallow depths. These soils have a coarse-textured surface horizon and are highly prone to compaction and erosion by water. About 60% of the West African Savanna is covered by Alfisols. Soils of the drier climate (300–800 mm annual rainfall) are mostly Aridisols, Inceptisols, and Alfisols. These soils are also susceptible to accelerated erosion. Ultisols and Oxisols occur in high rainfall (> 1500 mm annual rainfall) regions of West Africa, i.e. the coastal areas of the Ivory Coast, Liberia, Sierra Leone, eastern Nigeria and Cameroon. Some soils of central Africa now classified as Oxisols may, in fact, be Ultisols. Oxisols are predominant soils of the equatorial zone with high annual rainfall exceeding 2000 mm. Productivity potential of soils of West Africa is low because of low inherent fertility, low plant-available water reserves, variable and erratic rainfall, and severe erosion and erosion-caused degradation.

How serious is soil erosion in West Africa?

Soil erosion is severe in most areas of West Africa, especially where the natural vegetation cover has been removed toward an attempt to introduce food crop farming. In general, soil erosion is the most severe in savanna regions with an annual rainfall of 500 to 1000 mm. In the forest region, soil erosion becomes severe when the natural vegetation cover is removed. There are few quantitative data available for the region where measurements of erosion have been made over a long enough period to establish the trends. Estimates of erosion presented below are based on the reports by Fournier (1967), Lal (1976), Roose (1977a,b, 1987), FAO/UNEP (1978), Armstrong *et al.* (1980), El-Swaify *et al.* (1982), Jege and Agu (1982), Miti *et al.* (1984), Walling (1984), Lootens and Lumbu (1986) and Kattan *et al.* (1987),

1. *Sahel*: Large tracts of the West African Sahel region, lying between 14° and 18° north of the equator, are covered by sand dunes and laterized rock outcrops. The region is prone to wind erosion and the potential rates may range from 10 to 200 t/ha/year (Fig. 1.1). Soil crusting and compaction are severe factors in the Sahel (Dugue, 1986; Hoogmoed, 1986; Valentin, 1986).

2. *Sub-Sahelian Region*: The region lying between 10° and 14° north of the equator is the most severely affected because of its susceptibility to erosion by both wind and water. Estimates of erosion range from 10 to 50 t/ha/year by wind (Fig. 1.1).

3. *Savanna*: The sub-humid or savanna region has a rainfall of 800 to 1200 mm which is mostly received in one growing season of four to six months. Water is the principal agent of erosion, although wind erosion also occurs in some regions (Fig. 1.1). In the northern savanna zone, water erosion rates range from 10 to 200 t/ha/year.

Fig. 1.1. Estimates of potential risks of erosion by wind and water in West Africa (adapted from FAO/UNEP, 1978).

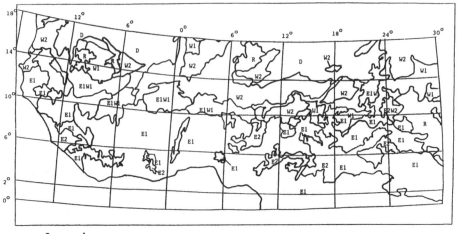

Legend
D = Dunes   E = Water Erosion   1 = 10–50 t/ha/yr
R = Rocks   W = Wind Erosion   2 = 50–200 t/ha/yr

4. *Forest*: Severe erosion in the forest region occurs only when the protective vegetation cover is removed for intensive crop production. Potential erosion rates are estimated at 10 to 200 t/ha/year (Fig. 1.1). Actual erosion rates can be as high as 100 t/ha/year on steep slopes managed for row-crop farming with plow-till method of seedbed preparation.

There are few quantitative measurements of wind erosion rates. In addition to the FAO/UNEP report, severe rates of wind erosion are also reported by Prospero and Carlson (1972) and Dregne (1983). Measurements of erosion rates by water, based on field plots established in different ecological regions, are shown in Table 1.1. Erosion rates from croplands range from 5 to 40 t/ha/year for the sub-Sahelian zone, 5 to 50 t/ha/year for the savanna, and 5 to 50 t/ha/year for the forest zone. Assuming that wind

Table 1.1. *Measured water erosion rates in different ecological regions*

Region	Country	Land use	Erosion rate (t/ha/year)	Reference
Sub-Sahelian	Senegal	cropland	5–30	Fournier (1967), Charreau and Nicou (1971), Roose (1977a,b)
Sub-Sahelian	Burkina Faso	cropland	5–35	WRI (1987)
Sub-Sahelian	Niger	watershed	40	WRI (1987)
Savanna	Ghana	bare-fallow	20	Mensah-Bonsu and Obeng (1979)
Savanna	Ghana	cropland	5–10	Mensah-Bonsu and Obeng (1979)
Savanna	Nigeria	cropland	10–20	Kowal (1972)
Savanna	Ivory Coast	cropland	10–50	Kalm (1977)
Forest	Ivory coast	cropland	10–50	Fournier (1967), Roose (1977a,b)
Forest	Ivory coast	bare-plowed	50–600	Fournier (1967), Roose (1977a,b)
Forest	Nigeria	bare-plowed	10–320	Lal (1976)
Forest	Nigeria	cropland	10–50	Lal (1976)
Forest	Nigeria	mixed	15	WRI (1987)
Forest	Ghana	bare-plowed	100–300	Mensah-Bonsu and Obeng (1979)
Forest	Ghana	cropland	5–20	Mensah-Bonsu and Obeng (1979)

erosion rates in the Sahel and sub-Sahelian regions may be as high as 20 t/ha/
year for the Sahel and 10 t/ha/year for the sub-Sahelian region, the combined
erosion rates for both wind and water for different ecological regions are
shown in Fig. 1.2. Annual average estimates of erosion rates correspond to
approximately 1 mm, 1.5 mm, 2.0 mm, and 3.0 mm for the forest, savanna,
sub-Sahelian, and Sahelian regions, respectively. These estimates do not
include deposition. The net erosion rates over the whole region may be 50 to
60% of those shown in Fig. 1.2.

What factors and policies are intensifying erosion?

Factors responsible for accelerated erosion include both biophysical and socio-economic. The most important among biophysical factors are
land use and management. Some soils are, no doubt, susceptible to erosion,
and climatic erosivity is also high. Field experiments have indicated that
erosion risks can be drastically curtailed by proper land use and judicious soil
management (Roose, 1977a; Lal, 1984a; Wright and Bonkoungou, 1985–86;
Hamel, 1986). The available research information supports the conclusion
that erosion is severe in those regions where land is used beyond its
capabilities and by those methods of soil and crop management that are

Fig. 1.2. Estimates of gross erosion rates by wind and water. These
estimates do not include deposition. The actual rates of erosion may be
50–60% of the gross rate.

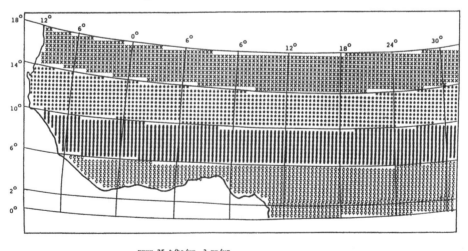

```
xxxx 35 t/ha/yr  3 mm/yr
**** 25 t/ha/yr  2 mm/yr
IIII 20 t/ha/yr  1.5 mm/yr
◊◊◊◊ 15 t/ha/yr  1 mm/yr
```

ecologically incompatible. Excessive grazing, uncontrolled burning, voluntary bush fires, and mechanized land clearing are important factors responsible for severe erosion in West Africa. Uncontrolled grazing and high stocking rate are major contributory factors for accelerated erosion in the sub-Sahelian and savanna regions.

Western Sahel and adjacent regions in West Africa witnessed five-fold increase in cattle population during the 25 years preceding the 1968 drought (Gallais, 1979). Uncontrolled and excessive grazing has depleted the protective cover and caused severe denudation of the landscape. Socio-economic factors also play an important role. Important social factors affecting soil erosion in regions of shifting cultivation and bush-fallow rotation are: (a) lack of awareness of the problem; (b) communal land rights; (c) insufficient resources to install conservation practices; (d) inefficient and inadequate extension agencies; and (e) subsistence farming.

Most resource-poor small farmers have to shift their land holdings every two or three years. They are often unaware of the erosion problem. Even if they are aware, farmers are reluctant to invest in conservation because they do not own the land and may not get the same parcel the following year. Any farmer would feel extremely reluctant to adopt conservation practices if she/he does not have permanent title to a piece of farmland. Soil erosion is in fact a serious problem in regions where land is communally rather than individually owned.

Subsistence farming in itself is an impediment to conservation. Resource-poor farmers often find it difficult to invest in conservation-effective practices even if they are aware of their benefits. Large family size and polygamous marriages further aggravate the situation by partitioning the individually owned small farms among many children. An uneconomic size of holding is difficult to manage properly. Conservation-effective measures require long-term investments for which the returns are usually not received for several years. Peasant farmers can neither afford nor are they interested in such long-term investments whose returns and benefits are hard to perceive.

There is a widespread and institutionalized tractor-hiring scheme available in most anglofone countries in West Africa. Plow-tillage (involving disc plowing and harrowing) is highly subsidized. Consequently, farmers get their fields plowed regardless of the slope steepness, soil characteristics, and cropping systems. Plow-tillage accelerates soil erosion especially when done at a wrong time of the year (Lal, 1984c). In contrast, conservation-tillage based on herbicides, is expensive. Furthermore, herbicides are often unavailable and are not subsidized. It seems, therefore, that governments in some ecological regions are subsidizing technologies that accelerate erosion.

There is also a question of the lack of appropriate technologies. It is widely recognized that capital-intensive technologies are not appropriate to the resource-poor farmers. Despite this realization, few attempts have been made to develop and transfer technologies suited to resource-poor and small farmers of West Africa. It is well known that most farmers are women, yet designs of farm tools and implements have so far ignored this important social factor.

What factors and policies improve soil conservation?

There are two practices that have potentially positive effects on soil conservation. One deals with subsidizing fertilizers. In some countries, such as Nigeria and Ghana, governments procure fertilizers and supply them to farmers at reduced prices. In theory, fertilizer application should ensure good crop growth and reduce risks of soil erosion. In practice, however, the system rarely works. The fertilizer is often not available on time. The distribution system breaks down and fertilizer bags are usually seen melting away in the rain at harbors and along road sides rather than being supplied to farms. It is also not uncommon that the wrong type of fertilizer is supplied. Continuous application of ammonium sulfate and other acidifying nitrogenous fertilizers may drastically reduce soil pH and adversely affect crop growth.

Another policy that has potentially beneficial effects on soil conservation is government encouragement to establish plantations and export crops. Perennial crops, including plantations of coffee, rubber, oil palm, and other tree crops, provide a protective vegetation cover. Tree crops are suited to the forest region and protect the soil from high-intensity rains. Land between the rows of trees is usually covered by low-growing legumes such as *Pueraria, Mucuna,* or *Centrosema.* Soil erosion from land used for tree crops is usually less than that from land used for food crops.

Attempts have also been made to develop legislation against burning, uncontrolled grazing, and cultivation of steep land. These legislative measures, however, have been ineffective.

How is soil erosion affecting water conservation on agricultural lands?

Siltation of water reservoirs has, in some regions, decreased the water-storage capacity of reservoirs for irrigation. In Ghana and Nigeria, for example, siltation of Lakes Volta and Kainji has, to some extent, reduced the irrigation potential of land in the vicinity of these schemes. In contrast to agriculture in tropical Asia, upland farming in West Africa is mostly rain fed.

There is little or no supplementary irrigation. Farm ponds and other hydrologic structures designed for supplementary irrigation are rarely used. Soil erosion on West African uplands, therefore, affects water conservation by reducing the onsite storage of soil water in the root zone. Accelerated erosion reduces onsite storage of rainfall by at least three processes. Firstly, crusted and compacted surface soil loses a substantial proportion of rainfall as surface runoff (Table 1.2). Consequently, potential reserves of plant-available water of the root zone are drastically decreased. Secondly, soil erosion is a selective process. It preferentially removes clay and other colloids leaving an inert coarse fraction behind. Lal (1976) observed that the clay and fine-earth fraction of the soil horizon decreased linearly with increasing amount of cumulative soil loss. The soil's available-water retention capacity is accordingly reduced because of the gradual depletion of clay, organic matter content, and soil colloids. Thirdly, soil erosion also reduces plant-available water reserves by decreasing the effective rooting depth.

Drastic effects of erosion on productivity of West African soils are related to severe drought stress. Most upland soils of the region not only have low available-water-holding capacity but it also decreases rapidly with increasing soil erosion (Lal, 1979, 1980). This is illustrated by the equation

$$\sigma = 38.7 - 0.02(E), r = -0.92*,$$

where E is cumulative soil erosion in t/ha and σ is soil moisture retention. Consequently, crops from eroded soils suffer more severely from drought than those from uneroded soils.

Depletion of soil organic matter and plant nutrient reserves

Most residual upland soils of West Africa are of low inherent fertility. Soil organic matter content is generally low and it declines rapidly with cultivation and accelerated soil erosion. Loss of organic matter causes deterioration in soil structure, resulting in soil compaction, decrease in water and nutrient retention capacity, reduction in infiltration rate, and further acceleration in runoff and erosion. Soil structure and pore size distribution, which are affected by soil organic matter content, also influence water retention and availability to crops. Both the permanent moisture content at the wilting point and field capacity are influenced by high organic matter content in a way that results in an increase in available-water capacity with increase in organic matter content (Table 1.3).

In addition to water, loss of soil organic matter also leads to reduction in effective cation exchange capacity and decrease in plant-available nutrients such as N, P, S, Zn, and Cu. Positive effects of organic matter content on

Table 1.2. *Effects of slope and soil management on runoff (mm) and soil loss (t/ha) in 1973. Rainfall: first season, 781 mm; second season, 416 mm*

Treatment	Runoff from slopes of:				Soil loss from slopes of:			
	1%	5%	10%	15%	1%	5%	10%	15%
(a) First season 1973								
1. Bare-fallow	315.7	347.3	311.0	316.5	7.5	80.4	152.9	155.3
2. Maize (mulched)	0	6.9	20.3	16.8	0	0	0.1	0
3. Maize (plowed)	55.7	158.7	52.4	89.9	1.2	8.2	4.4	23.6
4. Maize (no-tillage)	1.4	1.8	20.3	21	0	0.2	0.1	0.1
5. Cowpea (plowed)	19.8	81.2	51.4	46.1	0.6	5.6	3.2	7.6
(b) Second season 1973								
1. Bare-fallow	191.7	195.8	193.1	185.4	3.7	75.8	79.7	73.9
2. Maize (mulched)	0	4.0	9.0	7.5	0	0	0.1	0
3. Maize (plowed)	9.1	65.0	36.0	71.4	0.4	2.8	2.8	17.1
4. Cowpea (no-tillage)	6.0	6.5	10.2	10.7	0	0	0.1	0
5. Maize (plowed)	22.2	64.5	30.5	105.4	0.3	4.0	3.0	35.4

Table 1.3. *The effect of soil organic matter content and texture on soil water retention and plant-available water capacity.* C = *clay content* (%), S = *sand content* (%), X = *organic matter* (%)

	Country	Regression equation coefficient	Correlation	Reference
A. Field capacity (moisture content)	Zaire	$Y = 3.2 + 0.31C + 0.35S + 5.61X$	0.80**	Tran-Vinh-An and Nguba (1971)
	Nigeria	$Y = 9.3 + 4.42X$	0.68**	Lal (1979)
B. Wilting point (moisture content)	Zaire	$Y = 0.25 + 0.32C + 0.20S + 2.0X$	0.87**	Tran-Vinh-An and Nguba (1971)
C. Available-water capacity	Nigeria	$Y = 8.32 + 2.35X$	0.63**	Lal (1979)

fertility of West African soils are widely recognized (Lal and Kang, 1982). Soil erosion results in a preferential removal of soil organic matter content and other colloids. The enrichment ratio for organic matter content may be three to five (Lal, 1976). Soil organic matter content decreases linearly with increase in soil erosion (Lal, 1980). This is illustrated by the equation

$$\text{Organic carbon} (\%) = 1.79 - 0.002E, r = -0.71^{**},$$

where E is cumulative soil erosion in t/ha.

The quantity of organic matter loss due to soil erosion in West Africa may exceed 1 t/ha/year (1000 kg/ha/year) depending on soil properties, slope, and soil and crop management (Table 1.4). Similar results are reported from the Ivory Coast by Roose (1977). In addition to soil organic matter, there is also a substantial loss of N, P, K, Ca, Mg, and other essential nutrients. The quality of nutrients lost by erosion from cropland depends on many factors, e.g. growing season, rainfall characteristics, ambient soil nutrient status, terrain, and soil and crop management practices. The rates of erosion-induced nutrient losses are, therefore, hard to generalize.

In addition to loss of nutrients with eroded sediments, a substantial quantity of water-soluble nutrients is also transported in surface runoff. The quantity of nutrient loss in surface runoff is generally less than that in eroded sediments. The data in Table 1.5 show nutrient losses in water runoff for the same land from which the nutrient losses in sediments are reported in Table 1.4. Considering the fact that Africa uses on average less than 20 kg/ha/year of fertilizer, the loss of plant nutrients in eroded soil and water runoff are

Table 1.4. *Loss (in kg/ha/year) of soil organic matter content and plant nutrients in eroded soil from a field of 10% slope (Lal, 1976). T = trace*

| Treatment | Organic C | Total N | Bray-P | Exchangeable cations | | |
				Ca	Mg	K
Bare-fallow	3632	310	20	223	19	28
Maize–maize (mulch)	T	T	T	T	T	T
Maize–maize (plow-till)	168	14	1	10	4	1
Maize–cowpea (no-till)	T	T	T	T	T	T
Cowpea–maize	175	14	0.4	17	1	2

sufficiently high to deplete the soil's inherent fertility in a matter of a few years of intensive cultivation. As with organic matter content, depletion of overall soil fertility and of specific plant nutrients are directly related to the rate of soil erosion. The amount of fertilizer lost in erosion often exceeds that absorbed by crops. Some empirical relations between fertility depletion and soil erosion are shown in the equations below (Lal, 1980):

$$\text{Soil pH} = 5.6 - 0.02(E), r = -0.62^{**},$$
$$\text{Exchangeable calcium (meq/100 g)} = 5.3 - 0.003(E), r = -0.3,$$

where E is cumulative erosion in t/ha.

Soil erosion and crop productivity
Soil erosion affects crop yields both directly and indirectly. Direct effects of erosion on yields are related to damage to crop stand, and washing away or burial of young seedlings. Indirect effects are related to depletion of soil fertility, degradation of soil structure, reduction in plant-available water reserves, and decrease in effective rooting depth.

Soil properties and crop yield
Field experiments conducted in Nigeria have shown that maize grain yield on an eroded Alfisol was related to soil properties such as soil color, amount of gravel on the soil surface, and depth to the gravel layer (Table 1.6). These properties are related to soil erosion. Lal (1980) observed significant correlations between soil properties and grain yields of maize (Table 1.7) and cowpea (Table 1.8). Grain yield was correlated positively with soil organic matter content, pH, water retention at 0.01 MPa suction

Table 1.5. *Nutrient loss (kg/ha/year) in water runoff from a field of 10% slope (Lal, 1976)*

Treatment	Nutrient loss				
	NO_3-N	PO_4-P	K	Ca	Mg
Bare-fallow	11.5	3.7	18.2	33.9	7.0
Maize–maize (mulch)	0.6	1.0	1.4	0.8	0.3
Maize–maize (plow-till)	1.6	0.6	4.3	5.8	1.3
Maize–cowpea (no-till)	0.6	0.2	2.5	1.1	0.3
Cowpea–maize (plow-till)	1.3	0.8	5.3	5.0	1.3

(field capacity), soil N, and exchangeable cations. In contrast, grain yields of both maize and cowpea were negatively correlated with the cumulative soil erosion. Multiple regression analyses performed on these data indicated that yields of maize and cowpea were significantly correlated with the degree of soil erosion and other soil properties as shown by the equations below (Lal, 1980):

$$Y_{maize} = 1.79 - 0.007(E) + 0.70(M_0) + 0.002(I_c),$$

and

Table 1.6. *Relation between soil properties indicative of the degree of soil erosion and growth and yield of maize grown on an eroded Alfisol in West Nigeria (unpublished data of Miller and Lal, 1985). + = positive correlation, − = negative correlation*

Variable	Above-ground biomass	Average height	Harvest index	Plants lodged	Plants stand.	Shelling (%)	Slope (%)
Soil color	−	−	−	−	−	−	+/−
Gravel (0–10 cm) (%)	−	−	−	−	−	−	+
Gravel on surface (%)	−	−	+	−	−	−	+
Depth to gravel	+	+	+	+	+	+	−
Surface soil thickness	+	+	+		+	+	−
Slope (%)	+	+	+	−	+		−

Table 1.7. *Coefficient of linear correlation and regression equation of maize grain yield (t/ha) with soil properties (Lal, 1980). x = soil erosion (t/ha)*

Independent variable (x)	Correlation coefficient	Regression equation
Moisture retention at zero suction	0.61**	$Y = -4.18 + 0.227x$
Moisture retention at 0.01 MPa suction	0.61**	$Y = -2.48 + 0.309x$
Soil pH (1:1)	0.56**	$Y = -6.23 + 2.034x$
Organic C (%)	0.75**	$Y = 0.28 + 2.875x$
Soil N (%)	0.67**	$Y = 1.11 + 25.20x$
Extractable K (meq/100 g)	0.55**	$Y = 2.06 + 4.929x$
Extractable Na (meq/100 g)	0.55**	$Y = 0.046 + 74.6x$

$$Y_{\text{cowpea}} = 0.10 - 0.005(E) + 0.044(\text{OC}) + 0.02(\text{AWC}),$$

where Y is the yield in t/ha, E is cumulative soil erosion in t/ha, M_0 is soil moisture retention at zero suction (%), I_c is infiltration capacity (mm/hour), OC is organic carbon content of the soil (%), and AWC is available-water capacity (%).

For shallow soils of low inherent fertility, grain yield usually declines exponentially with increasing cumulative soil loss (or decreasing rooting depth) (Table 1.9). The numerical value of the negative exponents depend on soil, micro-climate environment, and soil and crop management systems.

Table 1.8. *Coefficient of linear correlation and regression equation of cowpea grain yield (t/ha) with soil properties (Lal, 1980)*

Independent variable (x)	Correlation coefficient	Regression equation
Moisture retention at zero suction	0.58**	$Y = -0.585 + 0.023x$
Moisture retention at 0.01 MPa suction	0.60**	$Y = -0.442 + 0.032x$
Soil pH (1:1)	0.49**	$Y = -0.705 + 0.188x$
Organic C (%)	0.55**	$Y = -0.040 + 0.225x$
Soil N (%)	0.54**	$Y = -0.0005 + 2.157x$
Extractable K (meq/100 g)	0.53**	$Y = 0.0387 + 0.503x$
Extractable Na (meq/100 g)	0.45*	$Y = -0.0965 + 6.467x$
Available water (%)	0.46*	$Y = -0.71 + 0.0347x$

Note: Available water = moisture retained at 0.01 MPa minus moisture retained at 1.5 MPa suction

Table 1.9. *Relationship between grain yield of maize and cowpea and cumulative soil erosion*

Slope (%)	Regression equation	Correlation coefficient (r)
Maize		
1	$Y = 6.41\,e^{-0.017x}$	−0.99
5	$Y = 6.70\,e^{-0.003x}$	−0.99
10	$Y = 6.70\,e^{-0.003x}$	−0.89
15	$Y = 8.36\,e^{-0.004x}$	−0.86
Cowpea		
1	$Y = 0.43\,e^{-0.036x}$	−0.85
5	$Y = 0.64\,e^{-0.006x}$	−0.97
10	$Y = 0.49\,e^{-0.004x}$	−0.91
15	$Y = 0.29\,e^{-0.002x}$	−0.66

The data in Table 1.9 show that decline in yield of maize and cowpea ranged between 30 and 90 % with the loss of soil equivalent to merely 10 mm depth (Fig. 1.3). For shallow soils of West Africa, which are underlain by a restrictive gravelly or lateritic horizon, the topsoil thickness is often 15 to 20 cm. Soil fertility and plant nutrients are mostly confined to the topsoil layer. Because of the high enrichment ratio for clay and organic matter, most of the root zone can be devoid of its clay and humus contents within a short period of five to seven years.

Economic trade-offs among traditional versus improved technologies

Conservation-effective technologies for specific socio-economic and biophysical environments of West Africa include those based on principles of mulch farming, conservation-tillage, legume-based crop rotations, agroforestry, and mixed or relay cropping system. Improved cropping systems consist of multiple cropping based on sequential, mixed or relay cropping patterns. Crops that provide a slow canopy cover in the initial stages of development can be profitably grown in association with other crops, e.g. cassava/maize, maize/cowpea. *In situ* production of mulch materials can be achieved by planted fallows, e.g. *Stylosanthes, Pueraria, Centrosema, Psophocarpus,* and *Mucuna.* Agroforestry is another innovative option whereby

Fig. 1.3. Relationship between crop yield and cumulative soil erosion on an Alfisol in south-west Nigeria.

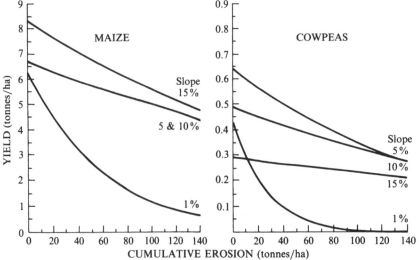

woody perennials are grown in association with food crop annuals. Some of these options have been discussed in detail in other reports (Lal, 1991). Effectiveness of these techniques may be improved by back-up engineering structures, e.g. terraces, diversion channels, drop structures, etc. Engineering techniques of water management are, however, not totally effective. Soil and crop management technologies encompass a wide range of cultural practices. Conservation-tillage, for example, may imply no-till with crop residue mulch on the soil surface in humid and sub-humid regions, plowing and soil inversion at the end of the rains in semi-arid regions, or tied-ridges on compacted soils of low permeability in the sub-Sahelian zone. Similarly variable are the appropriate crop combinations for mixed cropping. An appropriate practice of mixed or relay cropping may mean growing yam with egusi melon in humid regions, maize with cassava or cowpea in sub-humid environments, and sorghum with groundnut or pearl millet with cowpea in semi-arid and arid regions. There is a good correlation between soil erosion and unsuitable farming practices. The latter include all cultural practices that cause poor crop stand, stunted growth, low vigor and reduced canopy cover. Therefore, all cultural practices that ensure an early canopy cover are also useful erosion control measures. For example, ensuring a balanced fertilizer application and adequate pest (weeds, insects, and pathogen) control also provides erosion control through favorable canopy cover and a strong soil-binding root system. Agroforestry is an important component of improved technologies. Tree species used for agroforestry, however, differ among soils and agro-ecological regions.

Adopting these conservation-effective techniques does not necessarily involve additional costs. The capital investment for these techniques is comparatively less than that for engineering structures such as terraces and drop structures. A major impediment to widespread use of conservation-tillage, however, is the necessity to use systemic or selective herbicides. Herbicides are usually not available at prices that farmers can afford. The choice of appropriate herbicides is especially narrow for mixed and relay cropping systems where many species are grown simultaneously on a small patch of land.

Major constraints on adopting these improved technologies are the lack of adequate extension services and the insufficient number of trained personnel to conduct on-farm research and demonstrate the use of these improved technologies. If adopted on a wide scale, however, these technologies have both long-term and short-term payoffs. In addition to improving production by as much as 50%, adaptation of conservation-effective farming systems would reduce the pressure to bring new land under cultivation. This could have a significant impact on rates of deforestation. Severe ecological problems of pollution of water and wind, and of land degradation, would be drastically reduced. Some of the improved conservation technologies are listed in Tables 1.10, 1.11 and 1.12.

Policy recommendations

It is apparent that the food production potential of the presently used agricultural practices in sub-Saharan Africa is lower than those of improved and conservation-effective farming systems. The latter comprise a wide range of technologies including conservation-tillage, mulch farming, mixed and relay cropping, frequent use of legume covers in crop rotations, agroforestry, etc. Adoption of these technologies would reduce runoff and erosion, drastically curtail erosion-caused productivity decline, and facilitate

Table 1.10. *Possible techniques for decreasing runoff volume and peak runoff rate*

Soil	Strategy	Technique
Structurally unstable coarse-textured soils, in sub-humid regions	Prevent surface sealing and raindrop impact	Mulching, reduced tillage, cover crop
Soils with low-activity clays and coarse texture, in semi-arid to arid regions	Increase surface detention	Rough cloddy surface by plowing at the end of rainy season
Loamy to clayey-textured soils easily compacted	Improve infilatration	Vertical mulching, chiselling
Good-structured soils. Soils with well-aggregated, stable structure	Prolong draining time	Tied ridges, contoured ridge-and-furrow system
Vertisols: soils with expanding clay minerals in arid regions with short rainy season	Maintain soil surface at moisture potential above the hygroscopic coefficient and reduce heat of wetting	Mulching, soil inversion just prior to rain
Vertisols: soils with expanding clay minerals in semiarid or sub-humid regions with long rainy season	Dispose water safely and recycle for supplementary irrigation	Graded ridge-and-furrow system with grass waterways and storage tank, camber bed techniques

continuous cropping where shifting cultivation and bush-fallow systems now prevail.

There is an urgent need, therefore, for on-farm validation, adaptation, and transfer of these technologies on existing farmlands. Improving production efficiency on existing farmland by adopting conservation-effective technologies would be the most economical means to control erosion on the one hand and increase production on the other. National governments and international organizations should give the highest priority to strengthening

Table 1.11. *Possible strategies for soil surface management for sustained production on small-scale farms in tropical soils*

Structurally active soils	Structurally inert soils
(a) 1. Grain crop – cover crop rotation 2. No-till or minimum tillage method 3. Minimum inputs of chemical fertilizers	(a) Contour ridges (b) Tied-ridge system (c) Periodic subsoiling or chiselling to loosen the soil (d) Alley-cropping based on appropriate perennial species
(b) 1. Grain crop – alley-cropping 2. No-till or minimum tillage method 3. Minimum inputs of chemical fertilizers	
(c) 1. Integrating livestock with food crop production 2. Pasture in rotation with food crop production 3. Growing woody perennials to supplement feed during dry season	
(d) Agroforestry systems of growing food crops in association with perennial cash crops during the initial establishment phase. The food crops to be grown with system (a) above.	
(e) Strip-cropping with growing alternating strips of soil-conserving crops with grain crops. The grain crops to be grown with system (a) above.	

extension services in the region. In addition to financial help, international organizations could provide training opportunities for extension agents in the most practical means of technology validation and transfer. There is also a critical need for the training of personnel in practical aspects of soil conservation in the West African region. Universities and agricultural schools should include training in soil conservation as an important part of their curriculum.

Like extension agencies, soil conservation services are also non-existent in most countries in West Africa. A high priority should be given to the setting up of soil conservation districts/zones in each country. It is the district agents that should be charged with the responsibilities of transferring improved technologies and serving as the link between policy-makers and farmers. However, conservation districts will not function without major financial commitment by the governments concerned.

Governments and public institutions should periodically review their programs and alter their recommendations as per the new developments in agricultural science. Costs of some conservation measures should be borne by governments. It is appropriate, however, that governments subsidize the right type of technologies. Distributing seeds of improved crop varieties and of restorative cover crops and tree species, is an important task of the government. Government organizations must ensure that the right type of fertilizer is made available and at the appropriate time.

Table 1.12. *Possible strategies for soil surface management for sustained production on large-scale farms*

Structurally active soils	Structurally inert soils
(a) 1. Grain crop – cover crop rotation 2. No-till or minimum tillage method with periodic loosening of subsoil by chiselling or paraplow when needed 3. Planting trees or woody perennials on 1-m contours (b) 1. Integrating livestock with food crop production 2. No-till farming for grain crops following a lightly grazed pasture	(a) 1. Contour ridges following plowing 2. Terraces to control erosion (b) 1. Tied-ridge system 2. With or without terraces (c) Permanent ridge-furrow system following minimum tillage (d) Plowing at the end of rainy season and minimum tillage at seeding

There is an immediate need for restoring degraded lands. Research programs should be initiated to develop reclamation and restoration technologies for soils now suffering from impairment. The main soil-related constraints to an intensive land use and enhancement of productivity may be physical, chemical, hydrological, or biological. It is important to identify herbaceous and woody species with root systems and composition characteristics that favor deep penetration and support enhanced biological activity. Governments must give high priority to restoring degraded lands.

Conclusions

Soil erosion and erosion-caused land degradation are serious problems in West Africa. Low levels of crop yields observed in the region are partly due to severe levels of past erosion and resultant decline in soil productivity. Reduced yields on eroded lands are related to loss of water, reduction in rooting depth and plant-available water capacity, decrease in soil organic matter and deterioration in soil structure, and loss of plant nutrients.

Research information is available on improved technologies for soil and crop management and conservation-effective farming systems. Available technologies, however, are not being transferred. A successful program of technology transfer requires an integrated approach to improve logistic, structural, institutional, socio-economic, and cultural factors impeding progress.

Soil conservation technologies may require some inputs that are not available. These inputs may involve herbicides for conservation-tillage, fertilizers for improving soil fertility, seeds of appropriate legume species for growing as cover crops, agronomic packages of improved cropping systems to provide continuous ground cover and food security, and seedlings of new tree species useful for agroforestry systems. Governments should consider subsidizing these improved inputs.

There is a need to step up training in soil conservation and management, and in methods of technology transfer. Governments should set up conservation districts to bring home to the people the know-how to manage the precious soil and water resources prudently for the benefit of present and future generations.

The need to restore degraded lands cannot be over-emphasized. Productivity of vast tracts of lands now lying barren and useless can be restored. However, restorative measures are expensive. The costs of restoring these lands should be borne by governments.

2

Land degradation, famine, and land resource scenarios in Ethiopia

H. HURNI

Summary

Famines resulting from epidemic hunger must be seen as cultural catastrophes. Usually, but by no means exclusively, they are triggered by natural events. In Ethiopia, many recent famines emerged after severe droughts occurring predominantly in rainfed agro-ecological regions. Famines are cultural catastrophes, because the persistent local economic, social, political, and also ecological and land-use set-up does not respond to overcome the extreme shortage of food in the peasant sector.

The famine vulnerability of a country has to be sought in the land use, the human, and the natural elements of a geo-ecological system. In addition, international elements such as economic, political, and humanitarian influences must be included in the systematic approach, in order to understand famine vulnerability and the possibilities of reducing it. Such a geo-ecosystem approach is used to discuss famine vulnerability in Ethiopia in this chapter.

The role of ecology in the creation of famines must be seen in its long-term rather than its short-term impacts. Due to millenia-old traditional land management and use, land resources and productivity potentials have already been considerably reduced in many parts of Ethiopia through deforestation, soil erosion, and fertility decline. This contributes considerably to the present level of famine vulnerability. Long-term trends, on the other hand, give an even worse scenario indicating increased vulnerability in future.

Long-term ecological impacts of human and livestock populations on land resource utilization are modelled in this chapter in order to see long-term trends of land use in Ethiopia over the next 50 years. The model serves as a brain-storming tool for speculation on the effects of variable input levels in four sectors, namely family planning, environmental rehabilitation, agricultural development, and livestock development.

Selected scenarios are discussed in the model and with different sets of inputs, from 'pre-revolution' to 'conservation-based development'. It clearly shows up in the model that long-term sustainability of the land resources over the 50-year period can only be attained through a combined effort of environmental rehabilitation of the order of three to four times greater than at present, of agricultural development increased manifold, of a great emphasis on and high inputs into the livestock development sector, and, very importantly, of family planning inputs reducing growth rates to zero within the next 50 years, so that the Ethiopian population will not exceed 90–100 million people.

Ecological issues constitute a considerable portion of the present famine vulnerability in Ehiopia. Long-term trends have to be reversed through increased efforts in the rural development sector, including family planning programs. Total inputs to attain ecological sustainability over the next 50 years have to be increased by a factor of three to four times present inputs. New approaches and policies have to be centered on small communities with a strong local participation, increased training, awareness creation, education, appropriate technologies, and integrated conservation and development systems.

Introduction

The 1984/85 famine in Ethiopia affected almost ten million people, the highest figure ever experienced. As a consequence, the famine triggered the biggest global emergency operation ever carried out, overshadowed only by the fact that it was initiated three months too late. It also activated the largest support of international food aid to a single country up to the present time. The famine emerged from a significant production shortfall of about 20% in the agricultural sector, resulting in a decrease of the total annual production by about one million tonnes, mainly involving cereals and pulses. This shortfall was the result of a serious reduction of rainfall over large parts of Ethiopia in 1984. Despite the famine relief activities from November 1984 through 1985, several hundred thousand people may have died in this famine. Without emergency operations, 10 to 30 times more people would have died.

The 1984/85 famine in Ethiopia was not a natural disaster, but a cultural catastrophe following a natural event. Any famine at the present time is a cultural catastrophe, because it is foreseeable and could be avoided. Drought is not a causal factor for the creation of famines in Ethiopia, but may lead to serious shortfalls in production within the rainfed agricultural sector, which occupies over 95% of the peasant society. Food shortages, as a consequence of low production in vast areas, put such stress on the national economy that even a medium-term reorganization of food transports from surplus or storage regions to the affected areas was impossible on appropriate scales. Droughts, together with subsequent famines, were always overwhelming natural factors in the creation of disastrous situations. Other phenomena, like the rinderpest occurrence in 1889/90, also created famines in Ethiopia due to shortfalls of animal traction to plow arable land.

It is obvious that inherent causes of the recurrent famines in Ethiopia cannot be seen in natural factors alone, but predominantly in the subsistence agricultural system, the low economic development status, the lack of an adequate transport network and food storage system, and other reasons to be analyzed. Famine vulnerability, defined in Mesfin Wolde-Mariam (1984), will always be high in such a context. However, how can the risks of a famine be judged, and how could they be minimized? What are the long-term strategies, and what are the short-term preparedness and prevention measures (cf. Kent, 1988)?

Concerning long-term vulnerability trends, there are ecological issues which have to be seriously considered. Questions of carrying capacity at various input levels, environmental degradation due to deforestation and soil erosion, climatic changes and others, are extremely important to assess the current – and future – production level, which, in Ethiopia as in most other developing countries, is a major contributor to famine vulnerability.

This chapter intends to analyze ecological questions related to the creation of famines in Ethiopia (see Fig. 2.1). The term 'ecology' will be used in its widest sense, i.e. including man, land-use system, and nature in a complex interrelationship. The rates of ecological degradation in Ethiopia, primarily soil erosion from agricultural land, will be analyzed, and their impact on agricultural production shown. Long-term trends in population, land use, and livestock at present inputs will be assessed to show how ecology contributes to future shortcomings of production, to poverty, hunger, and eventually even to a higher vulnerability to famine. Needed inputs and their effects on land resources as well as on human and livestock populations will be modeled and discussed. Conclusions and some basic recommendations, finally, will be made concerning the ultimate goal to attain food self-sufficiency and long-term sustainable resource utilization within the next 50 years. Needed annual inputs of rural labor, foreign and local currency, and of government policies will be discussed. The ecological questions and issues this chapter addresses are summarized in the list overleaf:

Famine vulnerability:

What are the ecological components of famine vulnerability in Ethiopia?

Ecological degradation:

How does ecological degradation affect primary (agricultural) production – in the past, at present, and in future?

Sustainability:

What is needed to stabilize the ecological situation for the long-term use of land resources in Ethiopia – in terms of strategies, policies, participation, and external support?

Land resource utilization:

How will future trends in population and livestock affect land use, and how can the trends be influenced through inputs of the government, of external support, and of the Ethiopian peoples?

Scenarios for the next 50 years:

How will future land use within the next 50 years appear if different developments are envisaged?

Ecological stability:

What strategy has to be implemented in order to attain long-term ecological stability in Ethiopia, thereby reducing the vulnerability to famines?

Based on the long-term assumptions used in a model for land resource utilization in Ethiopia, it is concluded that the country, wisely using its resources of land, soil, and water, and making the best of its human potential, will be self-sufficient and sustainable within the next 50 years. To attain this goal, however, will require that some new programs and policies be introduced as soon as possible, and more external assistance and internal resource mobilization and reallocation will be needed than at present.

Vulnerability to famine

Anywhere in the world, droughts or other adverse impacts, such as animal or plant pests, frosts, storms, or political or social events, may considerably lower the agricultural production of a farmer, a community, a region, or a country. It is rather rare for most countries, however, that such impacts create famine situations. What is the peculiarity of a country like Ethiopia that makes such adverse impacts develop into crisis situations, affecting the national economy and millions of individuals?

The scheme provided in Fig. 2.1 offers a systematic explanation of famine vulnerability in Ethiopia. It is based on a geo-ecological and economic

Fig. 2.1. Systematic presentation of factors contributing to famine vulnerability in the Ethiopian context.

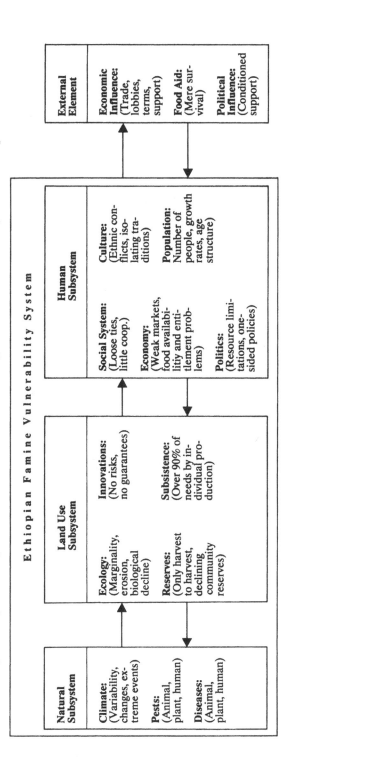

scheme developed by Messerli and Messerli (1978) for the Man and Biosphere Programme of UNESCO, has been applied to Ethiopian man–environment systems by Hurni and Messerli (1981), and is modified here to explain famine vulnerability.

The system in Fig. 2.1 is not intended to explain causal developments of vulnerability to famine in Ethiopia, but to show interrelationships between various subjects. The country is subdivided into the three subsystems: nature, land use, and man. There is an additional sector called 'external element', explaining influences from outside the country. System-inherent factors are listed within each subsystem. Inter-system enforcements are indicated by lines.

In the *Natural subsystem,* climate, pests, and diseases are listed as major inherent vulnerability forces inducing famines in Ethiopia. They are all *short-term,* i.e. a climatic instability may trigger a famine within a few months' time. The present trends of the individual factors in the natural subsystem are contrary, as while the trend of climate is towards more frequent drought and higher variabilities, especially in the north of Ethiopia, the impact of pests and diseases are currently decreasing due to the availability of medical treatments during the last 20–40 years. While a rinderpest epidemic could initiate the death of almost the total cattle population at the end of the last century, this would most certainly be stopped today with present medical care and services.

In the *Land-use subsystem,* worsening ecology, decreasing crop reserves, few innovations, and, most importantly, the subsistence agricultural economy are listed as major forces inducing famines in Ethiopia. They are factors with strong contributions to famine vulnerability. For example, land users on marginal land, such as on steep slopes or eroded soils, will subsist at a low production level, only barely guaranteeing survival as long as no adverse situations occur. With the slightest impact affecting harvest, hunger situations emerge. In cases where the community is in a situation similar to the individual, famine cannot be avoided, if assistance from outside the community does not arrive. In the Ethiopian context, reserves of food are usually sufficient until the next expected harvest, but cannot suffice for two periods. Innovations in agricultural development are difficult to implement in a society where risks cannot be taken by higher institutions in case of failure of the innovation. If more than 90% of the food needs are covered by home production, no market system will develop with a capacity large enough to supply food in case of need. Only an economy with a very strong exchange of consumer goods may have the capacity to supply the quantities of food required, if outside supply should replace the subsistence system. Induced changes and trends are all *medium- to long-term* and may be positive or negative at present development levels.

In the *Human subsystem,* it is important to notice that the social organization of individuals into communities is a rather recent phenomenon, the positive impact of which is not yet rooted enough to stimulate sufficient exchange of labor and goods in case of need. Inherent vulnerability forces are, here as well as in the land-use subsystem, rather *medium- to long-term,* and may also have positive or negative effects on famine vulnerability. The economic system, developed to mobilize food for townspeople, is too weak to also do the same for the rural subsistence population if need arises, since many more people may have to be fed. An analysis of the hypotheses of food availability or entitlement problems belongs to this aspect of economic vulnerability (cf. Amartya Sen, 1987). Ethnic diversity and conflicts, as well as isolating traditions, are no stimulations for further cooperation and economic exchange between regions. The mere number of people, increasing by almost 3% per year, would require agricultural development and production increases at least at the same rates. This is not the case at present. Political resource allocation and agricultural policies do not always stimulate better economic growth and production increases (cf. Brühne, 1988).

These inherent vulnerability factors of each subsystem are in close relationship with each other, and will enforce famine vulnerability in most cases. For example, drought in the natural subsystem is normally followed by pests and diseases. Subsistence agriculture in the land-use subsystem, coupled with ecological degradation, will normally lead to an extension of cultivated land into more marginal areas with even lower production potential. Population growth in the human subsystem, coupled with low economic activities, will increase social problems. All these examples induce a total vulnerability of each subsystem which is higher than the sum of its inherent factors.

In addition, there are *Intersystem enforcements* and *External impacts* which increase famine vulnerability to an even higher degree for Ethiopia as a whole. The inherent high vulnerability of subsistence agricultural land use is enforced by the climatic variability of the region. Population growth forces subsistence farmers to utilize the last remaining land resources for livestock and cultivated land in regions which are already ecologically damaged. External economic and political influence, finally, may also contribute to increased famine vulnerablitity. For example, if a country is interested in providing food aid only, but no development support, it will guarantee the survival of subsistence populations which are already surviving under highest vulnerability stress.

By their mere number and growth, they will experience an increase in vulnerability, since no improvements are induced and supported.

Ecological issues are mainly concentrated in the land-use subsystem of Fig. 2.1. There, the impact of deforestation, soil erosion, overgrazing, and biological degradation will have to be explained in the context of the other factors of the same subsystem, namely subsistence, food, and fuelwood reserves, and low innovative forces. However, natural as well as human factors will also have to be included, namely climatic change, growth of human and livestock populations, and the questions of the national economy and the social systems. Even external impacts are important, such as food aid, cash crops, and terms of trade. The future ecological vulnerability to famines in Ethiopia will certainly depend upon actions taken within the human subsystem and the external element. However, the land-use subsystem remains the central steering factor for solving these problems, at least during the next 20–50 years, or as long as agriculture remains the primary occupation of the Ethiopian people.

Man-made ecological degradation and rehabilitation

In Ethiopia, 88% of the human population, 60% of the livestock, and 90% of the agriculturally suitable area is concentrated in the highlands more than 1500 m above sea level (Constable, 1985). Suitable conditions of climate, ecology, soils, and environmental health made such concentration possible, which is exceptional for Africa. For example, the total population of Ethiopia, being over 45 million people, exceeds the total population of all other countries in the Sahel belt (IUCN, 1986). Favorable conditions attracted early human settlers to this largest mountain complex in Africa, and, gradually, most agriculturally suitable lands were occupied, including marginal land on slopes highly susceptible to soil erosion.

Land degradation in the Ethiopian highlands and mountains concerns processes of deforestation, soil erosion and biological soil deterioration, and overgrazing from most agricultural land-use types. These processes were initiated with the introduction of agriculture several thousand years ago; they are by no means a recent phenomenon. However, through population pressure in some areas, the processes are being accelerated at present. The present situation as regards degradation and rehabilitation in Ethiopia is summarized in the following list, then discussed in more detail below.

Deforestation:
Deforestation rates are still higher than re-afforestation, which covered about 300 000 ha with planted trees in the late 1970s and 1980s.

Soil erosion:

An average soil loss of 42 t/ha/year from cropland will remove the
   total soil of the present cropland within 100–150 years. The
   resulting annual production loss due to soil loss is between 1 and
   2%.

Biological Deterioration:

Biological soil deterioration adds a further 1% loss to annual
   production, because dung and crop residues are being burnt
   instead of brought back to the field.

Overgrazing:

A severe livestock crisis will emerge even before the cropland
   erosion crisis, since overgrazing exists almost everywhere, and
   may seriously affect agricultural production in the ox-plow
   system.

Rehabilitation:

Although rehabilitation activities in the form of afforestation and
   soil conservation are already impressive in Ethiopia, they have
   to be increased manyfold if stable and sustainable land-use
   systems are to be developed within the next generation.

Deforestation, in the beginning, was to clear land for cultivation and for
changing wildlife habitats into livestock areas. At a later stage, when forest
resources became more sparse, fuelwood requirements exceeded the natural
regeneration capacities of the last forest patches. Nowadays, Ethiopia's
forest resources are concentrated on very small land areas (3% of the
country), especially in the western parts of the highlands. In most other parts,
remaining trees are not even supplying enough fuelwood for home consumption. Wildlife using forests as a primary habitat has disappeared with the
forests. Contrary to common perception, it could be proved that deforestation dates back as much as several thousand years (Butzer, 1981; Hurni,
1987a). Growing trees with the help of man is a relatively recent phenomenon, which was introduced with the *Eucalyptus* tree about 100 years ago. At
that time, most regions of Ethiopia were already deforested to a large extent,
and the growing of *Eucalyptus* provided a source of fuelwood for the
developing towns.

Soil erosion caused by rainfall and runoff is a process as old as the history of
agriculture in Ethiopia. In the northern parts where agriculture was developed first, severe degradation, sometimes the complete removal of soil from
slopes, and its accumulation in valley bottoms has occurred. In regions like
Welo, Tigray, Gonder, and Eritrea, but also Harerge, Gojam, Welega,

Shewa, and Sidamo, shallow soils are frequent (Fig. 2.2), although they are used for grazing, while the deeper soils are cultivated. Shallow soils with less fertility and vegetative cover are more susceptible to erosion, their water retention and holding capacity decreasing with decreasing soil depth. As a consequence, less production will be obtained from such a soil, which in turn accelerates its further degradation. A vicious circle is initiated if long-term soil erosion is not stopped. The *loss of water,* on the other hand, is by itself threatening. Degraded soils and vegetation lead to direct runoff rates of up to

Fig. 2.2. Severity of soil degradation due to soil erosion in Ethiopia. 1, extreme (over 80% of the soils are only about 20 cm deep, and the rest about 100 cm); 2, very serious (60–80%); 3, high (40–60%); 4, medium (20–40%); 5, slight (less than 20%). Generalized from Barber (1984) based on Henricksen *et al.* (1983).

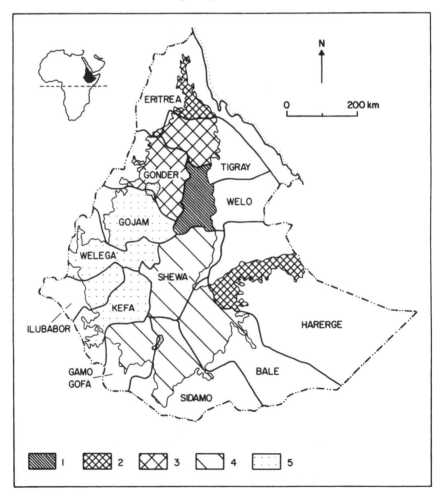

80% of the rainfall, while well-covered soils (long grasses, trees) will retain more than 90% of the rainfall.

On the other hand, a more balanced water regime of stable ecosystems could induce better growth of vegetation, supply springs and rivers during the dry season, and improve the irrigation potential outside the rainy season. Such favorable conditions, however, are rarely seen in Ethiopia today, except for some western parts with a forest cover of more than 25% of the area.

Current soil loss rates are highest from cultivated land (country average: 42 t/ha/year), where more than half of all field losses occur. Other land-use types produce only half as much total soil loss, although their areal extent is up to four times larger (see grazing land in Table 2.1). All in all, soil loss from slopes amounts to approximately 1493 million tonnes per year, with 45% from cropland alone, the latter having a significant impact on crop production. Highest soil loss rates on single fields are up to 300 t/ha/year, occurring in the western parts of the highlands, where rainfall erosivities are highest. Taking 60 cm of soil depth as a cropland average in the highlands, the annual loss of 4 mm of soil depth on average means that within 150 years most current cropland soils in Ethiopia would be eroded (except for gentle slopes below 1% gradients). From this, an annual reduction of soil productivity of 1–2% must be assumed, and has been verified in test areas (Hurni, 1987b).

Biological deterioration of the soil is an additional process common to Ethiopian agricultural systems. Lack of fuelwood, and the consumption of crop residues and dung as fuel, leads to deteriorating processes in the soil. As a consequence, organic matter decreases, soil fertility and productivity diminish, and agricultural production becomes less. Biological degradation may result in an annual decrease of soil productivity of up to 1%.

Table 2.1. *Estimated rates of soil loss on slopes in Ethiopia dependent on land cover (from Hurni, 1987b)*

Land cover type (from LUPRD)	Area (%)	Estimated soil loss	
		t/ha/year	Mt/year
Cropland	13.1	42	672
Perennial crops	1.7	8	17
Grazing and browsing land	51.0	5	312
Totally degraded	3.8	70	325
Currently uncultivable	18.7	5	114
Forests	3.6	1	4
Wood and bushland	8.1	5	49
Total country	100.0	12	1493

Overgrazing, finally, is extreme in many parts of the northern highlands, especially if coupled with sub-normal rainfall of high variability. The livestock feed problem in Ethiopia, an important factor for crop cultivation in the ox-plow systems, will eventually lead to severe constraints, thereby increasing famine vulnerability even much faster than the soil erosion processes. In general, the number of livestock per area unit is too high in almost all regions except Welega, Ilubabor, Kefa, and Bale.

Land degradation, to conclude, consists of processes whereby the productivity of the natural resources of soil, water, and vegetation is reduced. Particular attention is needed concerning the impoverishment or loss of vegetation (trees, grasses, agricultural plants), the diminishing of soil productivity (soil erosion, biological or chemical deterioration), and the loss of water quantity and quality (soil water, river runoff). Land degradation affects the agricultural production of the order of a 2% decrease rate per year.

Land rehabilitiation

The government of Ethiopia, with the support of international and bilateral agencies, has initiated a massive program for afforestation and soil conservation. The World Food Programme became involved in this sector of rural development in the mid-1970s and has since expanded its support continuously. Altogether, soil conservation and afforestation reached a total estimated input of US$ 50 million per year in 1987. Between 1976 and 1988, conservation and afforestation undertaken by the Ethiopian peasants amounted to some 800 000 km of soil and stone bunds for terrace formation on cropland, about 600 000 km of hillside terraces for afforestation of steep slopes, some 100 000 ha of closed areas for natural regeneration, and many other activities of land rehabilitation (Hurni, 1988, amended). Other activities of the Ministry of Agriculture, like irrigation development, soil fertility improvement, and land-use planning, contributed to the rehabilitation activities.

In view of the astonishing achievements, it is important to know the overall impact of the activities carried out so far. Problems of maintenance of the measures implemented, of perceptions and attitudes of the farmers, and, very importantly, of the negative impact of food-for-work schemes in areas where they are applied, are being studied in part. New approaches of participation amd mobilization are being tested in selected programs such as the Merhabete Integrated Conservation and Development Programme (Hurni 1987). Incentives other than food are offered to peasant communities, such as support for infrastructural development, water supply, milling, livestock, and agricultural development. They are offered in exchange for land rehabilitation works to be carried out by individuals on their land and by the peasant community on communal lands under the guidance of the extension workers of the Ministry of Agriculture.

It appears that at present input levels, despite the tremendous efforts, it will take another 70 years until all land in need of rehabilitation will have received a first treatment. As a consequence, only an increase of the order of two to four times the present inputs may result in the required effects within the next generation. Keeping the vast extent, ruggedness, and steep relief of the Ethiopian highlands in mind, it becomes clear that the task is almost impossible. Foreign assistance, now already covering 90% of the financial means and food supply, will have to be increased manifold, while the mobilization of all available local work power will be the central part of the program.

Ecological issues and famine vulnerability

Land rehabilitation forms an important part of any strategy to reduce vulnerability to famine in the medium and long term (cf. also Gopalakrishna Kumar, 1987). With no rehabilitation, the degradation rate would even increase due to inherent self-propelling acceleration of the processes involved. At the current input level of rehabilitation activities, the degradation rates may still persist and cannot even be reduced.

Given that all other factors of the vulnerability system (Fig. 1) are constant, degradation would lower the agricultural production by about 2% per year, thereby reducing food reserves, increasing hunger and poverty, and contributing to a higher vulnerability to famine. With present trends of population growth, livestock increase, cropland extension, and the relatively low level of government inputs, even into major programs like soil conservation and afforestation, it is likely that even within the next generation, a massive livestock and cropland crisis must be expected.

What inputs would be needed, on the other hand, to gain control over all adverse trends and developments, and to stimulate even more activities for long-term sustainable use of the land resources in Ethiopia? This question can be addressed with a land resource model specifically developed for Ethiopian conditions.

Population, land resources, and development inputs for sustainability

'Ethiopia could easily feed 200 million people'
'Land resources in Ethiopia are far from being exhausted'
'Ten times more land could be cultivated in Ethiopia'

Vis-à-vis such statements, the long-term prospects given in the last section concerning ecological issues appear very negative. Short-term policies, measures, and activities are being elaborated to reduce famine vulnerability in Ethiopia, since it calls for immediate action. Long-term prospects of resource utilization over several decades, however, should not be neglected. They may become even more important than the short-term issues. It is a global problem at the present time that resources are being greatly overutilized or abused, and fundamental changes have to be initiated worldwide.

For Ethiopia, there are some basic land resource questions which should be addressed and discussed as *long-term strategies* to reduce famine vulnerability:

* What environmental rehabilitation measures are possible in the present agricultural, social, political, and economic context?
* How much land can be cultivated with the present farming system and with realistic changes of the system within the next generation?
* What number of people can be fed with the land resources and with technologically feasible inputs at hand?
* How much can soil productivity be increased through management and inputs?
* How can the carrying capacity of grazing land be increased and the number of livestock be decreased, while still guaranteeing sufficient draught animals for plowing?
* What are the grazing resources and potentials of Ethiopia?

Although it seems impossible to address all questions together and to predict the future, any development policies attaining sustainable use of natural resources and food security will have to include the processes mentioned above. As a tool for brain-storming, a land resource model has been developed and is presented here (Fig. 2.3) for further feed-back and improvement. The model should not be seen as a tool for input planning, but rather as a means of re-thinking present processes, policies, and activities.

Various options of development inputs can be tested with the model presented here. It takes the basic resources of Ethiopia, namely people, land, and livestock as 'capital', and the current or future changes as 'interest' rates. 'Interest' rates can be modified according to inputs to be made by government and the people.

With the model, different scenarios of inputs may be tested and the changes in 'capital' resources observed. Of course, such a model is based on theoretical relationships. Its capacity to predict actual trends depends on the viability of the relationships and the basic assumptions used. The 'capital' resources of the model are *human population, land* (subdivided into total land, potential and actual grazing land, arable land, and currently cultivated land), and *livestock population*. Interrelationships between the various resources and the government on one hand, and the human population and the market-economy system on the other hand, have been simplified as much as possible. The numbered ones (1–7 in Fig. 2.3) are the functional relationships used for modeling the resource 'capitals'.

The seven impact functions used in the model are:

1. Impacts of family planning at various input levels on human population.
2. Relationship between population and cultivated land.
3. Impacts of environmental rehabilitation at various input levels on cultivated land.
4. Impacts of agricultural development at various input levels on cultivated land.
5. Relationship between human and livestock population.
6. Impacts of livestock development at various input levels on livestock populations.
7. Relationship between livestock population and grazing land.

Fig. 2.3. Land resource model for Ethiopia. For explanations see text. Functional relationships 1–7 are explained in the Appendix.

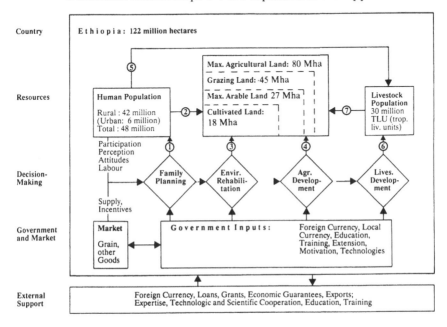

A description of the functional relationship of the seven impact functions dependent on various input levels is given in the Appendix.

There are some *basic assumptions* used in the land resource model, which have to be stated here before the model is applied:

1. The farming systems are assumed to change only moderately into modern mechanized systems, due to the fact that accessibility in Ethiopia will have to be developed over many more decades to come, and that steep slopes or narrow terraces will hardly be cultivated with tractors. It was assumed that within the next 50 years about 35% of the farms will be mechanized.

2. The attainment of food self-sufficiency is assumed to remain an overall goal of the government.

3. No dramatic. innovations for agricultural development are assumed to be available within the next decades, except for improvements of present technologies.

4. Perceptions, attitudes, resistance to change, and innovation barriers are assumed to take an order of one generation, until adaptations are made by the rural population.

5. No major loss of human or livestock population is assumed to occur during the modeling process. Food shortage years are assumed to be buffered by reserves, better famine preparedness and prevention, and eventual foreign food aid in case of need.

6. The following initial 'capital' resources were given (data from LUPRD, and EHRS, 1985);

   * Present rural population (1990) is assumed to be 45 million people.
   * The present cropland used for cultivation (or fallow) is 18 million hectares (ha).
   * The land actually grazed is 45 million hectares.
   * The present livestock population (1990) is around 30 million 'Tropical Livestock Units', TLU, equivalent to about 70 million head of cattle, sheep, goats, horses, mules, donkeys and camels.
   * The potential arable land is 27 million hectares.
   * The grazable land is 53 million hectares.
   * Arable and grazable land together give a total of 80 million hectares (66% of Ethiopia) the rest being deserts and semi-deserts. The total area of Ethiopia is 122 million hectares.

The inputs into the model

In the main sectors following below, inputs can be made onto the land resources model (see also Appendix). For each sector, the inputs can be varied at four levels, namely: none, low, medium, or high inputs.

The four sectors considered are:

1. *Family planning:*

   All aspects of family planning must be included in a program, such as extension, education, old age (retirement) security, availability and propagation of contraceptives, health services, mother and child care, awareness creation, education, and training. Of course, the impact will be determined according to the input level into this sector. With no significant input (= actual trend), population growth will remain at a 2.9% annual rate.

2. *Environmental rehabilitation:*

   Afforestation as well as soil conservation and improvement are central activities, based on extension, training, mass mobilization, incentives, organization, technical know-how, and technological inputs according to input levels. At the present (low) input (= actual trend), it will only be possible to avoid an increase in degradation rates, i.e. the rates of production loss per year will remain constant at a level of 2%, requiring an extension of cultivated land on the same order.

3. *Agricultural development:*

   Crop improvements, soil fertility improvements with biological and artificial fertilizer, extension services, trials and research, training, education, institutional setup, transport networks, and marketing services are needed to promote agricultural development, especially to attain increases in crop production. In sum, a 'green revolution' package including ecological improvements, social and land-use security, and an optimized allocation of resources will be needed. At the present (low) input level (= actual trend), only slight long-term increases of up to 1% annual growth in production can be expected. Such increases would in turn allow a higher concentration of production on plots becoming smaller at the same rate of change.

4. *Livestock development:*

   Improvements and local adaptations of breeds, improvement of grazing land through livestock management, controlled or rotational grazing, a better utilization of crop residues for animal feeding, and adaptation of farm implements for a more rational use of animal draught are elements in the livestock development strategy. Social questions, cultural values of animals, questions of livestock ownership and herd management, training, education, and awareness creation have to be included in an optimal program. At the present (low) input level, such improvements would help to reduce livestock populations, resulting in about a 1% decrease of respective livestock numbers at the time of implementation due to better livestock quality.

Magnitude of inputs into the model
For all input levels described above, respective costs of rural labor and financial inputs (foreign and local) were estimated. They are based on actual inputs and foreign aid for different input levels, on experiences of environmental rehabilitation and agricultural development in other countries, and on mere guesswork for unknown future inputs. In Table 2.2, some indications of the magnitude of inputs are given for the four input sectors and the four different input levels. For any input option selected in the model, approximate costs may be calculated using Table 2.2.

Land resource scenarios for the next 50 years
The model presented in the previous section is used here to get some ideas of how land resources in Ethiopia as a whole might develop if different input levels into the rural sector are made. The four input sectors, namely family planning, environment, agriculture, and livestock, with four possible input levels in each sector, would give a potential number of $4^4 = 256$ scenarios, assuming that no changes in inputs are made over the 50-year period.

Here, only five scenarios will be presented (Fig. 2.4). A first scenario (a) is assumed to be 'pre-revolution', i.e. Ethiopia about 20 years ago, when very little inputs and efforts were made by the then government into the agricultural sector. The next scenario (b) is assumed to be 'present day', when considerable efforts have been made already, but where constraints still exist. Two further scenarios (c) and (d), try to put emphasis on 'environment' and 'green revolution', i.e. strategies heading forward in one direction mainly, in order to solve degradation or production problems with high priority. In a final scenario (e), an attempt was made to see how, with minimum inputs, it would be possible to become ecologically sustainable over the 50-year period.

Fig. 2.4(a).

SCENARIO: "PRE-REVOLUTION"

Short Description: Very little attention is given to the rural sector, which comprises the majority of the population. The few programmes concerning agriculture and rehabilitation are located near towns, the funds are mainly absorbed by adminstration. Rural labour for reclamation is not mobilized. Naturally, inputs in such a system are also very low, it is a typical "exploitative" system.

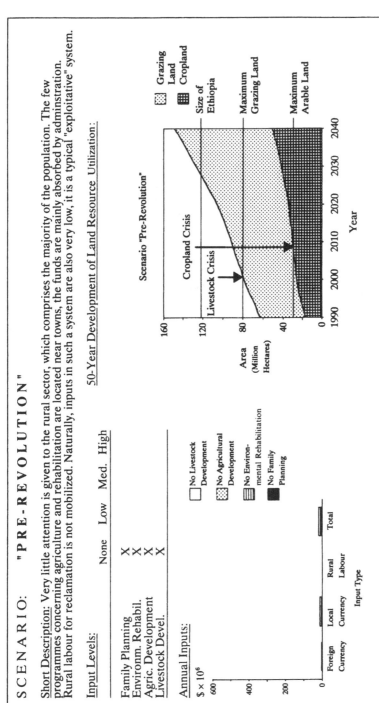

50-Year Development of Land Resource Utilization:

Input Levels:

	None	Low	Med.	High
Family Planning	X			
Environm. Rehabil.	X			
Agric. Development	X			
Livestock Devel.	X			

Annual Inputs:

Legend:
- No Livestock Development
- No Agricultural Development
- No Environmental Rehabilitation
- No Family Planning

Chart legend:
- Grazing Land
- Cropland
- Size of Ethiopia
- Maximum Grazing Land
- Maximum Arable Land

Conclusion: Such a model, leaving population and livestock growth at present rates and environmental degradation accelerating, quickly results in a complete absorption of land resources. First is a severe livestock crisis due to grassland degradation, then, after only 20 years, a cropland crisis due to complete utilization of all cultivable land.

Ecological Vulnerability to Famine: In the short period of 10-20 years, famine vulnerability will drastically increase due to the lack of any development and ecological stabilization. Especially in the livestock sector, problems of animal feed will emerge very strongly.

Fig. 2.4(b).

SCENARIO: "PRESENT DAY"

Short Description: A considerable effort is being made in the agricultural sector. Inputs are many times higher than in the "pre-revolution" Scenario. Agricultural organizations exist and services are being offered. However, considering the magnitude of the problems and the number of rural farmers at subsistence level, the impacts are still much less than needed.

Input Levels:

	None	Low	Med.	High
Family Planning	X			
Environm. Rehabil.			X	
Agric. Development			X	
Livestock Devel.			X	

50-Year Development of Land Resource Utilization:

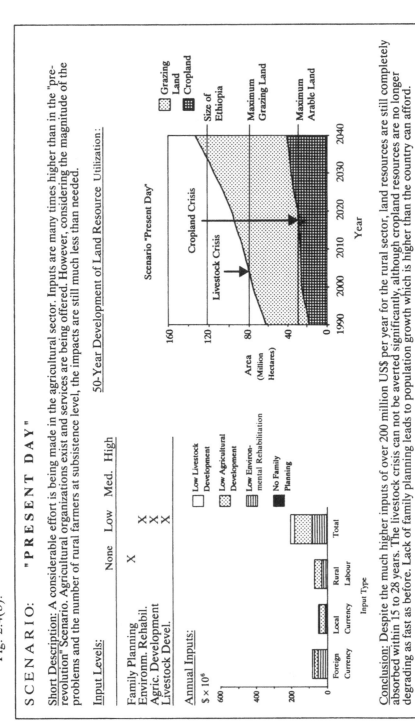

Conclusion: Despite the much higher inputs of over 200 million US$ per year for the rural sector, land resources are still completely absorbed within 15 to 28 years. The livestock crisis can not be averted significantly, although cropland resources are no longer degrading as fast as before. Lack of family planning leads to population growth which is higher than the country can afford.

Ecological Vulnerability to Famine: A considerable increase must be expected due to the livestock sector. Lack of properly fed draught animals will reduce production from cropland, and livestock reserves for periods of food shortage will become smaller in the near future.

Fig. 2.4(c).

SCENARIO: "ENVIRONMENTALIST"

Short Description: High emphasis is given to the ecological sector, while the other sectors remain at levels as low as today. Terraces are formed on cropland, forests are planted, and soil erosion and runoff is almost totally controlled within the next 50 years. Environmental rehabilitation receives annual inputs on the order of 200 million US $, 4 times more than at present.

Input Levels:

	None	Low	Med.	High
Family Planning	X			
Environm. Rehabil.				X
Agric. Development			X	
Livestock Devel.			X	

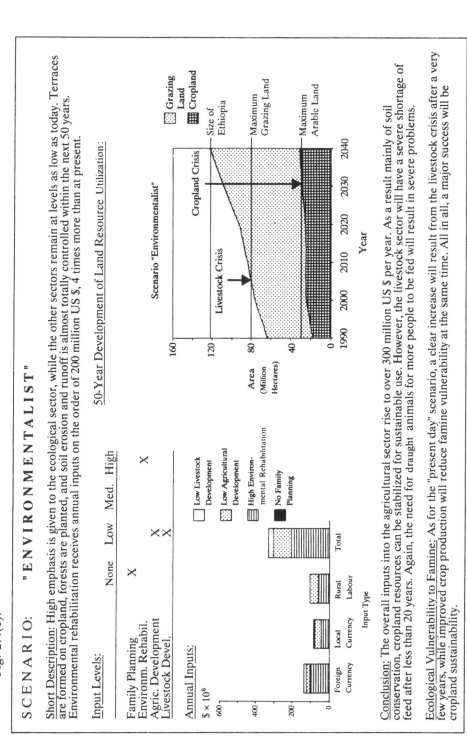

50-Year Development of Land Resource Utilization:

Annual Inputs:

Conclusion: The overall inputs into the agricultural sector rise to over 300 million US $ per year. As a result mainly of soil conservation, cropland resources can be stabilized for sustainable use. However, the livestock sector will have a severe shortage of feed after less than 20 years. Again, the need for draught animals for more people to be fed will result in severe problems.

Ecological Vulnerability to Famine: As for the "present day" scenario, a clear increase will result from the livestock crisis after a very few years, while improved crop production will reduce famine vulnerability at the same time. All in all, a major success will be cropland sustainability.

Fig. 2.4(d).

SCENARIO: "GREEN REVOLUTION"

Short Description: High emphasis is given here to the agricultural development sector, while the others remain at the present input level. Fertilizer programmes, improved seeds and high-yielding crop varieties are introduced and made available to a majority of the peasants. Inputs are similar to the "environmentalist" scenario. However, degradation as well as growth of population and livestock remain at common high rates.

Input Levels:

	None	Low	Med.	High
Family Planning	X			
Environm. Rehabil.		X		
Agric. Development				X
Livestock Devel.		X		

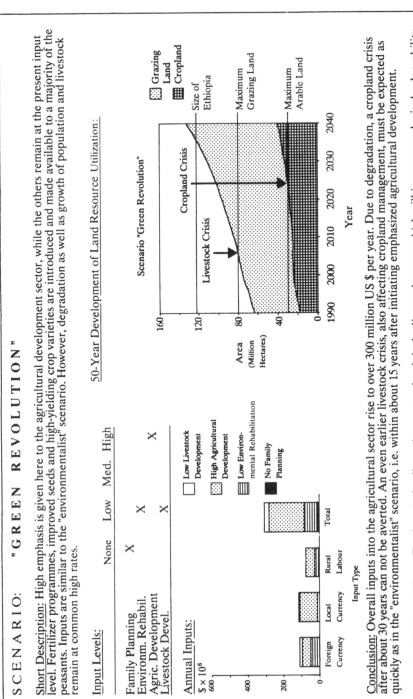

50-Year Development of Land Resource Utilization:

Conclusion: Overall inputs into the agricultural sector rise to over 300 million US $ per year. Due to degradation, a cropland crisis after about 30 years can not be averted. An even earlier livestock crisis, also affecting cropland management, must be expected as quickly as in the "environmentalist" scenario, i.e. within about 15 years after initiating emphasized agricultural development.

Ecological Vulnerability to Famine: As in all previous scenarios, it is the livestock sector which will increase ecological vulnerability to famine after a very few years, although higher cropland production will improve the situation, at least until degradation will irreversibly take its toll.

Fig. 2.4(e).

SCENARIO: "CONSERVATION-BASED DEVELOPMENT"

Short Description: In this scenario an attempt was made to attain the goal of sustainable use of land resources over at least 50 years. Besides medium inputs into agricultural development and environmental rehabilitation, high emphasis has to be given to livestock development. In addition, this positive scenario is only feasible if an input into the family planning sector is made.

Input Levels:

	None	Low	Med.	High
Family Planning			X	
Environm. Rehabil.			X	
Agric. Development			X	
Livestock Devel.				X

Annual Inputs:

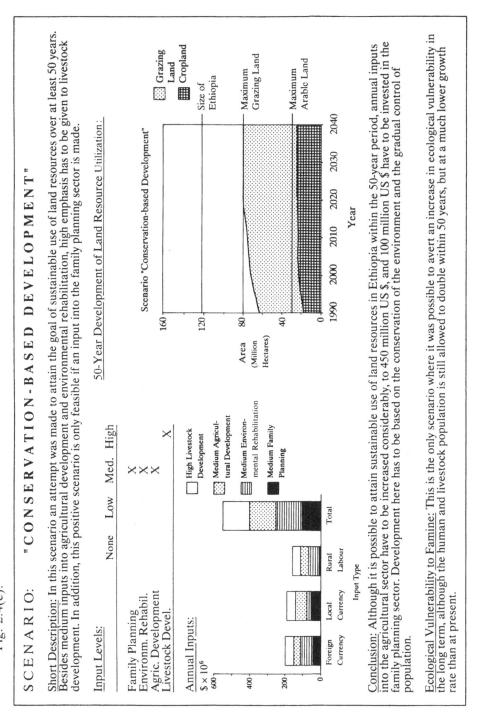

50-Year Development of Land Resource Utilization:

Conclusion: Although it is possible to attain sustainable use of land resources in Ethiopia within the 50-year period, annual inputs into the agricultural sector have to be increased considerably, to 450 million US $, and 100 million US $ have to be invested in the family planning sector. Development here has to be based on the conservation of the environment and the gradual control of population.

Ecological Vulnerability to Famine: This is the only scenario where it was possible to avert an increase in ecological vulnerability in the long term, although the human and livestock population is still allowed to double within 50 years, but at a much lower growth rate than at present.

50   *H. Hurni*

Interpretation

None of the first four scenarios was able to attain long-term sustainable use of the land resources on a country-wide scale, despite the fact that two tried to emphasize either the environmental sector or the agricultural production sector. While the two considerably improved the cropland situation, either through stabilization of the degradation problem or through higher land and crop productivity, both scenarios were unable to provide solutions for the livestock sector. A livestock crisis, with serious consequences also for cropland management in the ox-plow farming systems,

Table 2.2. *Total annual inputs in millions of US$ as required to attain the rates of change for the different scenarios of the land resource model (see also the Appendix)*

Sector		Annual input level			
		None	Low	Medium	High
Family planning	FC	2	20	40	90
	LC	3	25	50	100
	RL	—	5	10	10
	Total	5*	50	100	200
Environmental rehabilitation	FC	1	45	70	100
	LC	2	5	25	40
	RL	—	25	55	60
	Total	3	75*	150	200
Agricultural development	FC	1	20	40	50
	LC	2	35	65	100
	RL	—	45	45	50
	Total	3	100*	150	200
Livestock development	FC	1	15	30	50
	LC	4	8	25	50
	RL	—	2	20	50
	Total	5	25*	75	150

Notes: FC = foreign currency (food-for-work, technologies, fertilizers, equipment, expertise)
LC = local currency (investments, recurrent costs for salaries, education, training, propagation)
RL = rural labor (campaigns, individual works, social works. All works are valued at 1 US $ per work-day. They should be contributed at no costs, attained through participation, incentives, changes of attitudes, motivation, awareness creation, mobilization)
TOTAL = total inputs required in *million US $ per year*. Those marked with * are present input levels (1988)

emerged in all four scenarios after some 10–30 years. An explanation of this problem can be found in the growth of both the human and the livestock populations, which are closely correlated. More people need more land for cultivation, and as a consequence, more cattle to safeguard land tillage operations.

Only the last scenario, 'the conservation-based development' is able to 'end' with a stable situation. Here, medium inputs are made in the family planning sector as opposed to none in the other four scenarios. With annual costs of about 100 million US $, projects of family planning are propagated through extension, education, social and retirement security, motivation, propagation of contraceptives, health services, mother and child care, and other activities. Through this, the annual population growth rate will decrease from its present level of 2.9 to 2.4% after the first 10 years, to 1.8% after 20, 1.2% after, 0.6% after 40 years, and end at zero growth after 50 years from the inception of the program.

The marked difference between the last scenario and the first four is exemplified in Table 2.3, where population and livestock numbers are given for all scenarios at the end of the 50-year period. It is noteworthy to realize that despite a doubling of population in the last scenario within the next 50 years, it will still be possible to produce sufficient animal feed and food on a country-wide scale. This is a stimulation by itself. It will have its costs, however, since annual inputs into the rural sector will have to be redirected and increased by a factor of 2.5 from about 200 million US $ at present to about 550 million in future. In view of the fact that Ethiopia receives international support of less that 10 US $ per person per year, the African average being 18 US $, strategies to stimulate an increase in foreign support have to be discussed, since an increase in the African average would suffice to finance the ambitious 'conservation-based development' scenario.

Table 2.3. *Population and livestock numbers for various scenarios at the end of the 50-year period (2040)*

Scenario	Rural population (million)	Livestock (million TLU)
'Pre-Revolution'	173	86
'Present Day'	173	79
'Environmentalist'	173	78
'Green Revolution'	173	79
'Conservation-based Development'	88	41

On the other hand, if population increases four-fold and livestock by a factor of three, this being the present trend, sustainability of land use cannot be guaranteed any more.

Not considered in the scenarios were regional differences, especially between the heavily degraded northeastern regions and the western ones, which have less damage, but are ecologically more fragile. The model could also be used for regional scenarios, if regional input data are used for rural population, land resources, and livestock numbers. Considerable differences may result for the various agro-ecological zones of Ethiopia.

Conclusions and recommendations

The impacts of ecology on famine in Ethiopia are manifold:

* Natural factors, like climatic extremes, pests, or diseases are usually the short-term factors triggering production shortfalls, eventually ending in famine situations.
* Land-use factors, especially the subsistence agricultural systems dominant in the Ethiopian context, are the main reasons why famines cannot be averted with the existing exchange system of basic goods, like markets, relief organizations, or transport networks. Long-term land degradation contributes to the present low productivity status of soils, to dwindling food reserves, and to economic deterioration over generations, where innovations are no longer possible due to the high risks of survival.
* Human factors, being cultural, social, political, and economic processes, are unable to overcome and reverse ecological weaknesses.
* External factors such as food aid and political and economic pressure, as well as intersystem enforcements, add to ecological issues in the creation of famines.

In the short-term analysis of famine preparedness, ecological issues must be considered as 'matters of fact' rather than processes which can be reversed easily and in a short time. In a long-term prevention strategy, however, the impacts of ecological processes on famine vulnerability must be seriously considered. Current trends of population growth, land resource utilization, environmental degradation, and livestock numbers, indicate very serious constraints, and a sharp increase in famine vulnerability due to ecological problems in the next 10–30 years. With the model developed in this chapter, current development inputs in the rural sector and their impact on land resources over the next 50 years can be observed and discussed.

Ecological impacts on famine vulnerability can be reduced only if a package of rural development is implemented in four main sectors, namely family planning, environmental rehabilitation, and agricultural and livestock development, and applied over the next 50 years at input levels much higher than at present. Different scenarios of inputs can be observed with the model. All inputs consist of foreign currency, local costs and rural labor, to be mobilized with adapted policies, participatory approaches, extension, training, and appropriate technologies. No single-focused scenario, where high inputs are made into one sector alone, will yield satisfactory benefits.

A best scenario attaining sustainability of land resource utilization over the whole 50-year period is the 'conservation-based development' scenario. There, family planning programs, in their broadest sense, lead to a stabiliz- ation of population growth after 50 years, when the Ethiopian population will have doubled. To attain this goal, a reduction of growth rates from currently 2.9% per year to 2.4% must be achieved after only 10 years. Environmental programs will have to be increased four-fold. Agricultural development inputs will have to double, and livestock development inputs be increased by a factor of six. The overall increase in the rural development sector will have to reach about 600 million US $ per year, or 13 US $ per person per year. Such inputs call for increased international assistance, and appropriate policies and strategies of the government to attain this goal. Ecological vulnerability to famines, with such an optimum scenario, will already be reduced after 10– 20 years and reach a very favorable level after the 50-year period. The present model, however, cannot be used to assess other factors contributing to famine vulnerability in Ethiopia, such as economic, social, and political factors, although some of them may be closely interrelated.

The scenarios used in the model are global scenarios for rural develop- ment, which will need regional interpretation and specific application to the different agro-ecological zones of the country. Long-term strategies to reduce famine vulnerability and to attain land-use sustainability will have to be regionalized as well, since initial inputs into the four sectors of rural development of the land resource model will greatly vary within Ethiopia, from region to region as well as within each region. Furthermore, rural development programs should wherever possible be biased toward ecologi- cal stabilization as a '*sine qua non*'. Pilot projects in the Gojam Region ('Learning from Anjeni') and in Gonder can be examples for wider appli- cation. There, conservation-based experiments were done with single Peasant Associations, using infrastructural and basic need improvements (clinics, springs) as stimulations for mobilizing farmers to carry out soil conservation and afforestation. Such pilot project ideas are currently being applied on a wider scale in the Merhabete area (Hurni, 1987b).

Current policies of rural development, including state farm promotion, resettlement, village development, and distribution of subsidies for agricultural development, but also the promotion of higher production of the individual as well as the cooperative groups, should be analyzed and discussed in light of the ecological sustainability to be attained in all Ethiopian ecological zones in the long term. It appears that the livestock sector, including the associated land degradation processes, will need special attention.

Recommendations

Ecological issues in the creation of famines should be considered as follows:

1. The complexity of factors related to famine vulnerability should be acknowledged and considered. Natural factors, land-use factors, human factors, external factors and their interrelationships make up a vulnerability system. Ecological issues concern the long-term trends, which are heading towards more vulnerable situations in future.

2. The most crucial ecological issue is the long-term sustainability of the utilization of Ethiopian land resources. Resources are absorbed to the last reserves, deteriorating due to soil erosion and fertility decline, and creating problems for livestock feed and fuelwood supply. Rural development must have, as an ultimate goal, the attainment of ecological stability in the long term, while ensuring adequate production.

3. High emphasis must be given to the rural development sector, namely programs of family planning, environmental rehabilitation, and agricultural and livestock development, as main components. Overall inputs into these sectors must be increased manifold in order to reach the goals within a given time frame of 50 years for attaining stable situations.

4. Without family planning programs, any other development activity is doomed to fail in the long term. While a doubling of the population and a moderate increase of livestock may still be sustainable with other efforts performed, unlimited growth will lead to an excessive land resource overuse and an almost irreversible ecological disaster within a very few decades from today. Human resources must be utilized wisely, since people are promotors of any development activity.

5. Since massive external support will be needed to achieve foreign currency requirements, government policies and strategies must include the mobilization of such cooperation and take appropriate steps to invite a more active participation of the international community for conservation-based development in Ethiopia. If it is possible to receive foreign assistance at the African average at present, i.e. 18 US $ per year, all programs outlined above, supported by massive inputs of both the rural population as well as the Ethiopian government, can be achieved at needed scales of input in human, technological, and financial resources.

6. The national economy should be geared to stimulate people to become engaged in the second sector (handicraft, support services for agricultural development, industrial inputs into the agricultural sector, other industries), in order to allow some rationalization in the first (agricultural) sector.

7. For rural development, the combination of conservation activities with development packages, such as the provision of basic needs or infrastructures, has emerged as a most appropriate and efficient tool to stimulate rural development. New approaches and policies should be centered on small communities with a strong local participation, increased training, awareness creation, education, appropriate technologies, and integrated conservation and development systems.

Appendix: Impact functions of the land resource model for Ethiopia (Fig. 2.3, p. 41)

1. Impacts of family planning at various input levels on human population

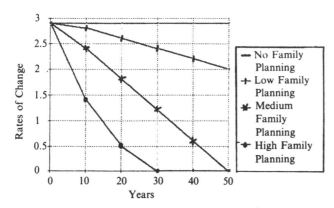

It is assumed that impacts of family planning at various levels will reduce the present population growth rate gradually within the 50-year period. Inputs will address programs in all issues concerning family planning, such as extension, education, retirement security, availability and propagation of contraceptives, health services, mother and child care, awareness creation, education, training.

* With no (present) inputs, the rates are likely to remain at 2.9% annual growth.
* With low inputs of about 50 million US $ per year, growth rates will only slowly decrease over the period, from 2.9 to 2.0% after 50 years.
* With medium inputs of almost 100 million US $ per year, rates are assumed to steadily fall to zero growth after 50 years.
* With high inputs of 200 million US $ per year, rates are assumed to fall sharply after the first 10 years, then to a lesser extent, and to reach zero growth after 30 years. This last version, however, will have to be associated with enforced birth control, and is therefore very unrealistic.

2. Relationship between population and cultivated land

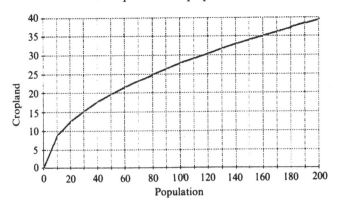

With the assumption that present farming systems will only moderately change into modern mechanized systems, the area of cropland used per person will only slightly decrease in future. However, due to dwindling cropland resources and increasing population, it is expected that farmers will automatically start to intensify agriculture and make higher inputs on smaller areas. This trend, while certain for normal cropland, is difficult to imagine on eroded land, the productivity of which is already at a low level. All in all, cropland and population are assumed to have a functional relationship following the root-power function

$$y = 2.77\, x^{1/2}$$

y being cropland in million hectares, and, *x* human population in million people.

3. Impact of environmental rehabilitation at various input levels on cultivated land

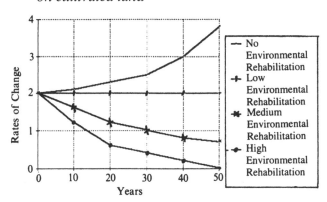

It is assumed that impacts of environmental rehabilitation at various levels will reduce the rate of erosion especially on cropland. At present, there is a decrease in productivity of 2–3% per year due to degradation processes. This means that a farmer will have to expand the area of cultivation by about the same amount, i.e. 2%, in order to attempt to maintain production at constant levels. Therefore, environmental rehabilitation is assumed to change the annual rate of cropland expansion according to the various input levels as follows:

* With no inputs, cropland will have to be expanded at increasing rates, since degradation is a self-propelling process accelerating its own rate. After 50 years without inputs, the rate of degradation, and subsequent cropland expansion, is assumed to rise to almost 4% per year.
* With the (present) relatively low inputs of about 75 million US $ per year, rates can be maintained at constant levels, but not yet be reduced, because of the magnitude of the problems.

   * With medium inputs of 150 million US $ per year, expansion rates may be decreasing to less than 1% after 50 years.
   * With high inputs of 200 million US $ per year, the rates of cropland expansion due to degradation are assumed to fall to zero, when sustainability of all cropland is attained.

4. Impacts of agricultural development at various input levels on cultivated land

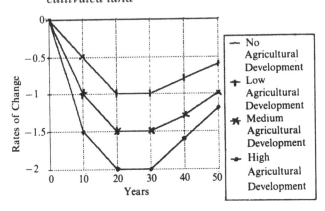

It is assumed that impacts of agricultural development will lead to an intensification of agricultural production for given areas. As a consequence, more will be produced on less area, where higher inputs of labor and fertilizer can be made. As a consequence, it is therefore expected that cropland areas will decrease slightly.

   * With no inputs, no impact on cropland size is expected.
   * With low (present) inputs of 100 million US $ per year, it is expected that the rate of land reduction will constantly rise until it reaches −1% decrease of land per year. After about 10 years of constant rates, however, the impact of agricultural inputs on land, such as fertilizer, is expected to become less, since soils are approaching maximum productivities. Similar inputs will not produce constant output increases from year to year, but slightly reduced increases after some 30 years of fertilizer inputs. Thus, after 50 years, annual production gains will be lower than after 30 years, and cropland reductions will be back to −0.6% annual reduction.
   * With medium annual inputs of 150 million US $, the same will happen at a slightly higher level, to −1.5% maximum annual

decrease of cropland size due to production improvements, and back to −1% after 50 years.
* With high annual inputs of 200 million US $, rates will stimulate production gains resulting in a reduction of cultivated land of −2% annually, lowered to −1.2% after 50 years.

5. Relationship between human and livestock population

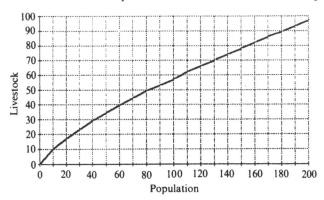

With the assumption that present farming systems will only moderately change into modern mechanized systems, the livestock number per person will only slightly decrease over the years. Since tractors cannot be utilized everywhere, especially not in mountainous areas, on steep slopes, and on small terraces, mechanization is not expected to replace traditional ox-plow systems on more than 35% of the farms after 50 years. As a consequence, the number of cattle for providing draught animals will have to be closely correlated to human population, although for more people, there will be a slight decrease of the rate.

All in all, livestock and population are assumed to have a functional relationship following the root-power function

$$y = 1.818 \, x^{3/4}$$

y being livestock number in million TLU, and x human population in million people.

6. *Impacts of livestock development at various input levels on livestock population*

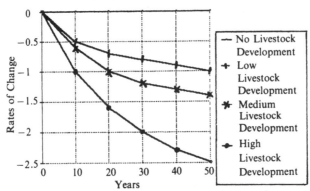

It is assumed that impacts of livestock development inputs will reduce the need of having as many animals as without inputs. Better animal feed, better breeds, more adapted oxen for plowing, better health services for cattle, grazing management, and other development issues will lead to a reduction of livestock number over the 50-year period. Due to the high value and use of livestock in the farming systems, however, reductions are expected to remain relatively low, although steadily increasing. On the other hand, more people will need more livestock, a fact which is expressed in relationship (5).

* With no inputs, livestock numbers will not be influenced.
* With low (present) annual inputs of 25 million US $, rates of decrease in livestock numbers are expected to quickly emerge and slowly become bigger, and reach −1% at the end of the 50-year period.
* With medium annual inputs of 75 million US $, rates will end at −1.4% annual decrease in livestock number.
* With high annual inputs of 150 million US $, rates are assumed to reach −2% after the 50-year period.

7. Relationship between livestock population and grazing land

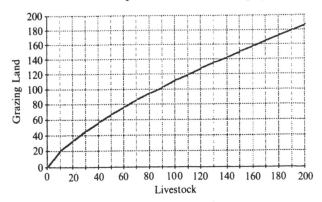

With the assumption that grassland improvement on heavily degraded soils is a difficult goal to achieve, rate of number of livestock per area of grassland will only slightly decrease over the 50-year period. Some improvements of forage, grazing management, and cut and carry systems, however, can be expected, so that more livestock can be fed on smaller areas after some time. All in all, livestock and grazing land are assumed to have a functional relationship following the root-power function

$$y = 3.5\, x^{3/4}$$

y being grazing land in million hectares, and x livestock number in million TLU.

3

Soil erosion and conservation in China

WEN DAZHONG

Introduction

Natural resources and environment are the bases for human social and agricultural development. China has a vast territory, a large population, and abundant natural resources. The total land area of China is 960 million hectares, which accounts for 1/15 of the total world land area. However, the average land area per person in China is only one-third of that in the world overall. The unusable land area, including deserts, glaciers, tundra, exposed rock, and other area, accounts for 28% of the total land area in China. The desertified (wind-eroded) and potentially desertified land areas account for 3.5%, and the water-eroded land areas account for 15.6%. The total cultivated land area is only 130 million hectares, or 13.5% of the total land area of China (Liu Yingqiu, 1988). About 6.7 million hectares, or 5% of total cultivated land, have become deserts. About 8 million hectares of land, or 6%, are too saline for cultivation. About 42 million hectares of China's cultivated land, or one-third of the total cultivated land, are undergoing serious water and wind erosion. Only a quarter of the cultivated land is well managed and highly productive (Peng Fude, 1987). Obviously, protecting land resources and improving land productivity are essential to feeding more than a thousand million people and for economic development in China. Clearly, controlling soil erosion and improving eroded lands are important tasks for effective management of land resources in China.

This chapter discusses wind and water erosion problems in China and considers current and future policies for effectively controlling erosion through soil conservation.

Soil erosion in China

China's vast mountain-land areas and plateaus are suffering serious soil erosion. According to statistics from the early 1950s (TMB, 1984) one-sixth of the soil in China was eroded, or 150 million hectares. The Chinese people have made great efforts to control soil erosion since the 1950s, and notable achievements have been made in soil conservation. Recently, 46 million hectares with serious erosion have been placed under a program of conservation (Liu Yingqiu, 1988). However, some land areas are still being eroded (Shi Deming, 1985a; Wang Yongan, 1987a; Zhang Zhongzhi, 1988)

The major erosion regions include the Loess Plateau, the southern erosion region, the northern erosion region, and the northeastern rolling land erosion region (Fig. 3.1).

The Loess Plateau

China's Loess Plateau is the most widely distributed loess area on earth. This loess is a deep deposit of paleosol, and the plateau is most systematically and completely developed for crops (Zhu Xianmo, 1986). The

Fig. 3.1. The major soil erosion regions in China. Scale 1 : 30 000 000

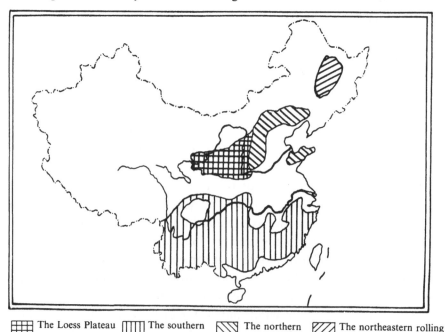

The Loess Plateau erosion region   The southern erosion region   The northern erosion region   The northeastern rolling land erosion region

plateau is located in the middle reaches of the Yellow River and extends over seven degrees of latitude (35–41° N) and 13 degrees of longitude (102–114° E), with a total area of 53 million hectares and a population of 70 million. The plateau formed after a large amount of thick loess was deposited on the vast area. The plateau area is 1000–1400 m above sea level and has a loess layer thickness of about 100 m. The grabens and valley basins are at an altitude of 400–500 m. The loess in the region was transported mostly by winds over a period of about three million years (Zhu Xianmo, 1986).

The land types on the Loess Plateau can be classified into four groups: mountains, hilly lands, high flat table lands, and stream and river valleys. The vegetation is very sparse, and the land surfaces are usually bare. The annual precipitation in the region is 400–600 mm, and most occurs during the summer season. Rainfall is intense and of short duration with large raindrops. This contributes to the serious erosion problem in the loess region. In addition, poor land-use practices over a long time period have accelerated loess erosion. Serious soil erosion in this region affects about 43 million hectares. Soil erosion rates are classified into four levels (Table 3.1). The area with an erosion rate of more than 100 t/ha accounts for 17% of the total loess. The total soil loss of the region in recent years is estimated to be about 2200 million tonnes annually. Three-quarters of the total lost soil is transported to the lower reaches of the Yellow River (Tang Keli, 1985; Jing Ke, 1986). Thus, the annual average erosion rate in the Loess Plateau region is 51 t/ha.

The total cultivated land area in the Loess Plateau region is about 13.3 million hectares or about 25% (NADC, 1981). Only about 10% of the area is flat, and the remaining cultivated area is on sloping land (Zhang Maosheng, 1987). The average erosion rate on the cultivated land is about 60 t/ha (Shi Deming, 1985b); i.e. about one-third of the total soil erosion loss in the Loess Plateau region occurs on cultivated land.

Table 3.1. *Soil erosion in the Loess Plateau region (based on Jing Ke, 1986)*

Erosion level	Area (Mha)	Annual soil loss rate (t/ha)
I	7.2	>100
II	10.8	50–100
III	5.7	20–50
IV	19.3	<20

The southern region

The southern region is located in the tropical and subtropical zones south of the Yangtze River valley excluding the Yangtze River delta, the Pearl River delta, the Beyang Lake plain, the Dongting Lake plain, and the Shichuan Basin. The total area of the region is about 160 million hectares with a rural population of 200 million. The total cultivated land area is about 16.6 million hectares, and more than 60% of the cultivated land is paddy rice. Recently, the annual grain production in the region has been about 100 million tonnes, accounting for one-fifth of the total cereal grain production in China. In addition, about 50% of the total forest area of China is in this region (NADC, 1981).

The annual precipitation in the region is 1000–1600 mm, and 50–60% of the precipitation occurs between April and June. More than 6 million hectares of cultivated sloping land and barren steep lands are the sources of most of the soil erosion. The soil layers of the steep land of the region are very thin and are easily eroded to expose parent rock materials. Soil erosion in this region has increased rapidly during the last decades. In the 1950s, serious erosion in this region covered 46 million hectares (TMB, 1984). However, this area has increased to 69 million hectares, or 50%, in the 1980s (Shi Deming, 1985b). The total soil loss of the region is about 2500 million tonnes annually, including 2400 tonnes from the Yangtze River valley (Wu Youzheng and Pan Jinlin, 1986; Wang Yongan, 1987b) and about 100 million tonnes from the Pearl River valley and other river valleys (TMB, 1984). The annual average erosion rate in the region is 36 t/ha.

Erosion in the western part of the region (the upper reaches of the Yangtze River) covers about 35.2 million hectares with a total annual soil loss of 1600 million tonnes (Yang Yansheng *et al.*, 1987). Thus, the annual average soil erosion rate in the region is 45 t/ha.

The northern region

The northern region of China is located in the warm-temperate zone including mountain-land areas of Hebei, Shandong, and Lianoning provinces and some areas adjacent to them. The annual precipitation in the region is 500–700 mm. The soil on the steep slopes is very thin and mixed with stones. The degree of forest cover is less than 10%. Summer torrential rains cause serious soil erosion on sloping land. Several sources (NADC, 1981; HCH, 1984; Zhu Junfeng, 1985) estimate that soil erosion in this region covers about 23 million hectares including about 10 million hectares of cultivated land. The soil erosion rate is estimated to be about 20 t/ha annually but may reach as high as 50 t/ha (IFS, 1985; Tang Defu, 1985; Zhou Guangyu, 1986). Total soil loss for the region is about 500 million tonnes annually.

The northeastern rolling land region

The northeastern rolling land region is located in Jilin and Heilong-jian provinces. Annual precipitation is 500–600 mm, and 65% of the rain falls during the summer season. In general, the slopes of the rolling land are less than 4°, seldom more than 8°. Fertile, thick, black soil with an impermeable subsoil covers most of the region. When the soil is saturated with moisture, soil erosion is severe. The region covers about 13 million hectares, two-thirds of which, or 8.7 million hectares, is cultivated land. Soil erosion extends over 2.5 million hectares, which includes two million hectares of cultivated land (Zhu Junfeng, 1985). Although the slopes are not very steep, soil erosion rates are relatively high. The annual erosion rate ranges from 50 to 70 t/ha (Tang Defu, 1985; Fu Dexing, 1986). The total soil loss for this region is about 150 million tonnes, 80% of which is from cultivated land.

Overall water erosion rates

Total soil erosion of the four regions covers about 138 million hectares including 30 million hectares of cultivated land. Total soil loss is 5300 million tonnes (Table 3.2). If additional seriously eroded areas outside these four regions were included, the total area affected in China would be at least 150 million hectares, including more than 35 million hectares of cultivated land, accounting for 27% of China's total cultivated land area. Total soil loss in China is calculated to be more than 5500 million tonnes, which accounts for an estimated 20% of total world soil loss.

Table 3.2. *The major soil erosion regions in China*

Region	Total area (Mha)	Cropland (Mha)	Human population (millions)	Eroded area (Mha)	Eroded cropland (Mha)	Total soil loss (Mt/yr)
The Loess Plateau	53	13.3	70	43.0	12	2200
The southern region	160	16.6	200	69.0	6	2500
The northern region	28	13.3	65	23.0	10	500
The northeastern region	13	8.7	20	2.5	2	150

Wind erosion and desertification

A large area of lighter and looser soils, including 14.7 million hectares of semi-fixed sandy soil, 14.6 million hectares of fixed sandy soil, 44.7 million hectares of shifting sand, and other soils, covers the semi-arid and arid regions along the Great Wall area in northeastern, northern, and north-western China (Hou Guangjun and Gao Huimin, 1982). The dry climate with strong winds in these regions causes serious erosion. In general, these regions experience 20 to 100 days with strong winds (> 8 m/s) annually (Zhu Junfeng, 1985). Poor resource management aggravates wind erosion and causes desertification.

According to some investigations (NADC, 1981; Liu Shuangjin *et al.*, 1987), the historically formed desertified area in China totals 12 million hectares, the modern desertified area formed in the last five decades is 5.6 million hectares, and the current potential desertified area is about 15.8 million hectares. Thus, the total area of the three kinds of desertification is 33.4 million hectares, accounting for 10.3% of the total area in these regions, and 3.4% of the total area of China. About 4 million hectares of crop fields, 5 million hectares of grasslands, and 35 million people in 212 counties are threatened with serious desertification (TMB, 1984; Liu Shuangjin *et al.*, 1987).

The desertified area has expanded rapidly in recent years. The average rate of desertification has been 100 000 ha/yr during the last five decades, but 180 000 ha/yr during the last two decades (TMB, 1984). If the annual expanding rate of desertification is kept at 3.5%, another seven million hectares will be desertified by the year 2000 (Liu Shuangjin *et al.*, 1987).

It is calculated that 25.4% of the total modern desertified area in China was caused by grassland reclamation, 28.3% by overgrazing, 31.8% by fuelwood collection, 9% by unsuitable use of water resources and construction, and only 5.5% by sand dune extension (Liu Shuangjin *et al.*, 1987).

Desertification reduces land productivity. About 60% of crop fields in these regions lose 0.7–1.0 cm of the fertile topsoil layer by wind erosion annually. The cereal grain yields of these fields are less than 1000 kg/ha, and the desertified grassland yields only 0.8 kg/ha of animal products annually (Zhu Junfeng, 1985). The serious wind erosion and desertification also threaten the highway and railway traffic and other economic activities in these regions. In 1978, the State Council decided to build a forest system to protect these regions from wind erosion and desertification. The project is called the 'Three North' protective forest system, and is also known as the 'Green Great Wall.' During the first phase of the project from 1978 to 1985, the government gave 1700 million yuan to the project, and six million hectares of plantations were established. Crop field shelter-belt systems were established in these regions, protecting eight million hectares of crop fields from wind erosion (BTN, 1986).

The harmfulness of soil erosion in China

Erosion harms soil productivity, the environment, agriculture, the national economy, and human well-being.

Land degradation

Serious erosion in China has degraded fertile topsoil, damaged crop and natural vegetation, reduced land productivity, and degraded the environment. It is estimated that about 1 cm of topsoil is lost by erosion in the Loess Plateau annually (Zhu Xianmo, 1984a). However, to accumulate 1 cm of loess in the plateau requires more than 100 years. Thus, erosion rates in the region are more than 100 times the reformation rate (Zhu Xianmo, 1984a). About 50% of the loess hilly area in the Loess Plateau has lost its A or A+B topsoil layer (Shi Deming, 1985b). Erosion has removed most of the fertile, black topsoil in the plateau (Zhu Xianmo, 1984b). Soil organic matter of the topsoil of some plateau areas has decreased from 3 to 0.3%, total nitrogen content from 0.2–0.3 to 0.03%, and available phosphorus content from 50 to 5 ppm by erosion (Zhu Xianmo, 1984b). The available trace element contents in the topsoils have been reduced as follows: Cu from 1.5 to 0.6 ppm, Zn from 0.6–0.8 to 0.2 ppm, Mn from 15 to 5 ppm, Fe from 12 to 2 ppm, B from 0.5 to 0.2 ppm, and Mo from 0.15 to 0.04 ppm. It is estimated that the contents of available Zn, Mo and B in topsoil of two-thirds of the plateau area and available Mn and Fe in topsoils of one-third of the plateau area have been critically reduced (Yu Chunzhu, 1983). Loss of trace elements is harmful not only to crops and other plants but also to human health (Zhu Xianmo, 1984b).

The average nutrient contents in the sediments of the Yellow River and its tributaries are as follows: organic matter, 0.5%; total N, 0.04%; available P, 15 ppm; and available K, 100 ppm; these are similar to their contents in topsoils of cultivated land in the Loess Plateau (Tang Keli *et al.*, 1987). It is estimated from the above data that the annual loss of organic matter, including total N, and available P and K from the eroded areas of the Loess Plateau, is about 11 million tonnes, 0.88 million tonnes, 0.03 million tonnes, and 0.22 million tonnes, respectively. About 20–40% of the total area in the southern region has lost its A topsoil layer (Shi Deming, 1985b). Organic matter, nitrogen, and phosphorus in the topsoil in the eroded region have been reduced to 10%, 5%, and 2% of their original contents in the soils, respectively (Shi Deming, 1985b). About 0.3–0.5 cm of topsoil layers in the erosion area are lost annually. This suggests that all of the topsoil in most of the eroded area will be lost in 50–100 years (Shi Deming, 1987). The contents of organic matter, total N, and available P and K in the topsoils on the eroded cultivated land of the regions are 0.5%, 0.05–0.07%, 5 ppm, and 80 ppm, respectively (Tian Xinyuan, 1988).

The northeastern rolling land region of China is considered to be one of the country's most fertile agricultural regions and is an important base for Chinese cereal grain production. In general, the fertile black topsoil layer in the region is 60–70 cm thick. The region has been cultivated for less than 100 years. However, serious erosion has made the fertile black topsoil layer thinner each year. After 60–70 years of cultivation, the black topsoil layer on sloping land has been reduced to only 20–30 cm. About 0.5 cm of the topsoil on sloping land has been lost annually. Since the land was first cultivated, the amounts of organic matter and total N in topsoils on sloping cropland in the region have been reduced from 11% and 0.6% to 2–3% and 0.2%, respectively (Tang Defu, 1985).

According to the above information, it is estimated that 5500 million tonnes of soil is lost annually in China due to erosion. This soil carries with it 27.5 million tonnes of organic matter, 5.5 million tonnes of nitrogen, 0.06 million tonnes of available phosphorus, and 0.5 million tonnes of available potassium from the 150 million hectares. The annual total of N, P, and K lost equals 46%, 2%, and 63%, respectively, of the total N, P, and K applied annually in Chinese agriculture.

An average of 138 kg of organic matter, 37 kg of nitrogen, 0.4 kg of available phosphorus, and 3.3 kg of available potassium is lost from each hectare of eroded cropland in China each year. According to the results of the national network of fertilizer tests in China (FTN, 1983, 1986), in general, 1 kg of nitrogen fertilizer could increase cereal grain production by at least 10 kg. Thus, it is calculated that the annual nutrient losses per hectare of eroded cropland would cause a reduction of about 400 kg/ha of cereal grains. The 35 million hectares of total cropland in China with serious erosion reduces the annual cereal grain harvest by about 13 million tonnes, which is about 3% of annual grain production in China. The greater nutrient loss and decreased nutrient input in these eroded croplands cause land productivity to decline year by year (Yu Chunzhu *et al.*, 1985). Soil erosion increases water loss from the eroded land. About 5–10% of total rainfall is lost from sloping cropland on the Loess Plateau (Yu Chunzhu *et al.*, 1985). The annual water loss from sloping cropland in the northeastern rolling land region is more than 1000 m^3/ha (Tang Defu, 1985). This water loss reduces crop production, especially in the low rainfall regions of northern China.

Serious gully erosion cuts the eroded land into small, irregular plots. It is estimated that gully erosion causes 0.06% of total Loess Plateau land loss each year (Zhu Xianmo, 1984b). The average gully densities are 0.32 km/km^2 on the eroded area of the northeastern rolling land region (Tang Defu, 1985), 2–3 km/km^2 in the southern area, and 6–7 km/km^2 in the Loess Plateau area (Jiang Dechi and Zhu Xianmo, 1962).

Environmental problems from erosion sediments

About 40% of the total soil eroded from the land, or about 2000 million tonnes of soil, is carried to the mouths of the rivers in China. The remaining 3500 million tonnes of sediments are deposited in lakes, rivers, and various water conservation facilities (TMB, 1984; Fang Zhengshan, 1987).

The Yellow River is 5464 km long and carries 48 000 million cubic meters of total annual runoff. The river passes through the Loess Plateau and carries about 1600 million tonnes of sediments eroded from the plateau to the mouth of the river. These sediments are collected through 35 major tributaries and more than 1000 streams. The highly concentrated sediments give the river the highest silt content of any river in the world. The average silt content in the river water is 38 kg/m^3, 80 times that of the Yangtze River, and 20 times that of the Nile River in Egypt. During periods of flooding, silt content in the Yellow River can rise to more than 650 kg/m^3 (Zhang Gueliang, 1987). About 1200 million tonnes of silt are carried to the mouth of the Yellow River, and 400 million tonnes of silt are deposited on the 400-km-long riverbed, which causes the riverbed to rise about 10 cm each year. The riverbed is currently about 6–10 m higher than the surface of the surrounding land, giving it the nickname 'the suspended river' (Huang Yong, 1987). Several studies report about 1500 floods resulting from the river breaking through its dikes during the last 2000 years. These floods affected about 25 million hectares of land area (TMB, 1984). The large amount of silt deposits on the riverbed have reduced the flood discharge capacities of the river. It is difficult to keep the flood discharge capacity as high as 22 000 m^3/s (Huang Yong, 1987). The large amount of sediment on the riverbed is seriously threatening the security of agricultural production, five oil fields, seven railway lines, and 100 million people in the North China Plain (Huang Yong, 1987).

Cities find it difficult and costly to use river water in the lower portion of the Yellow River. For example, a project was set up in 1981 to divert 200 million cubic meters per year of water from the Yellow River to Tianjin City, and it cost 200 million yuan/yr to clean up sediments in the diversion project (Tian Houmo, 1987). In order to keep the river from flooding, the river dikes in the lower reaches of the Yellow River have to be rebuilt and maintained annually. About 1400 million yuan have been spent to maintain the river dikes since the 1950s (Tian Houmo, 1985).

The large amount of sediment at the mouth of the Yellow River forms 27 km^2 of new land area on the coast and extends the coast about 300 m each year (Zhang Gueliang, 1987). This extension will influence the oil fields and other industries in the area (Huang Yong, 1987).

About 520 million tonnes of silt are deposited in reservoirs and other water conservation constructions in the Loess Plateau annually (Chen Zhilin, 1986). About 30% of total water-storage capacity and 48% of total irrigation capacity of the reservoirs in the Loess Plateau have been lost due to sediment deposits during the period from 1949 to 1975 (Fang Zhengshan, 1987). The Sanmenxia reservoir, the largest reservoir along the Yellow River, was constructed in the Loess Plateau region in 1959, at a cost of about 900 million yuan ($400 million). To construct the reservoir, more than 70 000 ha of land had to be flooded and 250 000 people had to relocate (Tian Houmo, 1987). However, 44% of its water-storage capacity, or 3400 million cubic meters, was lost during the brief period from 1959 to 1966 because of the large amount of sediments (TMB, 1984). The annual economic loss caused by sediments in the reservoir is nearly three times the annual water conservation support to the region (Zhuo Dakang, 1987).

The Yangtze, which is the longest river in China, is 6300 km long with a trillion (10^{12}) cubic meters of annual runoff and 180 million hectares of valley area (Wu Youzheng and Pan Jinlin, 1986; Wang Yongan, 1987b). This river system passes through the southern region, collecting 2400 million tonnes of soil sediments. About 680 million tonnes of sediments are deposited at the mouth of the river. The remaining deposits are in the river system, lakes, and reservoirs (Wu Youzheng and Pan Jinlin, 1986; Yang Yansheng *et al.*, 1987). For example, the large Dongting lake in the middle area of the Yangtze River has an annual input of 130 million tonnes of silt. About 70% of this silt is deposited on the lakebed and raises it about 3.5 cm annually. From 1949 to 1977, the water area, storage capacity, and navigable section of the lake have been reduced by 37%, 39%, and 31%, respectively (TMB, 1984; Wu Youzheng and Pan Jinlin, 1986). It is also estimated that about a thousand million tonnes of silt are deposited in reservoirs on the Yangtze River system annually, and about 390 million cubic meters of water-storage capacity are lost in the 20 largest reservoirs in the upper area of the Yangtze River annually because of sediment deposits. This reduces the total storage capacity about 1% per year (Wu Youzheng and Pan Jinlin, 1986). The waterway transportation distance of the Yangtze River system has been reduced about 40% because of sedimentation since the 1960s (Wang Zhan and Chen

Chuanguo, 1982). The silt content in some tributaries of the upper area of the Yangtze River system is more than 40 kg/m^3, which is higher than that of the Yellow River (Wang Zhan and Chen Chuanguo, 1982). Some scientists consider that the Yangtze River will become the second most hazardous river after the Yellow River in China (Wang Zhan and Chen Chuanguo, 1982).

Some rivers in the northern region also have serious problems with sediments. The annual average silt content in the Yongding River of this region is about 61 kg/m^3, which is 1.6 times the level in the Yellow River. About 600 million tonnes of sediments are deposited annually in the Guanting Reservoir near Beijing, which has reduced the reservoir storage capacity about 25% since the 1960s (TMB, 1984).

Population impacts and soil erosion

Some natural factors influence soil erosion intensities in erosion areas. For example, the Loess Plateau has sloping land with sparse vegetation, extremely loose soil, and intense rainfall of short duration. These natural factors have caused some soil erosion in the plateau for many centuries (Zhu Shiguang, 1987). If humans alter environmental conditions, then erosion intensities may be increased. The southern region of China is located in tropical and subtropical zones with high rainfall and luxuriant vegetation. If the vegetation were damaged, the steep slopes plus the intensive rainfall would cause serious erosion there. Generally, increased human activity because of rapidly growing populations will intensify soil erosion.

Based on historical information, an association between population numbers and soil erosion has been demonstrated for the Loess Plateau during the last 3000 years (Fig. 3.2). About 3000 years ago, the annual total erosion in the Loess Plateau was about 1100 million tonnes. This can be considered natural erosion without human impact. During the period from 1020 BC to AD 1194, the annual total soil erosion in the region increased to 1200 million tonnes. This slow increase was caused by both human activities and natural erosion. The annual total erosion in the plateau increased to almost 1700 million tonnes in the early part of the 20th century and to 2200 million tonnes during the last three decades (Jing Ke and Chen Yongzhong, 1983; Zhu Shiguang, 1987). These data indicate that erosion rates in the region increased only 7.9% during the more than 2000 years from 1020 BC to AD 1194, but that soil erosion rates increased almost 50% during the 20th century. Undoubtedly, the rapid increase in the erosion rates in this century are directly related to rapid population growth (Fig. 3.2). The human population in the Loess Plateau region has increased 37-fold since the Han dynasty (AD 25–AD 220) (Wu Zhenfeng, 1981; Jing Ke, 1986), and population growth rates in the region are still high. The total population in the region was 35.6 million in 1957 and had increased to 72.6 million by 1981, a doubling of the population in just 24 years (Tian Houmo, 1985).

More food requires more cultivated land

A growing population requires more total food, and increased food needs are met by increasing cropland productivity and expanding the cropland area. New improvements in agricultural technology in some regions of China have been slow. Thus, the people have had to rely on expanding the cropland area to increase food production. Every flat and fertile piece of land is used to plant crops in the mountainous regions. The inhabitants have even extended crop cultivation to natural pastures and forests on sloping land. Once the pastures and the forests are destroyed and cultivated, soil erosion increases. Expanding crop cultivation increased the soil erosion problem. When soil erosion eventually reduced crop yields on the marginal lands, the people had to destroy more pasture and forests, a vicious circle.

It is estimated that forests covered about 40% of the Loess Plateau during the period from the Qin dynasty (221 BC–207 BC) to the Northern and Southern dynasties (AD 420–AD 581). Forest cover declined to 33% in the

Fig. 3.2. The relationship between soil erosion intensity and human population growth in the Loess Plateau region of China (based on Wu Zhenfeng, 1981; Zhu Shiguang, 1987).

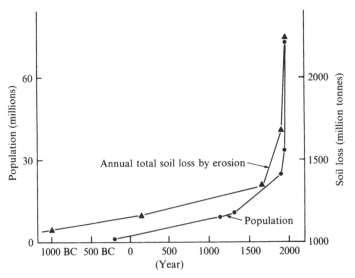

period from the Tang dynasty (AD 618–AD 907) to the Song dynasty (960–1279), and 15% in the period from the Ming dynasty (1368–1644) to the Qing dynasty (1644–1911). Forest cover in the region now is only 6.1% (Li Kaiming, 1984; Xu Zhemin *et al.*, 1985).

Forest cover in the middle area of the Yangtze River decreased from more than 50% before to 30% after the Ming dynasty (IGSP, 1987). Current forest cover is now less than 20% (Wang Yongan, 1987a). Before the extensive land cultivation about 100 years ago, forest cover in the northeastern rolling land region was about 15%, and little or no soil erosion occurred. Now, 70–80% of the total area of the region is cultivated, and about 85% of the original natural vegetation has been destroyed (Wen Dazhong *et al.*, 1978; Fu Dexing, 1986). It is estimated that western Liaoning, which is located in the northern region, had 30% forest cover about 200 years ago, but this forest cover was only 2–3% in the 1960s (Lei Chidi, 1983; Wen Dazhong, 1986).

Before the 1940s, no accurate statistics were available for evaluating the national forest resources in China. From 1973 to 1976, the second national forest investigation reported that the total forest area was only 121.68 million hectares, which accounts for 12.7% of the total national land area (TMB, 1984). In the period 1977–1981, the third national forest investigation reported that the total forest area had been reduced to 115.27 million hectares, or 12% of the total national land area. The forest areas in some major soil erosion regions are reported to be reduced about 20% (TMB, 1984). The total forest area in China is estimated (mid 1980s) to be only about 85 million hectares, only 8.9% of the total national land area (Han Yichi, 1987). Obviously, 36.42 million hectares of forest area, or 30% of total forest area, have been lost during a decade.

At the same time, the cultivated land area in China increased from 99 million hectares in the 1960s to 130 million hectares in the early 1980s (TMB, 1984; Liu Yingqiu, 1988). Most of the increased cropland area came from clearing forests, which were mostly distributed in the mountainous regions of China. Converting this type of forest land into marginal cropland causes serious soil erosion problems. Some experimental results indicated that water runoff rates were 30% greater than those of the forests, and soil erosion rates on sloping cropland areas were five to ten times greater than those of forests on the same sloping land (Jiang Dechi and Zhu Xianmo, 1962; Yu Chunzhu *et al.*, 1985).

More biomass burned as household fuel

Clearly, larger populations require greater quantities of household energy. The larger the biomass energy requirement, the more intense is the soil erosion problem. Total annual biomass production in China is estimated to be about 814 million tonnes, which provide about 3×10^{15} kcal of energy. About half of the total biomass is currently used as rural household fuel, and about 60% of all crop residues are burned as household fuel (Wang Changgui, 1985; Wen Dazhong, 1989). However, the household fuel supply in rural China is still less than 80% of the current needs. Because of the uneven distribution of biomass resources in rural areas, the household fuel in some areas, including the most serious soil erosion regions, meets only 40–50% of rural people's fuel needs (Zhang Qinghai, 1985; Wen Dazhong, 1987).

In these regions with serious soil erosion, crop and natural vegetation productivity is reduced. The people must collect nearly all the available biomass, including crop residues, leaves and branches of trees, grass roots, and manures for household fuel. Removal of this biomass further aggravates soil erosion. In the Loess Plateau, a 12-year-old child can collect about 50 kg of grass, roots, and shrubs in a day to meet his or her family's household fuel need for two or three days. However, collecting this biomass will destroy more than 200 m^2 of vegetation (Zhu Xianmo, 1984a). Burning large amounts of biomass reduces soil organic matter and makes soils more susceptible to erosion. The loss of organic matter in the loess has reduced soil bulk densities more than 1.2 g/cm^3 and the granular structure about 20%. Increasing loess water permeability will reduce erosion (Zhu Xianmo, 1984b).

In addition, the development of housing and other industries helps destroy vegetation and further intensifies soil erosion (Wang Guangren *et al.*, 1987; Gao Zhu and Zhou Lie, 1988).

Population impacts, soil erosion, and rural poverty
Serious erosion plus reduced biomass material and energy inputs in agriculture, forestry, and animal husbandry further reduce current low land productivity in rural areas. In the Loess Plateau region, for example, the annual grain yield is about 750 kg/ha, which is less that one-third of the national average yield. Now, one hectare of grassland can only feed one sheep for a year (Zhu Xianmo, 1984b; Zhu Junfeng, 1985; Tian Houmo, 1987). The annual average growth of timber volume in tree plantations is less than 1m^3/ha (Gao Qijiang, 1987). In the southern region, soil erosion on sloping land has reduced crop yields by 30–60% (Goa Zhu and Zhou Lie, 1988) and grass yields by 80% (Zheng Shaoxiang, 1988). The high population densities, serious soil erosion, and reduced land productivity are hindering economic development in these rural regions. During the period 1981–83, 225 counties in China were below the poverty level, i.e., with an annual

average per capita income of less than 200 yuan and a per capita grain production of only 200 kg per year. This accounts for 6.7% of the total national area and 8.1% of the total cropland area, and includes 86 million people. All of these counties are located in the serious erosion regions of China (Fei Xiaotong, 1986; Liu Xiang, 1986; Peng Fude, 1986b).

Soil conservation in China

Because of the long history of soil erosion in China, the Chinese people have accumulated much experience in soil conservation. The earliest record of soil conservation in China was 956 BC (Jiang Dechi and Zhu Xianmo, 1962). The major contribution of Chinese people to soil conservation was building terraced fields on sloping areas. Descriptions of the terraced fields were found in some literature on the Tang dynasty (AD 760) (Fang Zhengshan, 1960). Modern research work on soil conservation in China began in the 1930s (Guan Junwei and Wang Lixian, 1987), and large-scale soil conservation began in the 1950s. Soil conservation since the 1950s is examined below.

Government support and organizations

Soil conservation entered a new era in the 1950s when the Chinese government began to emphasize its importance. In 1952, the government issued a directive mobilizing the masses and promoting water and soil conservation. In 1957, the national program for erosion control was initiated. In 1963, the state council of China passed a resolution on water and soil conservation in the middle area of the Yellow River. In 1982, the state council issued 'regulations for erosion control,' in which general and specific policies on prevention, comprehensive management, research and education, and rewards and sanctions on erosion controls were stipulated. According to the regulations, the following policies for erosion control should be enacted: (1) erosion control should be an important component in rural development; (2) erosion control should be combined with the utilization and development of water and soil resources, promoting commercial production and well-being of the people in the regions; (3) the basic erosion control policy should devote equal attention to preventing and controlling erosion; (4) erosion controls should be combined with land-use policies; (5) there should be unified planning to provide suitable measures for the local community and environment to achieve comprehensive soil erosion controls; and (6) erosion control should eliminate disasters and contribute to the benefits and welfare of the people.

Under the leadership of the central government, local soil conservation organizations in the governments of provinces, cities, and counties have been established to implement the soil and water conservation regulations, to organize conservation plans, and to coordinate the local government departments of agriculture, forestry, water conservation, finance, and others to carry out successful soil conservation projects. Recently about 14 000 specialists have been working in these local soil and water conservation organizations. A total of 165 institutes and extension stations for erosion control have been established (Yang Zhenhuai, 1986). Several universities and colleges have established erosion control specializations to train soil and water conservation engineers. Based on the soil conservation experience during the last three decades, the central government has selected eight erosion areas with a total area of six million hectares as the focus of major efforts and has given these areas special financial and technical support for the conservation projects. This support is expected to have direct effects in reducing erosion in these areas during the next few years.

Some special work on erosion control has been carried out in China during the last decade. The special planning effort for erosion control for the Loess Plateau that was organized by the national planning committee is now finished. A map of national soil erosion in China is being constructed using remote sensing information (Yang Zhenhuai, 1986). The research project dealing with comprehensively controlling soil erosion in the Loess Plateau was chosen as one of the major national programs of science, technology, and development for the seventh national five-year plan. More than 600 scientists and engineers are involved in this project, and 11 experimental areas of small watershed control have been established (Jing Jiemin, 1988).

The Chinese government has provided significant financial support for erosion control. Annual expenditures by central and local governments for erosion control have totaled more than 160 million yuan. In addition, 30 million yuan of special support for the eight important control areas have been provided by the central government annually since 1983 (Yang Zhenhuai, 1986). Also, some departments of central and local governments have given financial support for various rural developmental programs in these erosion regions, which totaled more than 2000 million yuan annually. About a thousand million yuan of the total annual financial support are given to the Loess Plateau region for erosion control and some rural development programs. The total financial support in the Loess Plateau for these purposes has been more than 40 000 million yuan over the last three decades (Xie Lianhui, 1988).

Achievements in erosion control

According to statistics, about 30% of the total eroded area in China, or 46 million hectares, has been controlled since 1950. This area includes 25.7 million hectares of tree plantings, three million hectares of planted fodder grassland, and eight million hectares of terraced fields and dam-checked fields (Yang Zhenhuai, 1986). In the Loess Plateau region, about 10 million hectares of the eroded area, which accounts for 23.7% of the total erosion in the region, have been controlled since 1950. These controlled areas include 2.9 million hectares of terraced fields and strip-planted fields, 0.25 million hectares of dam-checked fields, 4.6 million hectares of tree plantings, 1.2 million hectares of established grasslands, and 0.7 million hectares of hills restricted only to trees and grassland (Bai Jinian, 1987). Erosion controls for small watersheds through a contract system of individual households have been launched along with the development of rural economic reform. Since 1980, about one-third of all households in the Loess Plateau region have contracted for erosion control of 1800 small watersheds with a total area of 5.3 million hectares and an annual control area of 0.6 million hectares (Wang Huayun, 1986).

According to some investigators, terraced fields could reduce the soil erosion rate by 80%, dam-checked fields could retain 45 000 tonnes of silt per hectare (Wang Huayun, 1984), and shrub plantings and grasslands on sloping areas could reduce soil erosion rates 70–80% (Chui Mengshen, 1985; Yao Huandou, 1985). In general, the erosion rates have been reduced about 50% in a comprehensive primary erosion control area in the Loess Plateau (Chen Yongzhong, 1988). It is estimated that various erosion control measures on the 46 million hectares of the controlled area could reduce soil loss by 600 million tonnes annually. In the Loess Plateau, about 600 million tonnes of soil are deposited annually in reservoirs. Thus, soil conservation measures could have major benefits for the Yellow River (Chen Zhilin, 1986; Wang Siangqun, 1986). These erosion controls have stabilized the Yellow River, which has not had a serious flood for three decades, even though the total soil erosion from the Loess Plateau has increased. This is very important for the safety of the people living near the lower reaches of the Yellow River.

This progress in soil conservation has not only controlled soil erosion, but has also produced several economic benefits. For example, the terraced field crop yields have doubled compared with non-terracing, and dam-checked fields have an even higher crop yield (more than 3000 kg/ha for grain) in some areas (Wang Huayun, 1984). According to some cost–benefit analyses of erosion controls in small watersheds, the ratio between total cost and total benefit for erosion controls is 1:2 or 3, for terraced fields about 1:4, for check-dams 1:1.2, for economic tree plantings 1:7, and for shrub plantings 1:10 (Zhang Jinhui, 1987). These erosion control benefits have accelerated economic development in the erosion area (Wang Siangqun, 1986; Xie Lianhui, 1988).

The development of erosion control techniques

Several erosion control methods have been developed in China. An array of soil and water conservation techniques can be used to control erosion in different crops and environments (Liu Wanquan, 1986). The major conservation techniques employed in China are outlined below.

1. Soil conservation for crop cultivation
   Some conservation cultivation techniques could greatly reduce soil erosion and increase crop yields on some lower slope croplands.

   (a) Contour ridge cultivation has been effectively employed for soil and water conservation on slopes of less than 7°. The erosion rates using ridge cultivation can be reduced 60 to 90%, and water runoff rates reduced 50%. Crop yields have been found to be 10–20% greater compared with downslope culti-vation (Hou Guangjun and Gao Huimin, 1982).

   (b) Contour ridge check requires that small banks be constructed every 2–3 m in furrows of contour ridges to retain soil and water runoff. This technique reduces erosion rates by about 80% and increases crop yields 7–30% on lands with an 8–20° slope (Jiang Dechi and Zhu Xianmo, 1962; Hou Guangjun and Gao Huimin, 1982).

   (c) Pit-cultivation requires that small holes 50 × 50 × 50 cm be dug with a distance of 50 cm between pits. Fertile soil and organic matter are placed in the holes and crops planted in the pits. This method is suitable for seriously eroded soils, and is often used to control erosion on extremely steep slopes and V-shaped gully walls.

   (d) Improving soils by deep plowing and using larger amounts of fertilizer is suitable for improving eroded soils, especially soil on newly terraced fields. However, deep plowing should be combined with other soil conservation measures to avoid losing any temporary benefits.

   (e) Crop rotations, intercropping and overcropping may be used for erosion control. These measures could increase plant cover on slopes to reduce soil erosion. Some agroforestry techniques may be employed, and the results are highly beneficial.

2. Planting trees, shrubs, and grasses
   Planting trees, shrubs, and grasses on eroded slopes and sloping fields unsuitable for food crops has good conservation effects and economic benefits. The often-used afforestation methods on eroded areas in China include higher edge terrace afforestation on 10–35° slopes (Fig. 3.3a), narrow terrace afforestation on 10–25° slopes (Fig. 3.3b), horizontal ditch afforestation on 25–30° slopes

Fig. 3.3. Afforestation techniques on eroded slopes in China: (a) higher edge terrace afforestation; (b) narrow terrace afforestation; (c) horizontal ditch afforestation; (d) fish-scale pit afforestation. Each picture shows a side view and an aerial view of an afforestation construction, respectively. (Based on Hou Guangjun and Gao Huimin, 1982.)

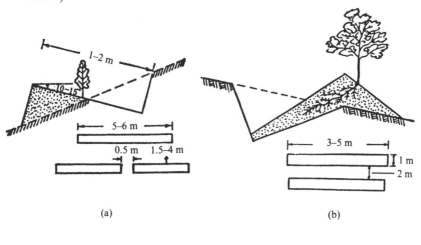

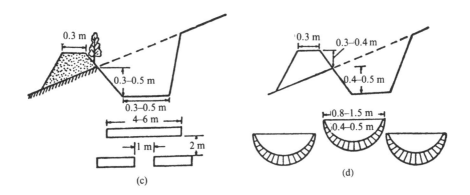

(Fig. 3.3c), and 'fish-scale' pit afforestation on steeper slopes (Fig. 3.3d). Afforestation by aircraft seeding has also been used in some remote areas in recent years. The methods of planting grasses on eroded slopes in China include contour ridge cultivation and strip interplanting fodder grasses combined with green manures and cereal grain crops. In addition, contour strip planting grasses and aircraft seeding of grasses are employed for soil conservation.

3. Engineering soil conservation measures

The most common engineering measures for erosion control in China include terracing, slope erosion control, and gully erosion control engineering. Terracing is the major and most effective measure for cropland erosion control on slopes of less than 25°. Terrace construction is expensive; 750–900 working days are required to construct terraces on one hectare employing only hand labor, and 15 working days where a 60-horsepower tractor is employed (Liu Wanquan, 1986).

Another erosion control is the construction of small contour dikes on gently sloping fields. Dikes 20–30 cm tall are constructed every 20–50 m, and contour ditches with a depth of 20–30 cm are established in front of the dike to control sediments and runoff.

Gully erosion control in the Loess Plateau region requires that check-dams be constructed across gullies to catch water and sediments. The captured silt forms the fertile new fields, which may have crop yields as high as three to five times the local average.

4. Comprehensive erosion control on small watersheds

Various erosion control measures should be used in combination in small watersheds for effective erosion control. In general, the area of a small watershed is less than 5000 ha. The current control standards for small watersheds in China are: controlling 70% of erosion; converting 80% of eroded bare land area into areas planted with trees and grass; establishing highly productive farms that are suitable for making the area self-sufficient in grains; and increasing people's incomes by *c.* 30–50% (Sun Jianxuan, 1985).

Recommendations

Although soil conservation in China since the 1950s has made great progress, soil erosion problems still exist. For improving erosion control, the following actions are needed:

Further investigation and understanding of the gravity of the soil erosion problem and the urgency for erosion control
Soil erosion is the most serious environmental problem in China today. Despite its achievements in soil and water conservation, China still has one of the most serious erosion problems in the world. Total soil loss in China accounts for one-fifth of the soil lost in the world. Erosion rates are still increasing more rapidly than control rates in the southern region and in the Loess Plateau region (Zhang Zhongzhi, 1988). Soil erosion has been a major hindrance to economic development in several regions. Unfortunately, few scientists and others in China understand the seriousness of this environmental problem. Thus, there is a need to conduct an active education program for all people, including policy-makers. Once the severity of the soil erosion problem is understood, full public support for action will exist.

Controlling population to lessen impact on agricultural and forest lands
The rapidly growing population in China is one reason for aggravated soil erosion. If the population continues to grow, the serious erosion situation will not be improved, no matter how many erosion controls are installed. Even though population control in China is highly successful, it will be impossible to limit the population to less than 1250 million by the year 2000 (Chen Yanhai, 1987). Unfortunately, population growth rates in the serious erosion regions are greater than the national average. The recent annual population growth rate in the Loess Plateau region, for example, is 2.2%, 50% greater than the national average (Tian Houmo, 1985; NSBC, 1987). It is impossible to control soil erosion without effective population control.

Integrating erosion control with economic development in regions with serious soil erosion
As more and more information is obtained on soil erosion, it is clear that a vicious circle exists between human welfare, population growth, and soil erosion. Everyone agrees that economic development is impossible without effective soil erosion control. But problems exist with tree planting and other erosion control projects that require several years to provide noticeable benefits. However, we have found that fast-growing shrub plantings and grasslands have greater acceptance than slow-growing tree plantings. The

reason is that local residents obtain benefits from shrubs and grasses in developing animal husbandry and household energy in a short time. Shrub plantings and grassland also have more rapid conservation effects than tree plantings.

There is a clear need to integrate some long-term benefits with short-term benefits, and to integrate the ecological benefits with economic benefits in soil and water conservation. Erosion controls should be considered an essential component of economic development. This integration would help to establish a sound relationship among humans, land, and economic development.

Support for erosion control

Most of the regions with serious erosion in China are also the poorest. Thus, it is necessary to support these regions not only for soil erosion control but also for economic development. Clearly, erosion control and economic development will not occur without some outside support. The evidence suggests that soil erosion harms not only the region where erosion takes place but also the more-developed regions that are downstream. Certainly, controlling 1 m^3 of silt on the lower part of the river costs about ten times more than if control were made on the cropland (Fang Zhengshan, 1985).

Implementing laws and regulations to protect and improve soils, land, vegetation, and environment.

In recent years, new legislation and regulations have been developed to protect the environment, forests, cropland, water and soil resources. The question is how to implement these laws to benefit China as a whole. Problems, for example, have developed in implementing some of these regulations in certain areas and stopping cropland expansion and forest destruction on steep slopes.

Developing rural household energy

The household biomass energy shortage in rural areas aggravates the damage to forests and pastures on slopes and to organic matter on cropland. Developing rural household energy will help prevent soil and water degradation and the start of a vicious circle. Development of fuelwood plantations will produce more household fuel and contribute to erosion control. Developing methane digesters and other equipment for solar energy will also help. Wind energy, for example, has advantages in some suitable areas.

Actively implementing erosion control policies

The basic policies of erosion control in China are sound. However, some problems in implementing such policies still exist. There is a need to pay more attention to river control, especially near the mouths of rivers. Attention should continue to be given to erosion control on croplands, especially the upper and middle river areas. More attention should be given to the use of engineering erosion control measures (Zhang Minghuan, 1984). A major effort is needed to protect existing natural vegetation; to plant trees, shrubs, and grasses; and to establish highly productive crops and terraced fields. Also there is a need to convert steep croplands into tree plantations and grasslands, to regulate land use for industrial construction, and to promote comprehensive small watershed erosion controls.

Strengthening research on soil erosion and conservation

Although China has a long history and experience in soil and water conservation, additional research is needed. Further research is needed on the distribution of soil erosion and its causes in China. Information is needed on the influence of soil erosion on crop and forest ecology and the environment, the sociology and economics of soil erosion, and the comprehensive utilization of natural resources in erosion regions. Investigation is needed on calculating potential erosion and on new techniques and measures for erosion control. This information will help promote soil and water conservation in China.

A little soil erosion occurs naturally, but it has increased with the growth in human numbers plus the clearing of land. Humans have intensified soil erosion and they should make efforts to control it. If humans can control their numbers and manage themselves, they can manage land and their living environment. Destroying the environment is always easier than controlling and restoring the ecosystem. Humans cannot survive if they destroy their land. It is to be hoped that the serious erosion in China and elsewhere in the world will eventually be controlled through continued cooperative efforts by Chinese people and people throughout the world.

4

A case study in Dingxi County, Gansu Province, China

L. MCLAUGHLIN

While examination of national erosion rates and patterns of erosion is enlightening and necessary, the full impact of soil erosion and the potential soil conservation to influence human lives cannot be fully appreciated without examining specific cases and specific regions. The benefits of soil conservation go far beyond the checking of environmental damage; they have a direct impact on the living standards of the individuals living in that environment. Particularly in developing countries, where efforts to maintain or increase rural living standards often result in environmental degradation, soil conservation has the potential and responsibility to become a force that contibutes to rural development. As has been noted by Wen Dazhong in Chapter 3, the Loess Plateau is one of the most severe erosion regions in China. The following discussion of soil conservation and its socio-economic effects in Dingxi County in the Loess Plateau examines the connections between the ecological benefits of soil conservation and the problems of rural development.

Dingxi County and the Loess Plateau

The Loess Plateau includes a total of 138 counties located in ten provinces including Gansu, Qinghai, and Ningxia. Around 80% of its total land area, 690 000 km^2, is affected by soil erosion. Although it is thought that the Loess Plateau was once largely forested, a history of intensive agriculture, over-harvesting of forests, and thoughtless development has left the Loess Plateau almost treeless. Population increase in more recent years has caused severe shortage of food and fuel and has even led to the digging up of grass and crop residues as people sought firewood and fuel, leading in turn to soil erosion and low agricultural yields. This cycle caused the Loess Plateau to become one of China's poorest regions.

Examples of deforestation can be found easily. In the Zhipanshan forestry area, the forested area decreased 20% between 1949 and 1972, and between 1960 and 1965 vegetational cover in the Ziwu Forest decreased 44.2%. Although severity of erosion varies from region to region, it is estimated that every year 0.5 to 1.6 cm of topsoil are lost per square meter (Ma Zijun, 1987).

Rainfall in the Loess Plateau is highest in the eastern part of the region, such as Shanxi, where yearly precipitation may reach 600 mm, and lowest in the west, where it may be less that 200 mm, as in parts of Ningxia. Rainfall tends to be concentrated in the summer months and severe thunderstorms are common. The high intensity of these storms and the runoff they create cause the deep gullies that are such a striking feature of the region.

Not all of Gansu is part of the Loess Plateau, but a significant portion of it is. In addition, almost all of the province is beset by some form of soil erosion. Of its 454 000 km^2 total area, the wind erosion area is 220 000 km^2 and the water erosion area 172 000 km^2, giving a total of 392 000 km^2 or 86.4% of the province. Gansu can be divided into five ecological regions: (1) the Long-dong Loess Plateau; (2) the Longzhong Loess Hills; (3) the Longnan Tushi Mountainous region; (4) the Gannan Plateau-prairie region, and (5) the Neilu Plateau Wind Erosion region (which includes the Hexi Corridor and the Southern Hexi Shishan region). All of the wind erosion occurs in this last region, and the first three regions have the most water erosion.

Broken down according to watershed, the Yellow River watershed accounts for 113 000 km^2 or 65.7% of the water-eroded area. The Yangtze River watershed accounts for 23 000 km^2, or 13% of the water-eroded area, and the Neilu watershed accounts for 36 000 km^2 or 20.9%. An estimated 644 000 000 tonnes of soil is eroded per year, 89% in the Yellow River watershed, 9% in the Yangtze watershed, and 2% in the Neilu watershed.

Dingxi County is located in the Longzhong Loess Plateau Hill region. Within that area, 36.9% of the land is used for agriculture, 8.1% is forested, 38.2% is barren slopes, and 16.9% is used for other purposes. It includes both arid and semi-arid areas, and account for 15.3% of the provincial area. Erosion affects 90% of the region, and each year more than 300 000 000 tonnes of soil are lost.

Dingxi's main agricultural constraint is lack of rain; the rainfall in the Longzhong Loess Hills ranges from 180 to 560 mm per year, but most counties average between 300 and 400 mm. Approximately 50% of the yearly precipitation falls between July and September. Drought is a serious and ever-present threat; in the years between 1956 and 1978, the seven counties that compose the Dingxi region had 18 drought years and only four years in which harvest was normal or above normal. During six years the harvest was

30–40% lower than average (Ma Zijun, 1987). The Dingxi area also lacks underground water resources. Even where there are underground water resources, it is difficult to undertake hydrology projects because of the area's many mountains and hills.

Dingxi County itself has an area of 3638.711 km^2 and an agricultural population of 340 000. Its altitude ranges from 2579 m down to 1700 m, with the southern part of the county mostly above 2300 m and the northern part mostly between 2000 m and 2100 m. It is characterized by its loess soil, which ranges from 20 to 100 m in depth. The bedrock below is usually either red clay, sandy clay, sandstone, or gravel.

Land use in Dingxi County heavily favors agriculture, with 46.82% used directly for farming, 5.8% used for forestry, and 19.93% used to grow grass. Such a distribution typifies the unbalanced economic structure found in the Loess Plateau. Although land is used mainly for agriculture, when analyzed on the basis of soil type and topography, prime agricultural land is extremely scarce, being only 3.3% of the agricultural land. On those lands, which are mostly flat and fertile and have access to water, grain yields are about 3750 kg/ha (500 jin/mu). More land is of fairly good quality, with a deep topsoil layer subject to only slight erosion. But this second category still accounts for only 25.11% of all agricultural land. Even though its dark loess soils and ortho-chernozems are fairly fertile, water supply is unreliable and droughts are frequent. Even in normal years, grain yields are only 1125–2250 kg/ha (150–300 jin/mu).

By far the largest amount of land 55.83% falls into the third category. Of only fair quality for agricultural use, these lands are subject to severe erosion and aridity. Although this type of soil is fairly easy to farm, its yields are usually only between 750–1125 kg/ha (100–150 jin/mu). The fourth category of land includes 9.39% of Dingxi's agricultural land, and is poor quality agricultural land with extreme soil erosion, aridity, and cold. This category of land yields only about 900 kg/ha (120 jin/mu). The remaining 6.37% of agricultural land is found on steep slopes, for the most part of greater the 25°. Erosion is extremely severe and a hardpan close to the surface makes farming very difficult.

The lack of good agricultural land in Dingxi is compounded by the climate. Rainfall is scarce, usually between 350 and 450 mm per year, but it is concentrated in the summer months and frequently comes in heavy storms that create runoff and erosion and render much of the rainfall ineffectual. More rain comes during the fall (autumn) than during the spring, and spring drought is a serious threat to Dingxi agriculture. Annual rainfall varies greatly; although average rainfall over the past 16 years was 415.09 mm, the standard variation was 90.39 mm. In addition, the growing season is short, only 141 days. In the northern part of the county the frost-free period is slightly longer, 146 days, but in the southern part, because of higher altitudes, it is only 112 days.

Conservation methods and their effects

Methods that attempt to control soil erosion in the Loess Plateau must also address the instability of rainfall and the problem of soil quality. Many different methods have been used for these purposes in northern China. These can be roughly divided into biological, engineering, and agricultural conservation methods. The basic biological methods of erosion control are afforestation and planting of grasslands, often on lands that were once used for agriculture. The call to 'return agricultural land to grass and trees' has been a major component of conservation strategies throughout the country, and the government prohibits the use of land with a slope of more than 25° for agriculture.

The grasses that have proven most effective in the Loess Plateau are *Medicago sativa, Onobrychis viciaefolia,* and *Astragales adsargeus.* Among the trees and bushes that have had significant use in the 'greening' of the Loess Plateau are *Caragana* sp. The species most often used is *Caragana microphylla,* or 'little ningtiao.' Ningtiao's adaptability and hardiness make it practically ideal for use in the Loess Plateau. It is highly resistant to cold, not showing any frost damage even at −35°C, and it grows extremely well under arid conditions. Even when yearly precipitation is only 100 mm, it grows normally, and, in fact, seems to do even better in drought years than in normal years (Guo Xiulin, 1986).

The effects of these biological measures are quite clear. Xifeng Soil and Water Conservation Station's experiments with planting strips of alfalfa show that, after ten years, yield in the areas where runoff had been trapped by the alfalfa increased by 9% to 20% (Huanghe Committee Xifeng Station, 1982). Research in Dongzhiyuan-Wangjia watershed showed that during storms alfalfa in the 40–50 m erosion furrow was able to catch 3 tonnes of silt. Research in Dingxi confirms that planting of strips of grass controls soil erosion and increases yield (Dingxi Area Conservation Station, 1983).

One drawback to planting grass is that in the year that the grass is planted, because ground cover is fairly low, erosion may actually be high. Research by Li Song (1987) in Longdong's Qingyang County illustrates this point. Research showed that vegetation coverage provided by different grasses was able to decrease the influence of water droplets on the soil and to maintain and protect soil structure. At the same time the grass's root structure improved the physical characteristics of the soil, enabling more water to seep into the soil and decreasing the amount of runoff. However, areas newly planted to grass had less vegetational coverage and more runoff (see Table 4.1).

The same phenomena occur when heavy rain follows the harvesting of grasses or bushes for firewood. Although grassland erosion is normally less than that of agricultural land, after grasses are harvested runoff may be significantly higher than that of cropland in full growth. Li's research also showed that while perennial grasses decrease loss of minerals to soil erosion, crops and newly planted grasslands show high losses. On average, the percentage of organic matter, total nitrogen, effective nitrogen, phosphorus, potassium, calcium, magnesium, zinc, copper, manganese, and fluorine contained in the soil carried away by runoff was less than that in the topsoil (0–5 cm) if perennial plants or trees were grown. For crops and first-year grasses, however, the percentages contained in runoff were actually higher than those in the first five centimeters of topsoil, with the exception of phosphorus, magnesium, and manganese. Phosphorus, magnesium, and manganese were 36.2%, 3%, and 9% lower in the runoff than in the topsoil (Li Song, 1987)

Aside from simply planting strips of grass, crop rotations with grass were also used to control soil erosion in the Loess Plateau. Research in Gansu's Tianshui (Chu Wenguang, 1981) compared the commonly used agricultural practices to an improved crop rotation sequence and to intercropping grass and crops at different slopes and found that the intercropping was more effective at controlling erosion than either the commonly used practices or the improved rotation, probably because it increased ground cover most effectively. The steeper the slope, the more obvious grass intercropping's superiority (see Table 4.2). On gentle slopes, the improved rotation and rising of ridges was most effective, but I will discuss that method in the agricultural methods section. Similar results to those of Chu were obtained by Mo Shiyu (1981). Two patterns of intercropping alfalfa were compared to the standard cropping pattern of the region. The two cropping patterns were:

    1. Lentil (*Dolichos lablab*)/alfalfa → alfalfa → alfalfa,
    2. Lentil/alfalfa → alfalfa/millet → alfalfa,

and the standard regional cropping pattern was:

    3. Lentil → winter wheat/buckwheat → corn/soybeans.

Although in the first year the intercropped grass pattern showed high erosion levels because the grass did not grow well, the second and third year proved

Table 4.1. *The relationship between percentage land cover by various crops and rate of soil erosion. (Adapted from Li Song, 1987)*

	1st-yr beans	1st-yr alfalfa	1st-yr Onobrychis	Natural grass	2nd-yr Onobrychis	Established alfalfa	Established apricot trees
Erosion (kg/m²)	0.80	0.91	0.86	none	slight	none	0.14
Land cover (%)	10	10	10	100	50	90	50

Table 4.2. *Soil conservation effect of grass-crop rotations. (Adapted from Chu Wenguang, 1981)*

		Slope (°)			
	Treatment	5	8.10	13.20	17
Runoff (m³/ha)	Std rotation	219.64 (100%)	1263.45 (100%)	185.59 (100%)	164.32 (100%)
	Improved rotation	3.37 (1.5%)	7.83 (3%)	45.38 (24.5%)	21.51 (13.2%)
	Grass intercrop	15.1 (6.9%)	22.17 (8.4%)	19.14 (10.3%)	9.79 (6%)
Erosion (t/ha)	Std rotation	3.4 (100%)	4.06 (100%)	6.34 (100%)	7.6 (100%)
	Improved rotation	0.08 (2.4%)	0.13 (3.2%)	1.34 (21.2%)	0.02 (8.2%)
	Grass intercrop	0.2 (5.9%)	0.54 (8.4%)	0.57 (9%)	0.76 (10%)

its ability to decrease both runoff and soil loss (see Table 4.3). Comparisons of the common agricultural method with planting pure grass on steep slopes over a 13-year period and a 3-year period also showed that grass was far superior to standard agricultural methods at decreasing runoff and soil erosion.

From the data in Table 4.3 it is obvious that planting grass is an effective way of controlling soil erosion. In addition, it is inexpensive and does not require much time compared to using other conservation measures on the same land area. Whereas building terraces requires 900 man-days per hectare and costs the government 30–450 yuan per hectare, planting grass only costs 45–75 yuan per hectare and takes farmers only 45 man-days per hectare. From the farmer's point of view, it is a major saving. Nor are the actual effects of planting grass or shrubs limited to controlling soil erosion. To farmers, its most important effect is probably its ability to lessen or eliminate entirely the fuel crisis farmers now face. According to research done in Guanxing Cha, Dingxi County (Gansu Province Hydrology Department, 1986), the average family uses 4138 kg of fuel per year and the average individual uses 300 kg per year. In recent years, because of abundant rainfall, grain yield has been fairly high and stable, but fuel has still been scarce. As a result, in areas where grass

Table 4.3. *A comparison between standard rotations and legume crop rotation for the control of erosion. (From Mo Shiyu, 1981)*

Year	Area	Crop	Runoff (m^3/ha)		Erosion (t/ha)	
1955	20	beans	48.0		3.12	
	21	beans and alfalfa	72.58		3.76	
	22	beans and alfalfa	69.68		2.99	
1956	20	winter wheat and oats	458.7		16.7	
	21	alfalfa	110.4		10.7	
	22	alfalfa and millet	73.48		5.09	
1957	20	corn and soybean	62.5		0	
	21	alfalfa	0		0	
	22	alfalfa	0		0	
Three-year average				(%)		(%)
	20		189.7	100	9.7	100
	21		60.89	32	4.82	50
	22		47.72	25	2.59	28

had not been planted, families had grain to eat but had nothing to cook it with.

Engineering methods of controlling soil erosion take many forms, but among the most important are level ditches, level terraces, and reverse-slope terraces. Actually, there are many other engineering projects that are major parts of development work and which are almost always done in conjuction with erosion control measures, but which do not actually have a direct effect on soil erosion. For example, dry wells do not directly affect soil erosion but have a very significant effect on rural lives because they provide the drinking water the rural people need.

Terraces are perhaps the most effective and widely used engineering measures. In general, it has been found that terraces in the Loess Plateau are able to retain between 20 and 40 m^3 more runoff than either grass or trees (Chen Zhangling, 1985). If terraces or ditches are level, they are able to effectively catch all the runoff (Anon., 1986a). In Dingxi's Anjiagou, on 4 June 1963, with a storm in which 101.4 mm of rain fell at a rate 231 mm/hr, level terraces and reverse-slope terraces basically retained all the rain, whereas in untreated slopes between 24.75 and 40.20 m^3 of runoff was carried depending on the amount of vegetation and the steepness of the slope (Ye Zhenou *et al.*, 1986). If, however, terraces are uneven or if animals have dug holes in the terraces or ditches, parts of them can be washed away by runoff coming from unterraced areas (Anon., 1986a). Where ridges are raised around the sloped fields, a large part of the eroded soil can be retained as well (Anon., 1986).

Although terraces and ditches are able to conserve soil and water, their effect is not limited to conservation. They have a positive effect on the entire ecological system, increasing soil moisture content and nutrients, and im-proving the soil's physical characteristics. In the Loess Plateau, where water shortage is the key threat to agriculture, soil moisture content is extremely important. Long-term research at Dingxi's Chankou Forestry Station showed that all forms of engineering control (including level terraces, reverse-slope terraces and level ditches significantly increased moisture content at soil depths of 0–1 m. Level terraces and reverse-slope terraces showed the largest increase, 2–5% more than other methods. Comparison made between different widths of reverse-slope terraces revealed that within 1 m depth, moisture content was greatest when terraces were 2 m wide, and least when 0.7 m; the 2-m wide terraces had 3% more moisture than the 1 m or 1.5 m terraces and 7% more than 0.7 m terraces (Chankou Forestry Station, 1986). Research in Dingxi confirmed that sloped land's water-use coefficient was 30% or more higher than that of terraced land, i.e. six years of research showed that whereas level terraces require approximately 2.19 mm of rain to produce 1 jin of grain, sloped land requires 3.31 mm (Ye Zhenou *et al.*, 1986).

In the long term, soil conservation methods have been shown to increase the amount of nutrients in the soil. Comparison at Chankou (1986) among level terraces, reverse-slope terraces, level furrows and untreated slope land, showed that reverse-slope terraces showed the highest increase in soil nutrients. The same research also showed, however, that during the first year all methods of treatment showed some decrease in soil fertility. The reason for this drop in fertility appears to be the loss of topsoil caused by soil preparation. By preserving topsoil when preparing terraces, level furrows, or other projects, soil fertility can be increased. Research in Dingxi (Chen Zhangling, 1985) showed that when topsoil was not kept organic matter decreased 7.1%, total nitrogen decreased 10.6%, water-stable aggregates decreased 42%, and soil aggregates decreased 50–68% compared with terraces in which topsoil was preserved. As a result yield decreased between 8 and 56%. Researchers in other areas have found similar results (Anon., 1986b).

The overall effect of increased nutrients and moisture content is a significant increase in yield. In areas where trees formerly could not be grown, level furrows enable trees, bushes, or grass to flourish. Terraces, both level and reverse-sloped, have significant effects on crops, as do level ditches when used for crops. The effect tends to be more in drought years than in wet years. When the yield obtained by millet and wheat planted in level ditches and on slopes was compared, in each case where direct controls were used, yield was at least two times higher in the ditches than in the control areas. Similar results were found in Huining and Qinan counties. Although both terraced and sloped land's yield varied from year to year because of rainfall variation, terraced land still gave consistently higher yield over a wide variety of crops (Anon., 1986; Yang Sheng and Lian Sheng, 1987).

After comparing soil moisture levels, soil fertility, the soil's physical characteristics, and yield after various conservation methods were implemented, researchers in Dingxi's Chankou Forestry Station concluded that reverse-slope terraces were the most appropriate method for use in Dingxi County because of their ability to conserve soil and water and concentrate soil nutrients (Chankou Forestry Station, 1986).

Farmers are quite aware of the superiority of terraces, whether reverse-sloped or level, and during land distribution fight to get more terraced land. One major drawback, however, is terracing's labor requirement: terracing one hectare of land requires approximately 900 man-days of labor, not including time for planting crops and for later maintenance. As a result, while farmers recognize terraces' superiority, they are not always willing to do the work to build them.

The final types of soil conservation techniques used are farming methods. In the 1950s, considerable emphasis was given to agricultural methods of controlling soil erosion, but in the 1960s and 1970s and on into the 1980s increasing attention was paid to terracing and conservation agriculture was to a large extent overlooked. In the last five to ten years, however, conservation agriculture has once again begun to come into its own. There are myriad of methods used in China today, but I will confine myself to some of the more widely practiced ones.

Perhaps the simplest farming method is contour cropping. Although this may seem an obvious and simple method, according to investigation in Shanxi, 60% of all sloped farmland and 'terraces' have croplines which are actually running downslope. This causes runoff to accumulate and form rills. During four storms in June and July of 1981, during which 152.2 mm of rain fell at a maximum strength of 25.2 mm/hr, sloped planting resulted in runoff 12 times that of evenly planted crops on 17° hills, and 4.2 times that of evenly planted crops on 20° hills (Zhang Maosheng, 1983).

A second important method commonly used in the Loess Plateau is 'deep plowing'. Research in Xifeng found that tilling 20–22 cm deep decreased runoff 10–47% and erosion 16% over soil plowed to 13–15 cm, probably because more rainfall could be absorbed by the soil. In fact, soil moisture increased about 43% and yield 13–29% (Huanghe Committee Xifeng Station, 1982). Research in Shandong found that plowing to a depth of 27–40 cm caused moisture content to increase 46%, but caused soil volume–weight to decrease 12.5%. Water infiltration increased 13.3 times and aeration also improved. Micro-organisms, particularly ammonifying bacteria, increased significantly as well (Liu De, 1983).

In general research has found that deeper plowing increases soil moisture content and that plowing three times ensures the highest soil moisture content; if soil is plowed more than three times, moisture content begins to decrease (Anon., 1986c). Chankou research (1986) added, however, that although in the first year the deeper the plowing the higher moisture content, in the second year fields plowed to different depths showed no significant difference in moisture content. Deep plowing needs to be combined with other methods, particularly increasing fertilizers, in order to ensure high yields. In addition, sometimes deep plowing will have a positive effect on summer crops but will have a negative effect on fall crops (Huanghe Committee Tianshui Station, 1983).

A third type of arid agriculture conservation method is the raising of ridges. There are many variations on this theme: level casting followed by various types of raised ridges. All involve raising ridges either before or after planting. Promoted as early as the 1950s, these methods are able to prevent soil from being washed away and increase yields, and are appropriate for slopes of less than 25°. Their main drawback is that when ridges are raised moisture can be lost; because evaporation is increased, these methods can increase aridity, and in certain years yields may actually decrease (Huanghe Committee Tianshui Station, 1983). Level casting followed by ridge raising eliminates this problem by raising ridges after seedlings have had a chance to begin growth. Care must be taken to raise ridges before major rains come, however, otherwise their effectiveness as a soil conservation method will be lost entirely.

There are many other methods developed to conserve soil and water in the Loess Plateau. In general, agricultural conservation methods usually require 15–30 extra man-days/hectare and can increase yield up to 100%. A partial list of these methods and their effects are given in Table 4.4. For the most part these methods are not as effective as terraces at controlling soil erosion, but in areas without terraces they are simple ways to decrease soil erosion and increase yield. Evaluating different methods of conservation agriculture, the Huanghe Committee Tianshui Station (1983) suggested that, overall, ridges were the most appropriate for the Loess Plateau, particularly when raising of ridges was done after plants had been given a chance to begin growth. Most authors agree that these methods should be used in combination with other

Table 4.4. *The effects of different soil conservation measures in the Loess Plateau (From Yang Songwang, 1983)*

Method	Runoff captured (%)	Decrease in erosion (%)	Yield increase (%)
Level terraces	89.2	96.3	—
Sloped terraces	36.6	7.7	—
Contour farming	64.2	79.9	25
Grain crop–legume intercrop	30–40	30–70	37.5
Deep plowing	76.3–92.3	82.6–86.2	11–12
Raised ridges	71–98	84–99	10–25
Crop–grass strip intercrop	60.5	65.7	50–100
Ridges raised after crop seeding	79	29–51	10–38.5
Grain–grass rotation	60.8	65–68	30–100

erosion control methods. Because of the large amount of labor required for terracing, agricultural conservation methods such as these provide a highly viable alternative.

Social and economic effects of soil conservation

It seems logical that the effects of various soil conservation methods at controlling erosion must surely in turn affect the living standards of the people dwelling in once-eroded areas. Yet the social and economic effects of soil conservation are often difficult to measure. One way to see the effect conservation has is to compare measures of living standards between areas where conservation has been carried out and areas where it has not. By comparing data both before and after conservation work began, some indication of the influence of erosion control can be found.

Fortunately, in China extensive soil conservation projects have been carried out and relatively complete economic data are available. The following is a comparison of the social and economic effects of soil conservation at three levels: the watershed, the village, and the family. Comparison is made between two watersheds that have carried out extensive soil conservation, namely Shijiacha and Guanxingcha, and one in which little work had been done, Shuangle. In Shijiacha primary emphasis had been placed on biological methods of conservation, whereas in Guanxingcha emphasis had been placed on engineering methods.

At the village level comparison is made between two adjacent villages, one within the Shijiacha watershed and one just outside the watershed border. In Daying, the village within Shijiacha, extensive work on soil conservation had been done, whereas in Xujiamen, the village outside the watershed boundary, very little had been done. Finally, analysis is made of data from 30 families within the Guanxingcha watershed. Unfortunately data was only available for the years 1983–85, so no comparison could be made with family living standards before the conservation project began.

The rural people of Dingxi have long faced an almost constant threat of starvation. Normally low agricultural yields compounded by frequent drought have made even basic subsistence difficult to maintain, and the government has often been forced to send grain and even water to Dingxi. One of the benefits of conservation is the effect of soil conservation on grain production. The data and results of analysis of variance for Daying and Xujiamen are shown in Table 4.5. As can be seen in the table, before conservation measures were implemented, no significant difference existed in per mu grain production between the two villages. Significant variation from year to year is evident, but this is to be expected as it reflects the yearly variation in weather that so strongly affects agricultural production. After conservation, however, the difference in production between the two locations is significant, and although yearly variation due to climate is still evident, it is not as striking as before conservation.

The same result is found at the watershed level. Although data were insufficient to compare grain production in the watersheds before conservation, Table 4.6 shows the data and analysis of variance of Shuangle and Shijiacha data after conservation. Because Guanxingcha grain production data was available only for 1983–5, it was not included in the analysis. As with grain production at the village level, both location and yearly weather conditions produce significant variation in grain production. That is, watersheds and villages that have carried out conservation measures have significantly higher grain production than areas without conservation, although grain production varies from year to year regardless of whether conservation methods are used.

Soil conservation also exerts a significant influence on income. As shown in Table 4.7, the two villages Daying and Xujiamen exhibited no significant variation in income before conservation, although, once again, there was significant yearly variation in income. As the incomes analyzed included only agriculture, animal husbandry, and forestry-related incomes, and because the economy in both villages was heavily dominated by agriculture, it is not surprising that income varied from year to year, just as agricultural production does. After conservation, however, the situation was quite different. Not only did location influence income, indicating that conservation significantly increased income, but yearly variation was no longer significant. That is, soil conservation significantly increased and stabilized rural incomes.

Although longer-term study would be needed to confirm the stabilizing effect of conservation on rural incomes, there are two possible causes for the stabilization observed. First, soil erosion has been shown to decrease soil's ability to retain water (Pimentel *et al.*, 1987). By reversing this trend, soil conservation measures enable plants to make better use of precipitation. As a result, production would be less affected by drought than in areas where conservation had not been carried out, and incomes would become more stable. A second possible explanation is that soil conservation was accompanied by a restructuring of the village economy, with the result that a smaller percentage of village income came from agriculture. Because animal husbandry and forestry are probably less dramatically affected by drought than agriculture, incomes would remain more stable.

Table 4.5. *Agricultural yield before and after soil conservation in Daying and Xujiamen*

(a) Agricultural production (kg/mu) before conservation

Year	Daying	Xujiamen	Total
1976	46.34	47.48	93.92
1977	61.76	63.42	125.18
1978	20.28	19.58	39.86
1979	26.51	21.81	48.32
1980	54.02	46.43	100.45
1981	20.25	19.44	39.69
Y	229.16	218.16	447.32
Y^2	10403.97	9669.16	40061.31

(b) Analysis of variance before conservation

Source	df	SS	MS
Location	1	10.08337	10.08337
Year	5	8356.0842	671.2168
Other	5	32.3663	6.4733
Total	11	3398.5339	

$F_{location} = 1.5577$ n.s.　　　　$F_{0.01(5,5)} = 10.97$
$F_{year} = 103.69^{**}$

(c) Agricultural production (kg/mu) after conservation

Year	Daying	Xujiamen	Total
1982	45.32	16.45	61.77
1983	115.93	77.8	193.73
1984	111.19	35.85	147.04
1985	145.91	81.18	227.09
1986	124.92	74.51	199.42
Y	543.27	285.79	829.06
Y^2	64751.618	19750.598	154309.8

Table 4.5. (*contd.*)

(d) *Analysis of variance after conservation*

Source	df	SS	MS
Location	1	6629.5957	6629.5957
Year	4	8420.8518	2105.2129
Other	4	717.7262	179.43
Total	9	3398.5339	

$F_{location} = 36.948**$ $F_{0.01(1,4)} = 21.20$
$F_{year} = 11.78*$ $F_{0.05(4,4)} = 6.39$

Table 4.6. *Grain production after soil conservation. Comparison of Shuangle and Shijiacha watersheds*

(a) *Grain production after conservation (kg/mu)*

Year	Shuangle	Shijiacha	Total
1982	20.51	29.05	49.56
1983	63.98	77.75	141.73
1984	46.02	73.5	119.52
1985	64.18	97.4	161.58
1986	51.50	81.2	132.7
Y	246.19	358.9	605.09
Y^2	13403.2633	28371.415	41774.7683

(b) *Analysis of variance*

Source	df	SS	MS
Location	1	1270.35441	1270.35441
Year	4	3659.61084	914.90271
Other	4	231.32224	57.83056
Total	9	5161.28749	

$F_{location} = 21.96665**$ $F_{0.01(1,4)} = 21.2$
$F_{year} = 15.8204*$ $F_{0.05(4,4)} = 6.39$

What is the nature of the relationship between income and soil conservation, and what influence do the different types of conservation measures
have on income? Statistical analysis of rainfall and percentage of land on
which conservation was being practiced did not produce a significant regression in Shijiacha, nor did analysis of after-conservation data from

Table 4.7. *Income and conservation. Analysis of Daying and Xujiamen
villages before and after conservation*

(a) Total income before conservation (yuan)

Year	Daying	Xujiamen	Total
1976	29606	29945	59551
1977	35276	34531	69807
1978	15967	10900	26867
1979	16428	13046	29474
1980	31523	26835	58358
1981	16560	10978.8	27538.8
Y	145360	126235.8	271595.8
Y^2	3045598414	3218752376	1.417×10^{10}

(b) Analysis of variance, before-conservation data

Source	df	SS	MS
Location	1	30477919	30477919
Year	5	939936785	187987357
Other	5	14983271	29966542

$F_{location} = 1.02$ n.s.
$F_{year} = 6.27^*$ $F_{0.05(5,5)} = 5.05$

(c) Income after conservation (yuan)

Year	Daying	Xujiamen	Total
1982	40199	10900	51099
1983	96378	39580	135958
1984	87580	21905	109485
1985	106860	40915	147775
1986	91309	34830	126139
Y	422326	148130	570456
Y^2	3.83×10^{10}	6.619×10^9	7.083×10^{10}

Xujiamen and Daying. When before-conservation data was added to the equation, the standard regression became

$$y = 0.05x_1 + 0.76x_2,$$
$$\text{with } F = 5.01^{**}, R^2 = 0.5975,$$

where x_1 is rainfall (mm), x_2 is the percentage of land on which conservation was being practiced, and y is the standardized village income. That is, when all other factors are held constant, a one unit increase in rainfall produced a 0.05-unit increase in income, whereas a one unit increase in conservation produced a 0.76-unit increase in income. R^2, which indicates the percentage of variation explained by the regression, was 0.5975.

The analyses of after-conservation data in Shijiacha and for the two villages, Daying and Xujiamen, although not statistically significant, both showed a higher influence of rainfall on income. The decrease in the influence after the addition of data from before conservation may indicate that the relationship between rainfall and income is different before and after conservation, or it may merely reflect the complexity of the relationship. Total annual precipitation is not actually a good measure even of agricultural production, as uneven distribution of rainfall may cause crops to experience spring drought even during a year with high precipitation, or to experience no drought despite low annual precipitation.

Taking the analysis one step further, separation of conservation methods into engineering methods (mainly terracing) and biological methods (planting grass or trees), produced the following two-factor standard regression equation for after-conservation data:

$$y = 0.976x_1 - 0.160x_2,$$
$$F = 8.98, R^2 = 0.72,$$

where x_1 is the percentage of terraced land, x_2 is the percentage of land

Table 4.7. (*contd.*)

(d) Analysis of variance, after-conservation data

Source	df	SS	MS
Location	1	7.518×10^9	7.518×10^9
Year	4	2.83×10^9	718392500
Other	4	2.016×10^9	504091250
Total	9	1.24×10^{10}	

$F_{location} = 14.95^*$
$F_{year} = 1.43$ n.s. $F_{0.05(1,4)} = 7.71$

using biological control measures, and y is the standardized village income ($\times 10\,000$ yuan). Considerably more variation is explained by this equation than the previous one, 72% as opposed to only 60%. When annual rainfall is added into the regression, it becomes

$$y = 0.354x_1 + 0.985x_2 - 0.283x_3,$$
$$F = 9.90^{**}, R^2 = 0.83,$$

where x_1 is the annual rainfall (mm), x_2 is the percentage of terraced land, x_3 is the percentage of land using biological conservation methods, and y is the standardized village income. Although the increase in explained variation obtained by adding rainfall into the equation is not significant in itself, it still raises the explained variation R^2 to 83%.

It is possible that the two villages have different regression equations, in which case their data should not, in fact, be pooled. Using only Daying data, I obtained the standard regression:

$$y = 0.145x_1 + 1.126x_2 - 0.300x_3,$$
$$F = 8.27^{**}, R^2 = ?,$$

where x_1 is annual rainfall (mm), x_2 is the percentage of terraced land, x_3 is the percentage of land using biological control measures, and y is the standardized annual village income. In this equation, the influence of biological conservation measures and terracing remained much the same as in the equation for the pooled Xujiamen–Daying data, but the influence of rainfall decreased. Here again it seems likely that this decrease is due to the unreliability of annual precipitation as a predictor of income. Since fewer data were used to calculate the regression, it is not surprising that the influence of rainfall appears smaller.

The negative relationship between income and biological conservation measures may at first seem perplexing, and may be considered a fluke. But the same negative relationship also held true in the Shijiacha watershed as a whole. At first this seems counter-intuitive. If one considers the nature of income calculation, however, the reason becomes clear. In calculating income, grain is valued at 0.60 yuan/kg, whereas grass, the primary biological measure, is valued at only 0.06 yuan/kg. Because during soil conservation the amount of agricultural land decreases as steep slopes are returned to grass, increasing biological conservation almost always means a change from planting grain to planting grass. As a result, total income decreases if biological methods of control are used alone. In actual practice, however, because biological methods are used in conjunction with terracing and other engineering methods of soil conservation, income increases after soil conservation methods are implemented. In addition, biological control methods have other direct advantages to farmers, which will be discussed later in this chapter.

One might reason that increased raising of livestock on land newly planted to grass should offset the loss of agricultural income, but during the years in question this was not the case. The soil conservation project in question coincided with the implementation of the individual responsibility system, and because prior to the implementation of the individual responsibility system livestock were raised by the collective, more livestock were actually raised per capita before than after implementation. Furthermore, livestock are often not allowed to graze newly planted grass or shrubs, so the correlation between number of livestock and amount of land planted to grass is not very good. Were animal husbandry to become a more significant part of Shijiacha's rural economy, we might find that biological conservation methods had a positive effect on income. But at the present stage of Shijiacha's economic growth, biological conservation has a negative effect on income. When the same analysis was carried out for Guanxingcha, however, the relationship found between biological conservation measures and income was positive, perhaps indicating that farmers in Guanxingcha have been better able to make use of land planted to grass or trees. The regression for Guanxingcha for conservation methods and income without rainfall was:

$$y = 0.70x_1 + 0.29x_2,$$
$$F = 77^{**}, R^2 = 0.97,$$

where x_1 is the percentage of terraced land, x_2 is the percentage of land using biological control measures, and y is the standardized annual village income. Why biological measures should have a positive effect on income in Guanxingcha while having a negative effect in Shijiacha is not clear. Further research is required to see if grass and trees are being used differently in the two areas.

The relationship between income and conservation, however, is only a partial indicator of conservation's effect on rural living standards in Dingxi. As mentioned earlier, the people of Dingxi have lacked not only grain, but also fuel and animal fodder. Before soil conservation, farmers would often be forced to dig up crop residues and grass roots in order to have sufficient fuel to cook their food and keep warm. The biological methods of soil conservation directly address the fuel and fodder shortages experienced by the rural people and therefore have a significant effect on the standard of living. Using the data from Guanxingcha, I developed a standard of living index. Although overly simplified, it does give some indication of the changes in the quality of living of the people of Dingxi.

The three main rural crises that soil conservation is able to address are those of food, fuel, and fodder. The standard of living index attempts to measure the fulfillment of food, fuel, and fodder needs by dividing the amount of food, fuel and fodder available per person by their respective per capita needs and summing these quantities. Because 500 jin/person/year (250 kg) is recognized as a minimum grain requirement for impoverished regions, that figure was used as the per capita grain requirement. For fuel, 600 jin/person/year (300 kg/person/year) was used, a figure derived from research done in Guanxingcha. Recent analysis has suggested that fuel consumption per capita may actually be as high as 600 kg/person. Finally, to determine animal fodder needs, the average number of sheep, pigs, and other large animals per capita per annum was calculated for Daying and Xujiamen from 1976 to 1986, then the result was multiplied by the amount of fodder required by each type of animal, as determined by investigation in Dingxi's Gejiacha (Ma Buzhen *et al.*, 1986).

The living standard index thus derived is:

$$LS = \text{per capita grain}/500 + \text{per capita fuel}/600 + \text{per capita fodder}/1340.$$

Calculating overall living standard in Guanxingcha using data available for the years 1969–73 and 1983–85 produced the following regression:

$$y = 0.370x_1 + 0.537x_2,$$
$$F = 11.16^{**}, R^2 = 0.8170,$$

where x_1 is the percentage of terraced land, x_2 is the percentage of land using biological conservation methods, and y is the standardized standard of living index. Thus, both terracing and biological methods of conservation have positive effects on the standard of living, but the effect of biological erosion control is actually greater than that of terracing.

Performing the same analysis using three years of data from 30 families in the Guanxingcha watershed did not produce a significant regression. One reason for this is that only after-conservation data were available. Possibly after conservation the differences between families were too small to be significant or were not linear. A second possibility is that after all families had implemented conservation practices, other factors, such as farming or husbandry skill, became primary determinants of living standard.

Finally, the increase in living standard as measured by the index used above may not actually be due to the conservation measures themselves. It may instead be the result of the changes in national economic policy during recent years, or could even be the result of the increased investment and attention shown by the local government and conservation experts during the duration of the project. In other countries, similar phenomena have been observed. At times the overall attention paid to a particular area, rather than any specific measures taken, can spur on the enthusiasm and confidence that allow the rural people to improve their living conditions. In such cases, however, the positive effects of the development project tend to decrease over time or disappear entirely with the conclusion of the project and the return of the experts to their own homes. Whether this is the case in Dingxi will become clear over time (McLaughlin, 1988).

Although Dingxi is just one example of soil conservation in China, it is clear that conservation can make a significant contribution to improving the lives of people living in eroded areas. While the social and economic benefits of soil conservation are difficult to measure precisely, the benefits are clearly evident. In fact they present one of the most compelling arguments for investment in soil conservation. Full recognition of the social and economic benefits of soil conservation will help raise the political will necessary for implementation of conservation work and, at the same time, attention given to the relationship between conservation and living standards will enable soil conservation work to contribute more fully to development.

5

Soil erosion and conservation in India (status and policies)

T.N. KHOSHOO AND K.G. TEJWANI

Introduction

The tropical and subtropical lands of India have been cultivated for centuries. A very rich civilization has flourished on these lands and they continue to produce food, fodder, fibre, and fuel; the entire countryside is dotted with ancient ponds and tanks – symbols of catchment management, water harvesting, and recycling in the semi-arid/arid areas. These bear ample testimony to the art and science that underlie soil and water conservation. Land degradation is a recent phenomenon primarily related to the historical and political ramifications of the latter half of the 19th century and early part of the 20th century.

The problems of severe soil erosion started, in the second part of the 19th century, with the establishment of British rule and distribution of forested lands to the people. For example, in Ajmer-Merwara (now part of Rajasthan State), all the forest and waste lands were handed over to the people by the British in 1850. The hills were rapidly denuded of timber and grazing was not controlled. The crops in this semi-arid/arid region were irrigated from tanks (ponds) formed by building embankments across ravines. Some of these tanks are indeed very old. The rainfall in this region, though scanty, comes in heavy showers. The water, rushing down in torrents, quickly eroded the hill slopes; the tanks filled up with silt and debris, and the embankments burst. In 1869, less than two decades after the forests were handed over to the people, the region was described as follows: 'The cattle had perished, the people had fled, large villages were entirely deserted and country was almost depopulated' (Anon., 1927).

In the early days of British occupation, the hill slopes along the West Coast of the Bombay Presidency (now Maharashtra State) were covered with extensive teak forests. By 1927, these magnificent and valuable trees had been cut down and the forests had disappeared for good. The denudation of the Eastern Ghats and the Deccan Plateau resulted in excessive erosion and gradual silting up of the rivers (Anon., 1927).

Hamilton (1935) has narrated the history of severe erosion in the Siwalik Hills of Hoshiarpur District in the Punjab. According to him, forests were strictly preserved for sport during the reign of the Mughals. Later, during the rule of Maharaja Ranjit Singh, *jagirs* (large land holdings) were granted to Rajput and other chiefs, who appropriated all the land for their use; the peasants who previously had sufficient land to live on, were compelled to seek their livelihood by grazing cattle and by cutting and selling wood from the forests. From this time onwards, the destruction of forests began, but the problem was not serious. In 1852, the Forest Land Settlement under the British rule was promulgated. As there were no records of land tenure, and many of the Rajput chiefs had been dispossessed owing to their hostility to the British, almost anyone who claimed possession of land was granted. The result was that the Siwalik Hill forests became the property of countless irresponsible landlords.

Under the British rule an era of 'prosperity and rapid progress' began. Railways developed, and cantonments were established; thus the demand for railway sleepers, charcoal, and firewood was greatly extended. The improvident landlords wasted no time in getting what they could out of their newly acquired property and the forests were quickly destroyed. Cattle herding, formerly a precarious source of livelihood, now became safe and profitable. Herds were grazed on the hills in increasing numbers, and goats in particular swarmed everywhere, eating up shrubs and bushes and preventing any worthwhile plant growth from developing.

Such history repeated itself again in the late 1940s and early 1950s. When India achieved independence in 1947, the *Zamindari* (i.e. ownership of large land holdings) system was abolished. The dispossessed landlords ruthlessly destroyed all the forests before they handed over the lands to the government. At present, hardly any area in India is free from the hazard of soil erosion, scarcity of water, or danger of flooding or waterlogging.

Nature and extent of soil erosion problems

Nature of soil erosion problems
The arid areas are subject to severe wind erosion and the desert is slowly engulfing the fertile land and endangering the lines of communication. The march of the desert can be gauged by observing not only the increase in the area under sand dunes, but also by the persistent and progressive

decrease in the yield of crops and loss of productivity of the land. People in the semi-arid areas are always under the threat of famine due to uncertain and insufficient rains. With the destruction of vegetation on the highly erosive soils and rocks in the Siwalik Hills (mentioned earlier), these hills spew sand and silt, damaging agricultural lands. The Himalayas, particularly the densely populated outer Himalayas, send down millions of tonnes of sediment. The 2.3 to 4 million hectares of ravines on the banks of the Yamuna, Chambal, Mahi, and other major rivers and their tributaries continue to destroy agricultural land, harbour dacoits, and engulf habitation. In the northeastern region of India, heavy rains, floods, stream-bank-cutting, and deposition of sand make farming a hazardous undertaking. In south and south-east India, erosion is widespread on the undulating lateritic soils. In black soils, the problem of surface erosion and subsurface drainage during the rainy season, and moisture conservation after the rainy season, need special attention. Road building, especially in the hilly areas, without proper consideration of disposal of runoff, stabilization of cut and fill slopes, etc., blocks drainage in flatter areas and causes landslides and landslips in hilly areas.

Extent of soil erosion problems

The first gross national estimate (in the 1950s) reported that 145 million hectares (Mha) of land, out of 328 Mha constituting the geographical area of India, were in need of soil and water conservation measures (MOA, 1968). In 1975, it was estimated (Bali, 1975) that 175 Mha were degraded (68 Mha critically and 106 Mha severely); in this case the emphasis was on land degradation due to water and wind erosion, salinity and alkalinity, etc. As far as land use is concerned, these 175 Mha were estimated to include 48 Mha of degraded forest and pasture lands, 87 Mha of cultivated lands, 23 Mha of fallow lands, and 17 Mha of culturable waste lands (Vohra, 1981). It is estimated that about 7.2 Mha of these lands were degraded due to salinity and alkalinity (Abrol and Bhumbla, 1973). The rest of the area was predominantly subject to water and wind erosion. The latest estimates of soil erosion and land degradation are given in Table 5.1.

The first gross national estimate made in the 1950s reported that about 6000 million tonnes (Mt) of soil were eroded by water every year in India (Kanwar: *vide* Vohra, 1981). This was subsequently verified (Tejwani and Rambabu, 1981; Narayana and Rambabu, 1983) by using the information on the land resources in different regions of India (Gupta *et al.*, 1970), the average annual erosion index (EI_{30}) values and iso-erodent map of India, and sediment data for 21 rivers of the Himalayan region and 15 rivers of the

non-Himalayan region (Gupta, 1975; Rao, 1975; Chaturvedi, 1978). Narayana and Rambabu (1983) concluded that, annually, 5334 Mt of soil (16.4
Mt/ha) are eroded: the country's rivers carry an approximate quantity of
2052 Mt of soil (6.26 t/ha); of this, nearly 480 Mt are deposited in various
reservoirs and 1572 Mt are carried out to sea. In other words, about 29% of
the total eroded soil is lost to the sea, 10% is deposited in reservoirs resulting
in a loss of 1–2% storage capacity per year, and 61% is transported from one
place to another. The amount of soil eroded by water in different land
resource regions of India is reported in Table 5.2.

Impacts of soil erosion
The consequences of soil erosion are all-pervasive and pernicious:
soil erosion adversely affects the functioning of natural ecosystems (ecological impacts), the production base (economic impacts), and the quality of life
of the people (social impacts). In the following assessment, the ecological,
economic, and social impacts have been taken into account simultaneously.

Loss of soil fertility and land resources
It is now a universally accepted fact that water erosion removes the
most fertile part of the soil (containing available plant nutrients and organic
matter), with the result that the physical condition of the soil deteriorates;
simultaneously the crop yields also decline. It is estimated that 5.37 to 8.4 Mt
of plant nutrients are lost every year due to soil erosion (MOA, 1985).
Experiments were conducted wherein when 2.5 cm of topsoil was removed
artificially, 14% decrease each was observed in the yield of maize grain and
stalk. When 7.5 cm of topsoil was removed, the maize grain and stalk yield
were reduced by 33 and 27%, respectively. In the agricultural fields in the

Table 5.1. *Areas subject to different forms of erosion in India
(MOA, 1985)*

Type of erosion	Area (Mha)
Water	113.3
Wind	38.7
Saline alkali soils (including coastal sandy soils)	8.0
Waterlogged soils	6.0
Ravines and gullies	4.0
Shifting cultivation	4.3
Riverine and torrents	2.7
Total	177.0

hilly areas (in the Himalayas) not only the topsoil, but also the subsoil has been lost from all the unirrigated bench terraces (these are usually outward sloping). Consequently, there are stones and pebbles scattered all over the slopes. The same is the case for outwards sloping bench terraces on which potatoes are grown in the Nilgiri Hills in South India. In this way, the entire production base is degraded, which leads to the formation of gullies, abandoning of agriculture, and ultimately the loss of the production base. It is reported that every year large areas of cultivated lands are engulfed by ravine lands in the states of Rajasthan, Gujarat, Uttar Pradesh, etc.

Loss of rooting depth

In Maharashtra State, over 70% of the cultivated land has been affected by soil erosion to varying degrees – about 32% of the land being severely eroded, where cultivation is no longer possible. In Sholapur district, over a 75-year period from 1870 to 1945, nearly 17% of the land with medium soil depth (more than 45 cm) deteriorated into shallow soils (less than 45 cm). Similarly in Akola, Buldana, and Yeotmal districts, the number of fields having less than 37.5 cm soil depth increased during the same period by 54, 16, and 8%, respectively (Tejwani, 1980a).

Loss of water resources

With accelerated rates of runoff there is very little opportunity for rainwater to infiltrate into an eroded soil. Due to erosion and thereby the loss of organic matter and rich topsoil, the subsoil with less organic matter holds less soil moisture. As a consequence of these two processes, there is less moisture in the soil profile for crops, owing to which the crops have poor growth and their yield is low. This phenomenon occurs at micro-level or on farm level. There is also a 'macro'-level dimension to the loss of water resources: the ground-water resources do not get recharged. As a result, two situations arise. Firstly, the period of the base flow of streams is reduced, and secondly, the ground-water-table drops. In the first situation, a larger magnitude of runoff is available over a relatively shorter period and the total time of flow of the stream is reduced. In the second situation, water is either available for a shorter period and/or greater effort is needed to pump it or fetch it. This phenomenon is a fact of life in the Himachal Pradesh and Uttar Pradesh Himalayas in India.

Sedimentation

Sedimentation studies of 21 major reservoirs in India (Gupta, 1980) have shown that the annual rate of siltation from a unit catchment has

Table 5.2. *Estimated soil erosion (in tonnes) per annum by water in differ- ent land resource regions of India. Land resource regions from Gupta* et al. *(1970), soil eroded from Narayana and Rambabu (1983)*

Land resource region	Area (km^2) (% of total)	Soil eroded (t/km^2)	Soil eroded in the region (Mt)	Remarks
A. North Himalaya Snow clad	116 000 (3.56)	—	—	
B. North Himalaya Alpine Grass and meadow	98 250 (3.02)	15.93	1.6	
C. North Himalaya Forest	131 750 (4.08)	287.0	28.4	In 75% of medium erosion area
D. Punjab–Haryana Alluvial Plain	101 250 (3.1)	330.0	33.4	
E. Upper Gangetic Alluvial Plain	200 000 (6.15)	1410.0 50.0 3320.0	242.7 0.4 63.5	Agric. Forest Gullies
F. Lower Gangetic Alluvial Plain	145 500 (4.41)	940.0 287.0	132.3 1.4	Agric. Forest
G. NE Himalaya Alpine grass and meadow	16 000 (0.50)	50.0	0.6	In 75% of medium erosion area
H. NE Himalaya Forest	161 000 (4.97)	287.0 4095.0	23.2 163.8	Agric. Forest In 75% of medium erosion area
I. Assam Valley	88 500 (2.73)	2815.0	249.1	
J. Rajasthan Desert	191 000 (5.87)	—	—	No water erosion
K. Rann of Kutch	46 500 (1.44)	—	—	No water erosion
L. Gujarat Alluvial Plain	62 750 (1.94)	480.0 3320.0	28.2 13.3	Agric. Ravines
M. Mixed yellow, red, and black soil	115 750 (3.58)	300.0 100.0	53.4 2.7	Agric. Forest
N. Black soil	673 500 (20.73)	6448.0 —	3375.5 37.8	Agric. Forest average value

Table 5.2. (*contd.*)

Land resource region	Area (km^2) (% of total)	Soil eroded (t/km^2)	Soil eroded in the region (Mt)	Remarks
O. Eastern red soil	573 500 (17.65)	346.0 —	141.2 37.8	Agric. Forest average value
P. Ganges Delta	25 250 (0.77)	—	—	Agric.
Q. Western Coastal	61 000 (1.88)	3930.0	234.8	
R. Red soil	347 750 (10.72)	359.0 —	95.0 16.9	Agric. Forest average value
S. Eastern Coastal	93 500 (2.9)	3930.0	367.5	
Roadside erosion	10 000 (km in Himalaya)	1969.0	19.9	
Total			5364.4	

been 40 to 2166% more than was assumed at the time of the project design (it has been lower in the case of only one reservoir). Using the average of 21 reservoirs, the actual sediment inflow has been about 200% more than the design inflow. Nizamsagar reservoir, which is the oldest in India (1931), had lost 52.1% capacity by 1967 (CBIP, 1981). Most of the existing reservoirs were planned with provision of dead storage designed to store the incoming silt with a trap efficiency determined separately for each reservoir. It was assumed that the entire sedimentation would take place below the dead storage level and the designed live storage would be available for utilization throughout the projected life of the reservoir. These assumptions have not been realized, since observations have shown that (a) the siltation is not confined to dead storage only, and (b) the quantum of siltation in the live storage is equal to or more than that in the dead storage (CBIP, 1981; Sinha, 1984a; Table 5.3). The encroachment on the live storage capacity has affected the functioning of the reservoirs and has also affected some important aspects of design such as the economic aspects of fixing the dead storage level, outlet still level, opening of the penstock, etc. (Murthy, 1980).

Apart from the above, any irrigation system which conveys silt-laden waters, runs the risk of causing a malfunction in the entire irrigation system. The malfunctioning of the Kosi Canal System is a good example of this malady (Sinha, 1984b). Observations on sediment deposition in the main Kosi Canal have shown that in the earlier years of operation of this project during the 1960s, as much as 20% of the canal capacity was filled with sediment, resulting in reduction of water conveyance capacity. Thereafter, desilting of the Kosi canals became an annual feature entailing heavy recurring costs. During the 20 years of canal operation, a number of canal siphons and ditches have choked. Frequent closure of the canals has become a serious constraint on supply of irrigation water.

Floods

As a result of severe water erosion and gullying in the hilly regions, the volume of runoff and peak discharge for any given rainfall increases greatly in small watersheds. Siltation of streams and channels reduces their depth and water-carrying capacity, and increases their width. Interpretation of satellite images has clearly shown that some of the Himalayan torrents and streams have widened by 106% and rivers by 36% over a period of 7 years

Table 5.3. *Sedimentation pattern of reservoirs (CBIP, 1981; Sinha, 1984a)*

Reservoir	No. of years	Distrib. of sediment (%)		Av. annual loss of storage (%)	
		Dead storage	Live storage	Dead storage	Live storage
Mayurakshi	11 (1954–65)	33.2	66.8	—	—
	8 (1965–73)	41.3	58.7	—	—
	19 (1954–73)	38.1	61.9	—	—
Maithon	1963	50.0	50.0	—	—
	1965	50.0	50.0	—	—
	1971	41.8	58.2	—	—
Ganchisager	14	72.0	28.0	1.94	0.82
Girna	13	—	—	2.6	0.55
Hirakud	22	—	—	1.36	0.41

(Gupta, 1981). Another example of this phenomenon is the area covered by the hill torrents which increased from 19 282 ha in 1852 to 32 023 ha in 1884, to 37 630 ha in 1897, and to 60 000 ha in 1936 (Tejwani, 1980a).

Channels and streams having their catchment in the degraded hills over-flow their banks more frequently and cause flooding of the most productive land, thereby wreaking havoc on crops, livestock, property, and the people. On an average, since 1953, floods have inundated 4.9 Mha of land annually in Assam, Bihar, Uttar Pradesh, and West Bengal, including 2.1 Mha of cropped area. In addition to the direct damage, there are indirect losses resulting from the disruption of rail and road traffic and dislocation of normal life and business, adversely affecting industrial production and trade.

Power generation

The adverse impacts of soil erosion and siltation of reservoirs used for hydropower generation manifest themselves in three ways. Firstly, the total volume of water available for power generation is less: hence the power generated is less. Secondly, an eroded catchment yields water much more quickly than an uneroded one and, therefore, the former has a less moder-ated and less sustained yield of water. Consequently, if a reservoir does not have the capacity to hold the total runoff, major flow losses occur. Thirdly, the silt-laden water clogs and abrades the turbines causing a faster rate of wear and tear, frequent shut-downs and higher maintenance costs.

Inland water transport

Siltation of streams and rivers and the consequent raising of their beds reduces the channel depth of water and results in restriction or complete loss of capacity to support inland water transport systems. The River Ganges is a good example of this.

Inland fish and fish breeding grounds

India has as many as 753 000 ha of land under water ponds, 1 447 000 ha under rivers, and 902 000 ha under brackish water. Such a vast inland water body resource has considerable potential for fish production (MOA, 1984). No direct evidence of an adverse impact of soil erosion in inland fish and fish breeding grounds has been reported so far. However, it is logical to conclude that siltation of tanks, streams, and rivers, variation in the depth and period of flow, and variation in water quality and water tempera-ture in tanks, riverine, or lacustrine environments adversely influence the breeding and growth of the fresh-water fish. This is also true of the brackish-water fish.

Factors and policies intensifying soil erosion

Misuse and mismanagement of land
The major factor which has contributed to the intensification of soil erosion is very extensive misuse and mismanagement of land in all agro-ecological and land resource regions of India. Man in his search for arable land, fodder, firewood, timber, minerals, power, etc., has removed the natural protective cover of vegetation, leading to accelerated erosion and sediment generation. For example, the average land-use values in 31 major river valley projects are reported to be 62% agriculture, 20% forest land (not necessarily forested), and 18% other land uses including grassland, habitation, water bodies, roads, etc. (Das *et al.*, 1981); on single catchment basis, Ramganga river catchment (307 600 ha) has as much as 35% land under agriculture (inclusive of fallows) and 48.5% land under forests. At micro-level, the land-use values in two small watersheds 393.69 and 509.33 ha in area in the Uttar Pradesh Himalayas are very revealing (Table 5.4): 21.6 and 50.8% land is under agriculture, 35.8 and 15.1% under forests, and 42.5 and 34.2% unfit for agriculture (but used as grazing land and for miscellaneous land uses, e.g. road, habitation, etc.) (CSWCRTI, 1981; GBPUAT, 1982).

Table 5.4. *Land use in two small watersheds*

Land use	Fakot* Area (ha)	Fakot* % of total	Kafra Bhaura** Area (ha)	Kafra Bhaura** % of total
Cropped land	79.29	21.6	258.6	50.8
Waste land unfit for agric incl. pasture	157.06	42.5	173.7	34.2
Orchard	0.44	0.1	—	—
Forest	132.55	35.8	76.5	15.1
no canopy	34.06	25.7	28.6	37.5
thin forest	15.56	11.7		
moderate dense forest	82.93	62.6	38.3	50.0
dense forest	—	—	9.6	12.5

* CSWCRTI, 1981; ** GBPUAT, 1982

Out of 132.55 ha of forest land in Fakot watershed, not a single hectare is under dense forest, as many as 34.06 ha (26%) have no canopy, and 15.56 (12%) are thin forest. On the whole, in both these watersheds about 40% of the forest land is degraded, a figure which compares with the value of degraded forest land in the entire Uttar Pradesh Himalayas.

Strictly speaking, agricultural land use should be determined by the capability class of the land. A very large area of land is under agriculture, but the land is not used according to its capability. For example, in the two small watersheds mentioned above (Fakot and Kafra Bhaura), there is no capability Class I or II land, and only 2.0 and 25.0% land in Classes III and IV (combined) is suited for agriculture, in the watersheds, respectively. However, as compared to the above two values, as much as 21.6 and 50.8% of land, respectively, is under actual cultivation (Table 5.4). Thus, a lot of land which capability-wise is unsuited, is nevertheless used for agriculture: in other words, the land is misused. It is obvious that once crops are grown on land unsuited for agriculture, the land available for forests and other non-agricultural uses is reduced. In the case of Fakot and Kafra Bhaura watersheds, 98 and 75% of land, respectively, should be under forest, pasture, and other non-agricultural uses, but only 35.8 and 15.1%, respectively, is actually under forest. In addition, the land is also mismanaged as no measures of soil and water conservation have been adopted. For example, in Fakot watershed, only 14% of the cultivated land is bench terraced (due to the fact that it is irrigated land).

Respectively, 42.5 and 34.2% of lands are classed as waste land unfit for cultivation in Fakot and Kafra Bhaura watersheds (Table 5.4). Most of this land is used for grazing. Being common lands, they are no one's responsibility, and are therefore grossly misused and mismanaged. Generally, there is a tendency on the part of the people to encroach on them, and grab as much land as they can get away with. Otherwise, they are overgrazed and overexploited by lopping and cutting of trees. Generally, they are infested with obnoxious shrubs (for example, *Lantana camara, Berberis lysium, Rhus cotinus, Euphorbia royleana,* etc. at an elevation of 1500 m in the Tehri district of Uttar Pradesh) and the grass cover contains a much greater proportion of the annuals than perennials. In the Fakot watershed, *Lantana camara* was recorded covering 33% of the grazing lands.

Grass yields are obviously low. No yield data from large plots are available. In the Fakot watershed, yield of air-dry grass for 1 square metre quadrats was 24.2 g in June and 312.9 g in September (i.e. after the monsoon). The 157.06 ha of poor-quality grassland including scrubland is expected to support 553 head of livestock comprising 71% cattle, 27% goats, and 2% sheep. This livestock population is equivalent to 425 cattle units and the area available per cattle unit is as low as 0.37 ha which, of course, is not all grassland.

Farm size
        In addition to the mismanagement and misuse of the land, farm sizes are very small. Farms are fragmented and scattered and fields are very small. For example, in Uttar Pradesh Himalayas, 70% of the farm holdings are less than 1 ha and only 13% are more than 2 ha; in small watersheds in Almora district, 75–83% of holdings are less than 0.5 ha. The number of locations for a single farm are as many as 13.5, and the number of fields per farm holding could be as high as 30.4. The average size of a field could vary from a ridiculously low value of 0.01 ha to a high value of 0.058 ha (Ghildyal, 1981; Shah, 1981). This is true of most parts of India (MOA, 1985). Small farms are inherently handicapped.

Population
        Over the two decades 1951–1971, the population of India increased by 52% from 361 to 548 million (in 1981, the population was 658 million). During the same period (1951–71), the net land area per capita decreased from 0.9 ha to 0.6 ha. Although the cultivated land area increased from 119 to 140 Mha (an increase of 24%) between 1951 and 1971, the per capita availability of cultivated land for production of food, fuel, fibre, and other needs shrank from 0.33 to 0.292 ha. Per capita forest area in India was only 0.12 ha in 1971 as compared to the world average of 1.08 ha. Even assuming no further deforestation takes place, due to increase in population alone, the forest area per capita would be reduced to 0.075 ha by the year 2000. There were 285.8 million cattle units in 1966, which increased to 308.34 million in 1977 (MOA, 1985).

        It is generally stated that the more fragile hilly areas forming the catchments of reservoirs are sparsely populated. This is true if the population density is calculated on the basis of total land area. For example, the density of population in the entire Indian Himalayas is 72 persons per km^2 as compared to the all-India figure of 211 persons per km^2. However, if the population density is reckoned on the basis of area of cultivated land, it works out to be 1432 persons per km^2 in the entire Himalayas as compared to 483 persons per km^2 for the whole of India. It may also be reiterated that most of the so-called cultivated land in the hills is unsuited for agriculture. The human and animal population pressure is the single most critical factor that determines the rate of accelerated erosion.

Given the pressure of human and livestock population and the consequent misuse and mismanagement of land, the geomorphological, geological, and climatic factors play a secondary role in soil erosion, land degradation, and desertification; they only influence varying rates of erosion and sedimentation (Tejwani, 1984a). The projected rates of population growth of both humans and livestock should be a very serious cause of concern.

Developmental activities

Reservoir and road construction, and quarrying and mining activities are essential for economic development and growth. However, if these are undertaken in an unscientific manner, as is being currently done, they lead to deforestation, soil erosion, and land degradation on an extensive scale. This is particularly true of hilly areas. As much as 1.99 t of sediment are generated per annum by every 10-m length of road in the Himalayas (Narayana and Rambabu, 1983).

Landless labour

India has a large number of landless people, and it has been the policy of the government to distribute land – either belonging to the government or taken from landholders under land reform – at the rate of 1 ha per family. This policy is good in principle but extremely harmful in practice since the land which is distributed is invariably severely eroded and unfit for cultivation. It is not clear how much more erosion occurs due to cultivation of this type of land, but, presumably, if it is covered with bush and grass as grazing land, it will lose less soil than when it is opened up for cultivation.

Factors and policies improving soil conservation

Soil and water conservation development programmes

As early as the late 19th century and early 20th century, the problems of soil erosion received attention at the hands of the British Government in India, when they took notice of the problems of torrent erosion in Punjab in the 1860s and implemented schemes for gully control in the then United Provinces (now Uttar Pradesh) in 1920.

After independence, soil and water conservation programmes were incorporated as part of the First Five Year Plan (1951–56). This action was indeed very heartening. However, while soil and water conservation must be done for all land uses – agriculture, forestry, grasslands, etc. – emphasis has been laid mostly on agricultural lands. Forestry programmes were also initiated during the First Five Year Plan. It was realized by the end of the Second Five

Year Plan (1956–61) that there was an urgent need to pay attention to the reduction of sedimentation rates of reservoirs. Catchment treatment programmes were therefore started from the Third Five Year Plan (1961–66) onwards. Gradually, soil and water conservation programmes have evolved with emphasis on watershed management including all land uses and water resource development. Since the start of the Sixth Five Year Plan (1980–85) all soil and water conservation programmes and agricultural development programmes are firmly grounded on the concept of watershed management.

Structure of soil and water conservation programmes
The following is a historical account of soil and water conservation programmes since 1951.

First Five Year Plan, 1951–56: In the Ministry of Agriculture, Government of India, a Soil Conservation Division with multidisciplinary expertise (soils, agronomy, forestry, and engineering) was established. Moreover, a chain of soil conservation, research, demonstration, and training centres was established. Groundwork for undertaking soil and water conservation development programmes in the States was laid and work initiated.

Second Five Year Plan, 1956–61: Soil and water conservation development programmes were started and strengthened in all the States and Union Territories. The All-India Soil Survey and Land Use Organization was established, soil and water conservation research and training was strengthened and enlarged, and a model bill on Soil Conservation was circulated to the States and Union Territories for them to enact suitable legislation.

Third Five Year Plan, 1961–66: Considerable expansion of the soil and water conservation programmes took place, a centrally sponsored scheme for soil conservation in 13 river valley projects was started, and a survey of waste lands and ravine lands was initiated.

The years 1966–69 were a Plan Holiday.

Fourth Five Year Plan, 1969–74: The concept of integrated approach and programme planning on the basis of micro- and macro-watersheds was advocated for soil and water conservation. Hydrological and Sedimentation Data Collection Centres were established in the catchments of river valley projects.

Fifth Five Year Plan, 1974–79: Soil and water conservation on the basis of watershed management received wider acceptance;

pilot projects on control of shifting cultivation were initiated. State Land Use Boards or some alternative/equivalent bodies were established in all States and Union Territories.

Sixth Five Year Plan, 1980–85: A national policy was adopted to use watersheds as a unit of land and water resource development and conservation. A National Land Resource Conservation and Development Commission, and a National Land Use Board were established.

Physical achievements of the programmes

Soil and water conservation programmes were undertaken by the State Departments of Agriculture and Forests. Apart from bunding and terracing of agricultural fields in the earlier phases, at present there is an integrated multi-treatment on watershed basis. On the non-agricultural lands tree plantation programmes have been undertaken. Special attention has been paid to programmes like *nala* building, check-dams, gully plugging, land shaping, percolation tanks, water harvesting, water conveyance, etc. Up to 1987–88, a total land area of 29.34 Mha has been treated by the State sector programmes; in addition, 2.75 Mha have been treated under the central sector. Central sector/centrally sponsored schemes are initiated and funded fully by the Government of India and in some cases implemented by it. They serve as pilot projects to test solutions of different problems, to serve as catalytic agents, and to support essential activities of national importance. Three central sector/centrally sponsored major schemes are currently in operation. These are:

Delineation of priority watersheds: The term 'priority' watershed means that a watershed in a river valley project (RVP) or Flood Prone River Catchment (FPC) is in urgent need of treatment either to reduce sediment or runoff. The total catchment area of 31 RVP and FPC is 78 574 000 ha (Das *et al.*, 1981). Out of this, up to 1983–84, an area of 55 280 000 ha had been covered.

Control of shifting cultivation: This scheme was initiated in 1977–78 in six States (Andhra Pradesh, Assam, Meghalaya, Nagaland, Orissa, and Tripura). Up to 1978–79, an area of 1800 ha had been developed. After this, the scheme was transferred to the States. Two Union Territories – Arunachal Pradesh and Mizoram – had developed 2500 ha up to 1983–84.

Soil and water conservation in RVP: This scheme was initiated in 1961. At present, it operates in 31 RVP. Up to 1984–85, a total of 717 sub-watersheds had been worked on – out of which 216 have been completed and 501 continue to be treated.

Financial outlays of the programmes

The total expenditure on soil conservation programmes in India, since 1951 and up to 1988, was 17 389 million rupees (Table 5.5). It is noteworthy that the outlays and expenditure have been progressively rising with every Plan; this is indicative of a firm and positive commitment of the country to conserve its land and water resources.

While reviewing these all-India financial outlays, it may be worthwhile to study the profile of a State. Uttar Pradesh (UP) State has 56 districts, out of which eight are Himalayan hill districts. In these eight districts, agriculture and allied sectors (which also include forestry and horticulture) accounted for 23.7 and 26.5% of the expenditure during the Fifth and Sixth Five Year Plans, respectively.

Soil conservation and watershed management programmes are implemented by the Agriculture and Forest Departments. During the Fifth and Sixth Five Year Plans, sums of 65.2 and 327.9 million rupees, respectively, were spent on these two programmes which represented 12.3 and 18.7% of the expenditure on agriculture and allied sectors (Uttar Pradesh Government, 1985). These are large sums of money and a substantial proportion of activities in the agriculture and allied sectors.

Table 5.5. *Expenditure (in million rupees) on soil conservation programmes in India from the First Five Year Plan (1951–56) to the Seventh Five Year Plan (1985–90)*

Period	State Sector	Central Sector	Total
I Plan (1951–56)	15.3	0.7	16.0
II Plan (1956–61)	226.4	7.2	233.6
III Plan (1961–66)	655.2	113.1	768.3
Annual Plan (1966–69)	766.7	101.1	867.8
IV Plan (1969–74)	1374.5	254.4	1628.9
V Plan (1974–79)	1904.0	543.0	2447.0
Annual Plan (1979–80)	555.8	123.1	678.9
VI Plan (1980–85)	4253.7	1228.6	5482.3
VII Plan (1985–90)	3884.8*	1382.2*	5267.0*
Total	13636.4	3753.4	17389.8

* Up to 1988

In addition, the irrigation (and power) sector also undertakes watershed restoration. For example, to help people, development of springs, construction of water tanks, preparation of water conveyance channels, etc., under the minor irrigation programme, are implemented privately with the support of the government or solely by the government. Most of these activities help in soil conservation. For example, springs depend on water conservation, and development of irrigation potential normally leads to improvement of bench terraces. During the Fifth and Sixth Five Year Plans total sums of 135.9 and 447.7 million rupees, respectively, were spent on minor irrigation programmes including wells, pumpsets, high drums, and tanks. This led to a creation of a potential for irrigating 35 810 and 68 840 ha in the two plans, respectively; furthermore, State Government created a potential of irrigating 22 650 and 33 620 ha, respectively, in the two plans, and 6033 km of irrigation channels were constructed during the Fifth Five Year Plan.

On the other hand, certain programmes, such as road construction, contain elements of soil erosion and watershed degradation. Road construction activity, though vital for the overall development of the hilly regions, at present contributes to very severe erosion. In fact, unscientifically constructed roads have been a mixed blessing. The dimension of road construction in the hills is an indicator of the dimension of soil erosion. For example, sums of 659.2 and 1451 million rupees were spent on roads and bridges in the eight hill districts of UP during the Fifth and Sixth Five Year Plans, respectively. During the Fifth Plan (1974–79) and Annual Plan (1980), 3282 km of roads were constructed. One can be certain that these are sore spots contributing to severe erosion and high sediment generation, particularly because they are constructed with little or no consideration for environmental stability of the areas in question.

Forestry: physical achievements and financial investments

Forestry plays a multifarious role in watersheds, which is both protective and productive. From 1951 to 1980, India had invested 128 690 million rupees in agriculture and allied programmes – which included 34.1% on agriculture and land reforms, 18.8% on minor irrigation, 5.3% on soil and water conservation, and only 3.9% on forestry. In the earlier plans, it was the intention of the planners to create infrastructure for increasing food production and it was easy to rationalize the priorities of investment. However, keeping in view the fact that, compared with the 47% of land under agriculture, 23% of land is under forests, and that India not only needs firewood and timber but also long-range ecological security, the investments in the forestry sector have been very low. Long-range ecological security includes climate stabilization, and conservation of soil, water, and biodiversity. It is indeed remarkable to note that up to 1980, India invested more in soil and water conservation (6489 M rupees) than in the forestry sector (4839 M rupees). It was only for the first time in the Sixth Five Year Plan (1980–85) that a quantum jump was provided in investment in the forestry sector (6925 M rupees); also, it was for the first time in the Sixth Five Year Plan that the outlay in the forestry sector exceeded that in soil and water conservation.

In spite of these efforts from 1951 to 1980, while 3.6 Mha were afforested, as many as 4.1 Mha under forests were lost to other land uses. In the Sixth and Seventh Plans outlays in the forestry sector were enhanced significantly, but the result of this enhancement is not readily visible.

Success stories and benefits of soil conservation

It is a well-recognized fact that soil and water conservation and afforestation are long-range and long-term benefit programmes. That they conserve soil and water and restore ecological balance is a well-known fact. However, in almost all countries, and especially in developing countries which have many other pressing problems to solve, the political system always looks at short-term gains. Politicians have to be assured about the associated direct productive benefits even from the long-term plans. The success stories and benefits of soil conservation are numerous and have been demonstrated at various scales, e.g. river basins, reservoirs, catchments, small watersheds, etc. At river basin scale, the success of watershed management in the Damodar Valley Corporation in flood control, provision of irrigation water, supply of power, promotion of industrial growth, navigation, etc. is most remarkable.

Prolonging the life of reservoirs

Soil conservation and watershed management programmes in the catchments of reservoirs has a primary concern of sediment control to prolong the life of reservoirs and increase the efficiency of the system. It is reported that the sedimentation rates in Bhakra, Machkund, Maithan, Panchet, and Hirakud catchments have decreased (Table 5.6) as the area treated in their catchments has progressively increased (Das *et al.*, 1981). While these reductions in the rates of sedimentation are welcome, they are not satisfactory because (a) the designed inflows of sediment were underestimated, (b) the sediment is being deposited in live storage also (even though it was presumed that it would be deposited only in dead storage), and (c) encroachment of live storage capacity has adversely affected the functioning of the reservoirs and also some economic and design aspects (Murthy, 1980).

Table 5.6. *Trends in sedimentation rates (SPR) for areas treated for soil conservation. (From Das et al. 1981)*

Name of the River Valley Project	Catchment area (km^2)	Year up to	SPR ha.m/100 km^2		Prog. area treated (10^3 ha)
I. Bhakra	56876 (18200 in India)	1982	8.04		—
		1966	6.52		32.67
		1968	6.32		53.85
		1972	6.08		77.86
		1976	5.99		81.43
		(V Plan: % reduction 25.5)			
II. Machkund	2222	1959	3.38		—
		1966	2.43		22.53
		1969	2.67		36.54
		1972	2.5		53.68
		1976	2.3		58.66
		(V Plan: % reduction 31.95)			
			Maithon-Panchet	Total for Damodar Valley Corporation	
III. Maithon, Panchet		1962	—	13.16	—
		1963	15.27	—	
		1964	—	12.21	81.96
		1965	13.98	—	
		1966	—	10.78	
		1966/7–69/70			140.64
		1971	13.02	—	160.65
		1974	—	9.92	195.86
		(1971–74: % reduction 14.32		24.62)	
IV. Hirakud	83390	1947	4.91		
		1952	4.03		
		1957	3.73*		
		1961	4.47**		
		1966	4.12		80.04
		1969	4.0		136.41
		1974	3.84		245.73
		1975	3.8		255.01
		1976	3.69		257.81
		1977	3.58		256.1
		(% reduction: 13.11)			

* Due to extreme dry years preceding
** Due to excessive rainy years preceding

However, it may be noted that none of these adverse results have affected the success of the soil conservation and watershed management programmes.

Direct benefits
Direct benefits of soil and water conservation are strictly marketable commodities such as food, fuel, fodder, fruits, fibre, clean water, etc. Information about the direct benefits is gradually emerging owing to growing concern about the nature and extent of the benefits. Some of the more outstanding success stories are given below:

- In a study of the Hirakud catchment, small sediment reduction structures were evaluated. These structures not only reduced the runoff and sediment (protective benefits), but also resulted in reclamation of the land below them and consequent increase in crop production as well as employment generation. The area treated was 2321 ha, the area reclaimed for cultivation was 91.4 ha, and the area irrigated with stored runoff was 546.8 ha (Das and Singh, 1981). The benefit–cost ratios ($b:c$) for production purposes were 1.13 and 1.66 over a time-span of 10 and 20 years, respectively. For protective purposes the $b:c$ ratios were 2.32 and 3.14 over a time-span of 10 and 20 years, respectively. As a result of these soil conservation works the employment opportunities also increased, as evidenced by a post-project:pre-project employment ratio of 2.97 and 2.38 for a time-frame of 10 and 20 years, respectively.
- In an experimental watershed of 370 ha in the Himalayas, where soil and water conservation measures were tested for all the land uses, the overall $b:c$ ratio was 1.5, while for individual land uses the $b:c$ ratios were 0.97 for irrigated land, 3.6 for rainfed agriculture, 2.9 for orchards, and 0.92 for fuel and fodder plantations (Seckler, 1981).
- In a submontane high rainfall area, an embankment-type unlined farm pond constructed for the purpose of runoff storage and recycling, as well as indirect benefits, had a direct benefit of making water available for irrigation. This had a $b:c$ ratio of 1.85 and 1.96 over a period of 30 and 50 years, respectively, if the water was applied before sowing the crop (Vimalkishore *et al.*, 1980).
- In the case of conservation and utilization of lands not suited for agriculture, silvi-pastoral systems (fodder and fuel plantations), evaluated experimentally, gave $b:c$ ratios varying from 1.05 to 3.43 depending upon the combination of trees and grasses (Mathur *et*

al., 1979). With respect to pure plantation of *Eucalyptus* in degraded and denuded land, a *b:c* ratio of 1.68 was reported for a nine-year-old plantation. Furthermore, this plantation had the scope for coppicing and was expected to give a high *b:c* ratio because of the absence of land clearing and plantation (Rambabu *et al.*, 1980). A number of case studies in other regions of India have been reported showing *b:c* ratios of direct benefits to be more than one for soil and water conservation practices on agricultural and non-agricultural lands (Tejwani and Rambabu, 1982).

Indirect benefits

Apart from the foregoing types of direct benefits, many indirect benefits accrue due to soil and water conservation measures. If soil is conserved, the land continues to produce crops indefinitely; if water is conserved, streams flow for a longer period, the water remains clean and is available for irrigation and drinking, floods are moderated, and rates of sedimentation are reduced. These indirect benefits have been experimentally demonstrated. Examples include:

– In a 20 ha watershed in the highly erodible Siwalik Hills, the rate of sediment was reduced from 80 t/ha/yr to 5–7 t/ha/yr within 5–6 years of treatment (Mishra *et al.*, 1975). In another study in the same area, as a result of closure, afforestation, and gully control measures, peak discharge and total runoff were reduced by 61% each, 6 years after the treatments.

– In a high rainfall degraded and denuded submontane area, when a plantation of *Eucalyptus* was undertaken after clearing of the watershed originally covered with brushwood, runoff was reduced by 28% (Mathur *et al.*, 1976). Similar positive, indirect benefits have also been reported from other regions of India (Tejwani and Rambabu, 1982).

Lessons learnt and new directions

With over 40 years of planning and implementation of soil and water conservation programmes, a number of lessons have been learnt and attempts should have been made to incorporate them, as far as practicable, in programme formulation and implementation. However, a closer scrutiny indicates that no attempt has been made either to undertake critical review of previous policies, programmes, practices, and performances, or to act upon the lessons learnt so as to prescribe future policies and programmes and evolve appropriate solutions/applications.

On the basis of the above observations, it is necessary to indicate a future course of action, keeping in view the lessons learnt in the past. In some cases, though the malady may be identified, it may not be possible to suggest a remedy, or, even if the remedy is identified, its application may not be possible at least in the near future due to diverse reasons – mostly social and economic. The issues could be divided into social, economic, biophysical, and political categories: these cannot be divided into water-tight compartments as they are interdependent. These issues may require action at various levels – national, inter-state, state, department, district, village, etc. They may be institutional or technological in nature. All the same, these have to be based on socio-economic considerations. For convenience, these issues are identified as related to policy, programmes, practices, research, training, and demonstration.

Policy issues

Awareness
Among the policy issues, awareness is perhaps the foremost prerequisite for any development programme. In the case of soil and water conservation, the governments (both Central and State) and sectoral departments, the farmers and users of the land, and people at large have to be aware, active, and responsive.

India has been rather fortunate in as much that government has been aware of the importance of looking into problems of soil erosion. For example, soil and water conservation programmes were included in the First Five Year Plan (1951–56) and have continued and been diversified since then. In fact, watershed management was recognized as a priority activity even before the First Five Year Plan, since it was included in the Damodar Valley Corporation Act in 1948. Today, at least on paper, no land–water development programme is approved unless it is planned on watershed basis. At the people's level, the 'Chipko' movement in the Himalayas is a classic example of people, in particular the women (who have to bear the brunt of farming and fetching fuel, fodder, and water), initiating a conservation movement (Kunwar, 1982). Although apparently only environmental, it has socio-economic overtones as well.

As a consequence of this awareness and the concern expressed at various levels from time to time (NCA, 1976; Tejwani, 1982), the government of India recognized the urgency of the matter and constituted in 1984 the National Land Use and Wastelands Development Council headed by the Prime Minister of India. This is the apex body under which are constituted two national boards, namely the National Land Use and Conservation Board under the Ministry of Agriculture and the National Wasteland Development Board under the Ministry of Environment, Forestry, and Wildlife. It is hoped that a national land-use policy, keeping in view the current and future needs and pressures on land, will soon be formulated for the whole of India.

Population

If a single issue that influences soil erosion or conservation were to be identified, then it would be population pressure – both human and animal. It is obvious that there is an upper limit for any land–water system to support increasing numbers of human beings and livestock, without affecting the ecological balance. Though there is considerable awareness about the serious threat posed by population pressure, due to various social, political, and religious considerations, the issue of population control is not being faced squarely. Unless population is controlled, soil and water conservation and even food production cannot be assured effectively. There can be no alternative to a firm policy on population control in relation to development and use of a limited non-renewable resource like land and renewable resources like water, forest, and crops. For long-term good, the government has to be prepared to share even a measure of unpopularity on this account (Khoshoo, 1986).

Land holdings and ownership

The issue of small size of land holdings was discussed earlier. In addition to this, there are issues of land encroachment and allocation. Most of the common land, used for grazing, has been 'grabbed' by powerful individuals, with the result that the pressure has increased on the remaining common land. In India, there is a declared policy of 'Land for the Landless'. This policy is good provided there is good land available for allocation. However, generally, the land allocated is either too steep, has too shallow a soil depth, or is subject to other severe limitations like being alkaline/saline. 'Land for the Landless' policies, though politically desirable, ultimately have adverse impact on proper land use.

Given this scenario, it is difficult for subsistence farmers with little land to manage their land efficiently and construct terraces, provide irrigation facilities, or go in for high inputs. The obvious approach is to consolidate the holdings. However, the situation is not as simple as it appears to be. There are many biophysical and legal difficulties and the issue is complex, intractable, and hence remains unresolved.

Development strategy

The structure and extent of soil and water conservation programmes have been discussed earlier. In the First and Second Five Year Plans (1951–61), the emphasis was on soil conservation of agricultural lands. It was during the Third Five Year Plan that a centrally sponsored scheme for soil conservation in the river valley projects was initiated. However, it was really in the Fourth Five Year Plan that the importance of soil conservation and watershed management in the hilly areas of India was fully recognized. The Himalayan watersheds are particularly fragile as compared to the non-Himalayan watersheds. The Himalayan rivers carry far more sediment and much more coarser material than the non-Himalayan rivers (Gupta, 1975). The hilly areas are also more densely populated on the basis of cultivated area than the plains. In addition, the hills have many biophysical and socio-economic constraints, such as isolation, poor infrastructural facilities, wide variation in the agro-ecological conditions within very short distances due to altitudinal variation, and conditions of slope and aspect. These constraints call for a specific development strategy. The political system and planning process, recognizing the importance of hilly areas, declared these as 'Backward Hill Areas for Special Consideration' (Planning Commission, 1981). The Fourth Five Year Plan recognized that there will be higher costs and poorer returns from investment in the hilly areas, and, therefore, more central assistance was allocated at more favourable terms to meet the specific needs and problems of the hills. It is noteworthy that, while laying down the development strategy for these backward hill areas, one of the major items was maintaining the ecological balance.

There is another aspect of developmental strategy in the hills, namely interrelationship between uplands and lowlands. How the land and water is used and managed upstream determines the quality and quantity of water and the duration for which it will be available downstream. If the uplander manages the catchment and uses the water wisely, not only will he or she get good quality and sufficient quantity of water for a longer time, but the lowlander also will reap similar benefits. In the prevailing situation, where everyone is poor and has to eke out existence from land, it is difficult to expect the uplanders to manage land and water both for their own good and that of the lowlanders. Under these circumstances, the economic security of the uplanders must be assured, and, as an outgrowth of this, the uplanders will have to be persuaded to realize that their life is also interwoven with that of the lowlanders. They should, therefore, perceive that they are sharing and equal partners in the development of the hills as well as the plains. This will not come automatically. The *sine qua non* for this is that the people in the plains must also perceive that their development and growth are intrinsically linked with those of the uplanders. Both the uplanders and lowlanders must realize that they have different roles but common goals and have to work together. As far as water and forest resources are concerned, their mainten- ance and conservation starts upstream, but many of the benefits flow downstream. It is unfortunate that this is not fully realized, particularly by people living in the plains. For instance, the well-being of the Gangetic Plains is irrevocably interwoven with the well-being of the Himalayas. In fact, there is a growing awareness that excessive demand created for goods and services in the plains is the major cause of destruction of the highlands. A strategy based on this realization will alone save the interdependent lowland – highland system (Khoshoo, 1986).

Fiscal policies
Soil conservation, watershed management, and afforestation cost money and pay returns immediately to neither the farmer nor the community. In fact, sometimes the crop yield per unit area may be less, since the bunds/risers/ bench terraces remove land from cultivation. Even so, investments in soil and water conservation must be made. The questions arise as to (a) who pays: the farmer, the community or the government?, (b) if a farmer or an individual has to pay, then for what type(s) of work and how much must be paid?, and (c) what are the obligations of the people who live in the plains towards supporting the upland watershed programmes? Do they pay directly? Fortunately, it has been realized and accepted in India that since soil and water conservation works also have many social benefits – indirect, offsite, and long-term – the farmers and beneficiaries need not be required to pay 100% for the works. Currently, the subsidies/incentives vary from 25 to 100%. With respect to the contribution of the plains to the restoration and maintenance of the upland watersheds, there is the excellent example of Colombia where a percentage income (4%) from power generation is transferred to the rehabilitation and development of watersheds. This mechanism needs to be studied and adopted.

Legislation
There is a distinct need for legislative support, because soil and water conservation programmes are concerned with land treatment and water disposal, which go beyond the boundaries of individual farms, require investments of funds (usually public) with and without incentives, require

maintenance of structure, and lead to sharing of long-term benefits (e.g. tree plantation, water in ponds).

In India, land, water, and forest are state subjects. In order to encourage State Governments to enact a legislation, the Government of India formulated the 'Model Soil Conservation Bill' in 1955 and circulated this to the States. By 1981, sixteen States and Union Territories had legislation relating to soil conservation (Jacob, 1981). In some States (e.g. Gujarat, Maharashtra, and Andhra Pradesh), soil conservation measures are being implemented under earlier legislations, e.g. Bombay Land Improvement Act (1953) and Madras Land Improvement Schemes Act (1949).

With rapidly changing socio-economic conditions as well as diversification of performances, soil and water conservation, watershed management, afforestation, and social/community/farm forestry programmes have acquired urgency. Hence, there is need for new legislative approaches and initiatives. Currently, forests, grazing lands, community, and revenue lands are used as common lands to varying degrees. The use of all these lands is determined more by arbitrary custodial and property approaches, than by their best sustainable use. Chaudhry (1987) suggested that in developing a land-use policy for these so-called 'uncultivable lands', three objectives should be kept in mind, namely: (i) halting deforestation and degradation of the production base; (ii) improvement in the productivity and production of these lands; and (iii) improvement in the quality of life of the rural poor in terms of subsistence needs and employment generation. A number of issues within each of the above category of lands have to be sorted out, namely, who should have custodial and property rights, and how the productivity and production may be enhanced on a sustained basis. It is expected that the Government of India will evolve policies for land use and conservation to meet the new challenges.

Programmes

Perspective planning

A reasonably good institutional infrastructure to implement the soil and water conservation, watershed management, and afforestation programmes has been developed. Mostly this has been achieved through the State Departments of Agriculture and Forestry and in some cases through the State Departments of Soil Conservation. However, if one looks into the planning and implementation processes, an element of arbitrariness is clearly discernible. Every five years, a five year plan is made, and every year, within the period of the plan, a part of the five year target is planned and implemented. The planning and implementation is done on the basis of what was done in the previous five year plan or in the previous year. For the next five year plan or the next year's programme, the quantity of work to be done or amount of money to be spent is increased by some percentage, which does not bear any relationship to either the dimension of the problem or any foreseeable time-frame to solve the problem. For example, during a period of 35 years (1951–85), in the whole of India 29.38 Mha (27.07 Mha under State Sector and 2.31 Mha under Central Sector) were treated for water and wind erosion; this represents 20% of the total area (150 Mha) subject to water and wind erosion. However, if one remembers that most of the structural works have a life of 15 years only, then in effect by 1985, work has been undertaken effectively only on 18.6 Mha.

It is time that, with every five year plan, the financial outlays are sufficiently increased, because due to escalating costs, the average cost of reclamation has also increased progressively from 53 rupees per hectare in 1951–56 to 877 rupees per hectare in 1981–84. In effect, this has resulted in slow progress of work.

Assuming that no new area is eroded or degraded (which is not true in any case), the planning process should aim to treat all the 150 Mha in the foreseeable time-frame. The yardstick should be that the pace of restoration is always more than the pace of degradation. It is difficult to make *ad hoc* suggestions for the time-frame. There is an urgent need for serious perspective planning both at the State and Central levels. Surprisingly, there is no monitoring and planning cell either at the Centre or at State levels.

Planning whole systems
It is now widely recognized that if a whole system is planned, designed, and implemented, it is less costly as well as more efficient and beneficial in the long run. Whatever the reason in the past to follow a particular approach in planning, designing, and implementation of the various components of land and water conservation systems (reservoirs, catchments, and command areas), there is no justification now to follow the same antiquated procedures. A hard look needs to be taken at the total system.

At present, a reservoir is planned and designed assuming that the inflow of water and silt will continue to be the same in quality and quantity as observed in the past. It is only subsequent to the project implementation that adverse effects on the life and stability of the reservoir and structures become apparent and then curative measures of catchment treatment are thought of. This is a very short-sighted planning and implementation process which has dangerous consequences.

With the present planning and implementation process, the country does not get the designed benefits both in terms of quantity and quality from the investments made. The country also loses more capital on account of the higher costs of operation, repair, and maintenance than originally provided for. Furthermore, dam sites are lost within a shorter life-span, and the projected *b:c* ratios are never attained. If, however, the costs are realistically allocated, the project may in fact have negative ratios. This unfortunate situation arises partly due to political and prestige considerations, and partly due to the eagerness of engineers in the business to only construct the dams and not to handle the whole system (catchments, dam site, and command area) (Khoshoo, 1986). Enough has been learnt during the last 40 years about dam construction. It is, therefore, imperative and most urgent that a firm decision be taken, that hereafter, as well as the dam site, the catchment, reservoir, and command area will be surveyed, planned, designed, con-structed, and managed, at least simultaneously. In fact, it may be wiser to plan these aspects even in advance of the reservoir construction (Tejwani, 1984b). This would have to be a policy decision.

It is reported that if soil conservation and watershed management practices are included as a part of a project from the planning and design stages, and 25% silt discharge is planned to be reduced, the percentage increase in the utilization of water can be 1.48–1.79 and 3.02–3.81 over a design period of 50 and 100 years, respectively. In another case study, it was observed that a pre-planned 25% reduction in sediment would help in reducing the height of the dam from 32.6 m to 32.0 m, thereby reducing construction costs by 5 to 6%. In addition, there would be less submergence, reduced costs of rehabili-tation, reduction in the loss of land resource and forests, and, due to the smaller surface area of the reservoir, there would be reduced evaporation loss (Sinha, 1984a,b). However, funds are often made available too late for the catchment treatment and the work is done in an indifferent manner. The adage 'a stitch in time saves nine' is very apt in such a situation.

Coordinated action

Soil conservation and watershed management programmes, as is obvious, deal with cultivable land, forest land, pasture land, and water resources. There is also a need for conservation works in quarrying and mining areas, roads, irrigation channels, construction of power lines, etc. It is common knowledge that each sectoral department dealing with land, water, forest, mining, roads, etc., works in its own limited sphere, without coordinating its work with others. It is also common knowledge that the activities of the Forest/Irrigation Departments cause damage to roads. Simultaneously, Roads and Power Departments cause damage to forest plantations, water courses and bench terraces undertaken by the Forestry, Irrigation, and Agriculture/Soil Conservation Departments. As everyone goes merrily about their own sectoral pursuits, without heed to one another's problems, more damage is caused. It is obvious that if soil conservation, water resource development, and watershed management are to be effective, cooperation and coordination between various departments is urgently needed.

Similarly, if soil and water conservation and watershed management are being carried out in a catchment by two or three departments (Agriculture/ Soil Conservation/Forestry) there is need for treating small watersheds (500 to 5000 ha) simultaneously so as to make these works beneficial. This is currently not being done, with the result that the benefits that normally accrue are neither visible nor realized. Again, there is, thus, an urgent need for coordination.

Practices

Plantations

If any project for tree plantation in soil conservation and watershed management is examined, two things stand out as sore spots. These are the unsuitability of the species and the low survival percentage. Though criteria about species selection may have become precise, the situation with regard to survival percentage is still very critical. The issue with respect to species is complex and has many location-specific socio-economic and biophysical aspects. Over the last 35 years, it has become sufficiently clear that foresters must plant multipurpose tree and shrub species specific to the particular agro-ecological situations. However, it is not justifiable to have 20–40% survival rate in the plantations. Often, it is reported that so many million/billion seedlings were *distributed*: such reports do not even mention how many were actually planted. Often one finds reports that between 1951 and 1980, 3.5 Mha of land were planted with trees, but nothing is reported about survival percentages and the number of hectares harvested/extracted. Therefore, public concern about poor survival rate is legitimate. Often failures are attributed to the 'failure of monsoon'. These are not isolated examples. There is need to do quality planting: for instance, no plantation is good in a rainfall area of over 1000 m if it does not give 80% survival rate at the end of the first year. In cases where the site conditions are not good for planting trees, it is better to fence the areas because at many places the Indian agro-ecological conditions are such that the area will be clothed with grasses, shrubs, and even trees and lead to a natural jungle.

Maintenance of works

Though investment is made on soil conservation structures, they are not maintained even for a year. The bunds and terraces, even under the best of circumstances, are weak and give way during the monsoon; they are seldom repaired. There is no institutionalized maintenance, and ultimately the programme loses its credibility. The lesson to be learnt is that, as with tree planting, the quality of workmanship in soil conservation works needs to be good. Furthermore, the farmer's stake in the soil conservation has to be high so that he or she becomes a willing participant.

Roads and mining

At present one of the major causes of soil erosion, sediment generation, instability of hill-slopes, man-made landslides, gullying, and forest degradation, is the vast network of roads and unscientific mining in the Himalayas and other hill regions of India. Most of these roads were made in a hurry, with scant respect to ecological considerations (Khoshoo, 1986). This may have been understandable in the 1960s and 1970s, however, as a result, in the 1980s, the maintenance of the vast road system involves an enormous amount of money which is not available. The techniques and technologies of alignment and construction of roads on a scientific basis and stabilizing the slopes are well known. The issues are not technological but socio-economic and political. Money is allotted for road construction without any consideration of or allocation of money for conservation and stabilization works. In the process, a road is made which causes more damage and costs more in maintenance, than if it had been made with adherence to technical criteria.

The same is true of mining. In the case of mining, the law provides for stabilization and rehabilitation of the area during and after the mining operations. In fact, some funds are collected from the royalties for the purpose but nothing is done and the law is also not enforced. In this case, private companies and government departments are equally guilty of causing damage. The lime quarrying by private companies in Mussoorie Hills and the surface mining of coal by government agencies in Chota Nagpur Plateau in Bihar show classic examples of damage caused by mining (Khoshoo, 1986).

These facts are widely known and have been discussed and debated in the press, in parliament and in public (Planning Commission, 1982). There is need for a firm commitment and firm action. The scientific and technological solutions are known.

People's participation

If an enquiry is made by a development/extension worker, a researcher, a policy-maker, or a social worker, as to what 'people's participation' means in actual terms; the invariable response will be that it implies a participatory role for the beneficiaries/farmers in soil and water conservation, and watershed management. Women's roles may also be specifically brought out. However, if the same enquiry is made by the beneficiaries/farmers/women, it becomes clear that, in their opinion, in addition to themselves, all the service personnel are also participants. Therefore, people's participation should involve both the service personnel and farmers/beneficiaries/women. Otherwise, the results are disastrous because service personnel do not take an interest, with the result that there is no technical input.

The issue of the participatory role of service personnel at all levels is very crucial to the successful initiation, design, functioning, and implementation of any development programme. Policy-makers, decision makers, directors, project managers, and technicians, all have to play specific and well-orchestrated roles to make these programmes successful.

Beneficiaries/farmers/women will not change a land use or a practice on which they are subsisting unless they perceive that the proposed change will benefit them directly and immediately. For instance, if the project provides for irrigation water coupled with conservation, they will be more willing to participate, in fact, they will provide labour, borrow money, and go out of the way to protect the system.

Research and training

Research

Soil and water conservation and watershed management as a mix of applied field techniques, need to be backed by a strong research base. India has a good record in this area as, even in the 1930s during the severe depression days, there was a Dry Farming Research Scheme at four centres in the country to develop soil and water conservation techniques for successful farming in semi-arid lands. Furthermore, within one year of independence, India passed the Damodar Valley Corporation Act, which also included research and demonstration activities. During the framing of the First

Five Year Plan (1951–56), it was obvious that while problems of soil erosion and associated degradation and destruction of the production base were broadly known, solutions backed by scientific investigations were not. Consequently, a chain of Soil Conservation Research Demonstration and Training Centres was established in the late 1950s. This process is continuing and new institutions and centres were established/added in the Fourth Five Year Plan (1969–74) and the Sixth Five Year Plan (1980–85) (Table 5.7). Apart from the research institutes directly concerned with soil and water conservation, there are a number of other research institutes which are concerned with the problems of land and water. These include the All-India Soil Survey and Land Use Organization (New Delhi); the National Bureau of Soil Survey and Land Use Planning (Nagpur); the Central Arid Zone Research Institute (Jodhpur); the Indian Grassland and Fodder Research Institute (Jhansi); and the Forest Research Institute and Colleges (Dehra Dun).

The chain of eight research centres of the Central Soil and Water Conservation Research and Training Institute, the centre of the Damodar Valley Corporation, the two centres of the State Governments, and the one centre run by the Government of India (in Nepal) are unique in Asia (Table 5.7). While all these research institutes and centres have contributed to the development of technological packages (Tejwani *et al.*, 1975; Tejwani, 1980b; Sharma *et al.*, 1980; Venkataraman *et al.*, 1980; Verma *et al.*, 1986), the full potential of their application in the field is yet to be achieved. A developing country like India should aim at a mix of basic and applied research, the former explaining the whys and wherefores of a process, and the latter indicating the way to modify a process(es) to bring about a desired practical change.

For soil and water conservation and watershed management, some of the research gaps identified (Tejwani, 1984a,b,c) are: availability of watershed data; techniques of delineating priority watersheds for sediment control; flood moderation, etc.; and storing and updating information about characteristics of watersheds and retrieval of such information for practical application.

While all the foregoing research institutes and their centres have developed individual techniques, transfer of these techniques to the field in the form of a package is a major lacuna. This lacuna was identified as early as 1972 by the Central Soil and Water Conservation Research and Training Institute, when it recommended that the 'institute should undertake 14 operational research projects'. This concept is now practised by many other institutes of the Indian Council of Agricultural Research (ICAR). Operational research should find much greater acceptance in soil and water conservation, and watershed management.

Another major shortcoming in research and development in soil and water conservation and watershed management is the lack of research on social and economic aspects. This area needs to be strengthened, because there are many participants, e.g. politicians, administrators, technocrats, project managers, supervisors, middle-level technicians, field-level workers, social workers, local leaders, and, last but not least, the beneficiaries/farmers/ women. Each one of these participants has the potential to contribute to the success or failure of the programmes. Sociological research and sociological input should be available at the project formulation stage and not as a post-mortem service.

Soil and water conservation programmes cost money. No country, and more specifically a developing agrarian society with limited resources, can afford to invest large funds on conservation alone, however desirable it may be. Ecological security must, therefore, be combined with economic security and the research should develop packages which can achieve both these goals. There is a need for clear understanding of direct/indirect advantages, onsite and offsite benefits, short-term and long-term gains, and benefits to individuals and to the society. Generation of data on benefit/cost relationships by the research institutes can help in the formulation of better policy and legislation, and fiscal management.

Training
Soil and water conservation and watershed management programmes need skilled technicians and professionals to plan and implement them. They require structural works which call for investment. If the job is not done properly, not only is the investment wasted, but often greater damage is done than would have been caused otherwise. Also, due to the failure of structures, often the whole programme gets a serious setback. It is, therefore, essential that the work can be done by trained, fully oriented, and committed personnel. The Central Soil and Water Conservation Research Institute (Tejwani, 1979a,b) has established regular training courses, and has trained thousands of personnel. In spite of this massive effort, due to ever-expanding soil and water conservation programmes, there still is lack of technical personnel (Tejwani, 1981), which is a major constraint in starting, implementing, and reaping benefits from soil and water conservation programmes (Tejwani, 1979a,b, 1981, 1986). The same is true in the case of

Table 5.7. *Research institutes concerned with soil conservation*

Research institute and location	Region	Year of establishment	Remarks
1. Central Soil and Water Conservation Research and Training Institute, Dehradun, and its chain centres	NW Himalaya	1954	Also responsible for coordination of research and training of all its centres. Under Indian council of Agricultural Research
– Chandigah	Submontane tracts in NW region in India, esp. Siwalik Hills	1954	
– Ootacumand	S India high hill areas with high rainfall	1954	
– Bellary	Semi-arid black soil region	1954	
– Kota	Ravine problems, esp. on banks of Chambal River	1954	
– Vasad	Ravine problems, esp. on banks of Mahi River	1955	
– Agra	Ravine problems, esp. on banks of Yamuna River	1955	
– Hyderabad	Semi-arid red soil region	1962	
– Datia	High rainfall black soil region	1986	
2. ICAR Agricultural Research Complex for NE Hill region, Shillong, and its chain of centres	High rainfall lateritic soils of Meghalaya	1975	Under Indian Council of Agricultural Research, India. Also responsible for coordination of all research at its centres
– Impal	High rainfall lateritic soils of Manipur	1975	
– Aizal	High rainfall lateritic soils of Mizoram	1975	Main problem is shifting cultivation in NE hills

– Jhamapani	High rainfall lateritic soils of Nagaland	1975	
– Agartala	High rainfall lateritic soils of Tripura	1975	
– Arunachal Pradesh	High rainfall E Himalaya region	1975	
– Gangtok	High rainfall E Himalaya	1975	Main problem is hill agriculture and watershed management
3. Soil Conservation Research Centre, Hazaribagh	Red soils, high rainfall, Chotanagpur Plateau	1949	Soil Conservation Directorate of Damodar Valley
4. Soil Conservation Research and Training Centre, Rehmankhera	Alluvial soils and ravine problems on banks of Gomti River		With cooperation of Uttar Pradesh State Government
5. Soil Conservation Research Centre, Sholapur	Semi-arid soils of Maharashtra		Maharashtra State
6. Soil Conservation Centre, Chhatra, Nepal	NE Himalaya with special reference to Kosi catchment	1956	Govt. of India research centre in collaboration with HMG of Nepal

forestry education/training (Tejwani, 1981; Khosla, 1983; Khoshoo, 1986). While a lot has been done, much more is required. Some estimates of the trained manpower needs have been provided (Tejwani, 1981). Training technical personnel is no doubt important; however, equally important are the special training and orientation courses for farmers, social workers, managers, and policy-makers.

There is need for change in curricula even in degree courses leading to B.Sc. (Agric.) and B.Sc. (Eng.). For example, engineers, who are concerned with reservoir construction and mining operations, though being competent in their respective specialities, have no concept of the needs and techniques of soil conservation and restorative ecology. All such gaps in the type, quality, and quantity of training need to be filled up on a priority basis.

Concluding remarks

Land, water, and forest, the three important environmental assets, fall under the jurisdiction of the States under the Indian constitution. Although some progress has been made in soil and water conservation, immediate concerted action is required. Today, on account of the escalating population of both human beings and livestock, India has one of the highest man-to-land and animal-to-land ratios in the world. The resultant poverty can be directly linked to the state of the land degradation on account of deforestation due to overcutting and ever-increasing firewood and fodder needs and the ensuing soil erosion (Khoshoo, 1982, 1989, 1990). Nearly 175 out of 329 Mha of land are degraded, to a varying extent. Due to deforestation alone, around 1.5 Mha of forests are lost every year (Khoshoo, 1989). In addition, another 1 Mha are lost due to unsustainable agricultural practices, faulty irrigation and waterlogging, water and wind erosion, and unscientific road building and mining activities. There follows a vicious cycle of droughts and floods leading to a vicious downward spiral of rural poverty. The poor over-use land resources as a survival strategy. Therefore, much of India's poverty is intimately connected to land degradation (Khoshoo, 1990). Agriculture and animal husbandry are the two chief vocations of 76% of the people in India living in 567 000 villages. These two activities, together with forestry, constitute the three land-based activities of the majority of the populace. There is need for cooperative and coordinated action in order to first halt, and then reverse, the process of large-scale soil degradation. There is also a need for comprehensive land management programmes using modern technologies. If the country fails in this front, many of the gains made in the industrial sector will soon wither. Sustainable agriculture will continue to be important to India's well-being for the foreseeable future. Mahatma

Gandhi's philosophy of *Antodya* (welfare of the weakest) leading to *Sarvodaya* (welfare to all) will remain central to all the anti-poverty schemes which are essentially based on healthy cropland, grassland, and woodland.

The most urgent task before the country is to evolve a holistic Soil and Water Conservation Policy. In fact, this is overdue, and has now become all the more important on account of the decision of the Government of India to establish the institution of *Panchayati Raj* (governance through village councils) at the grass-root level. Such a policy has to be comprehensive and should tie up all the related aspects (policies on land tenure, water, irrigation, forestry, grazing, livestock, etc.) into an integrated whole (Khoshoo, 1990). Such a policy should also delineate clearly the Central and State responsibilities, as also those of farmers involved in resource-rich industrial agriculture and resource-poor subsistence agriculture. There is also a need for a high-powered continuing committee to oversee and monitor implementation, and suggest changes and update the policy as and when required. Land and water are critical to the well-being of an agriculture-based country like India and must get top priority. These are among the key elements for long-range ecological security and also sustainable agriculture in India.

Such a comprehensive policy, as and when evolved, would need long-term commitment for adequate and assured funds. Herein lies the salvation of the teeming millions in India: a rich country of poor people (Khoshoo, 1990).

6

Soil erosion and conservation in Australia

K. EDWARDS

Introduction

In any discussion of soil erosion in Australia it is necessary to place soil erosion and its effects in context. Despite the size of the Australian continent, because of constraints of climate, terrain and soil, the area potentially suited for agriculture is limited to 77 million hectares (Nix, 1976). There are many different forms of degradation affecting this land resource including: soil erosion, salinity (including both irrigation and dryland salinity), declining soil pH, a decline in soil structure, and the invasion of grazing lands by inedible shrub species.

Most of this degradation can be linked to inappropriate land uses. There is considerable interaction between the various forms of degradation; for example, erosion and soil structural decline are frequently associated with the same factors. The costs of each form of degradation are hard to quantify but one estimate is a reduction in annual production of more than A\$200 million for the Murray–Darling Basin in southeastern Australia (Table 6.1). The Basin accounts for about half of the nation's rural production which is worth about A\$15 000 million annually (1985–86) (Hawtrey, 1987). This review concentrates on soil erosion and soil structural decline and their combined effects on production.

Rates of erosion

On the basis of a nationwide survey conducted in 1975, some 2.7 million square kilometres or 51% of the area of the country used for agricultural and pastoral purposes (5.2 million square kilometres) requires treatment for at least one form of degradation (Department of Environment, Housing and Community Development, 1978). The study dealt with the land resources of Australia in two parts; an arid zone (where low rainfall limits land use) and a non-arid zone. The arid zone covers approximately 70% of the continent and within it is a large area, some 2.7 million square kilometres, that is not used for agricultural or pastoral purposes and was not included in the survey.

In the arid zone, vegetation degradation has affected some 55% of the area surveyed (Table 6.2). This degradation is generally associated with soil erosion. In the non-arid zone, water, either alone or in combination with wind, is an important agent of degradation on three-quarters of the land requiring treatment. Wind erosion is mainly of significance in areas used for extensive cropping and there it affects about one-third of the land requiring treatment. Some of that land is also affected by water erosion (Table 6.2).

About 44% of the area requiring treatment can be adequately managed by using farming practices such as contour cultivation, stubble mulching, temporary destocking and crop–pasture rotation. The remaining 56% needs measures which involve the construction of works such as gully control structures, and contour banks, as well as associated practices. At June 1975 prices these works were estimated to cost A$675 million with annual maintenance costs of A$50 million (Department of Environment, Housing, and Community Development, 1978).

The survey provides estimates only of the areal extent of erosion and not of the magnitude of the erosion. Since the time of the survey, a major drought

Table 6.1. *Land degradation costs: estimated value of annual yield losses in the Murray–Darling Basin of New South Wales, Australia (from Murray–Darling Basin Ministerial Council, 1987)*

Form of degradation	A$ million/y
Largely non-reversible and cumulative:	
wind erosion of cropland	0.1
water erosion of cropland	5.0
Reversible but requires expensive infrastructure:	
shallow watertables (including saline) in irrigation areas	39.1
dryland salinization of cropland	0.4
Reversible through appropriate farm management:	
soil acidification of cropland	25.2
soil structure decline of cropland	144.8
Total	214.6

affected large areas of the country and resulted in further degradation, including severe wind erosion damage in areas that had not previously experienced this form of erosion.

Currently, areas with particular erosion problems are the extensive cropping areas of northern New South Wales and southern and central Queensland. Widespread erosion also occurs on land used for intensive agriculture in all states and especially New South Wales and Queensland (mainly on land used for sugar cane production).

Table 6.2. *Land requiring treatment for various forms of degradation in Australia at June 1975 (from Department of Environment, Housing, and Community Development, 1978)*

	Area (thousand km^2)	%
Arid zone		
Area not requiring treatment	1506	45
Area requiring treatment for:		
vegetation degradation and little erosion	950	28
vegetation degradation and some erosion	467	14
vegetation degradation and substantial erosion	284	8
vegetation degradation and severe erosion	148	4
dryland salinity – sometimes in combination with water erosion	1	<1
Area in use	3356	100
Non-arid zone		
Area not requiring treatment	987	55
Area requiring treatment for:		
water erosion	577	32
wind erosion	57	3
combined wind and water erosion	55	3
vegetation degradation	92	5
dryland salinity – sometimes in combination with water erosion	10	1
irrigation area salinity	9	<1
other	14	1
Area in use	1804	100
Other lands	2522	
Total	7682	

Data on the magnitude of soil losses are limited and are available largely only from the eastern states. A summary of available data is presented for small plots (Table 6.3) and small catchments (Table 6.4). Data for larger catchments have been summarized by Olive and Walker (1982).

Table 6.3. *Soil loss data from plot studies*

Land use	Location	Soil loss (t/ha/y)	Source
Bare			
	Ginninderra, ACT	44.0	Kinnell (1983)
	Cowra, NSW	31.3	Rosewell (1986)
	Gunnedah, NSW	87.0	Rosewell (1986)
	Inverell, NSW	51.4	Rosewell (1986)
	Wagga, NSW	59.4	Rosewell (1986)
Cropped			
	Cowra, NSW	0.1–11.9	Edwards (1987)
	Gunnedah, NSW	0.6–9.6	Edwards (1987)
	Inverell, NSW	0.0–14.6	Edwards (1987)
	Maluna Creek, NSW	16.0	Elliott *et al.* (1984)
	Wagga, NSW	0.2–2.9	Edwards (1987)
	Wellington, NSW	0.0–1.7	Edwards (1987)
	Nambour, QLD	7.0–36.4	Capelin and Truong (1985)
	QLD	70–150	Prove (1984)
Pasture			
	Cowra, NSW	0.0–1.3	Edwards (1987)
	Gunnedah, NSW	0.0–1.1	Edwards (1987)
	Hunter Valley, NSW	0.9–1.0	Elliott and Dight (1986)
	Inverell, NSW	0.0–1.9	Edwards (1987)
	Maluna Creek, NSW	0.0–0.3	Elliott *et al.* (1984)
	Scone, NSW	0.0–0.1	Edwards (1987)
	Wagga, NSW	0.0–0.3	Edwards (1987)
	Wellington, NSW	0.0–0.8	Edwards (1987)
	Nogoa, QLD	0.0–21.1	Ciesiolka (1987)
Bushland			
	Cattai, NSW	0.0–0.3	Humphreys and Mitchell (1983)
	Cordeaux, NSW	0.0–0.7	Humphreys and Mitchell (1983)
	Engadine, NSW	39.6–64.2	Atkinson (1984)
	Lidsdale, NSW	0.1–7.0	Humphreys and Mitchell (1983)
	Narrabeen, NSW	2.4–8.0	Blong *et al.* (1982)
	Shoalhaven, NSW	0.0–4.3	Williams (1973)
	NT	0.2–2.4	Williams (1973)
Mine sites			
	Hunter Valley, NSW	0.4–11.8	Elliott and Dight (1986)
	Jabiru, NT	20–102	Duggan (pers. comm.)

The soil losses presented in Tables 6.3 and 6.4 generally refer to the results recorded over a number of years; however, much higher losses have been recorded for shorter periods or for single events. For example: over a three-month period there were losses of 140 to 686 t/ha on potato lands in Western Australia (McFarlane, undated); losses of 350 t/ha resulting from a series of storms on cultivated land in New South Wales (Harte *et al.*, 1984); losses up to 350 t/ha were recorded for a single storm falling on freshly seeded soils in Western Australia (McFarlane and Ryder, 1986); and losses of the order of 380 t/ha were reported over one wet season in canefields in Queensland (Prove, 1984).

One form of erosion that is rarely quantified but is of major importance in southeastern Australia is gullying. Generally gullying is expressed in terms of length of gully per square kilometre but with no estimate of rates of sediment production. Crouch (1987) has presented a method for classifying gully sides and assessing erosion rates for each form. Rates up to 75 mm/y (> 1000 t/ha/y) were reported for a catchment in central New South Wales. The potential contribution to sediment yield for catchments with a high gully density is apparent.

Before summarizing the available data, three comments are necessary. One: a broad range of methods have been used by different researchers in assessing erosion rates and the results obtained are not directly comparable. It is necessary to study the methods and the period of observation to assess their value. Two: results have been obtained over different periods of time

Table 6.4. *Sediment yield from small catchments*

Land use	Location	Sediment yield (t/ha/y)	Source
Cropping			
	Capella, QLD	0.0–16.1	Sallaway (pers. comm.)
	Greenmount, QLD	2.1–61.0	Freebairn and Wockner (1986a,b)
	Greenwood, QLD	1.8–31.6	Freebairn and Wockner (1986a,b)
	Adelaide Hills, SA	2.3	Hartley *et al.* (1984)
Grazing			
	Ginnindera, ACT	0.0–0.4	Costin (1980)
	Hunter Valley, NSW	1.1–11.0	Lang (1984)
	New England, NSW	0.0–0.1	Field (1984)
	Wagga, NSW	0.0–1.7	Adamson (1974)
	Nogoa, QLD	1.6–113	Ciesiolka (1987)
	Collie, WA	1.5–2.0	Abawi and Stokes (1982)

and, as highlighted earlier, short periods of records can lead to biased results. Three: the catchment area on which observations have been made varies enormously, from two square metres to hundreds of thousands of hectares. While the same processes might operate on all sizes of catchments, the relative importance of the processes varies greatly. Overall, sediment yield per unit area decreases as catchment size increases because the number of sediment sinks also increases. However, for small catchments, erosion rates may actually increase with catchment size because of greater opportunities for concentrated flow.

General conclusions to be reached from the available data are:

1. Soil loss rates in Australia generally increase from south to north and from inland to coastal regions as a result of changes in rainfall erosivity. The pattern demonstrated by Freebairn (1982) is illustrative of this trend for bare-fallow cultivation of crops in eastern Australia (Fig. 6.1).

2. Land-use and management practices have a major effect on soil loss. Any practice that reduces the amount of protective ground cover increases the risk of high soil losses (compare, for example,

Fig. 6.1. Annual erosion loss (t/ha/y) recorded under 'conventional' cropping practices in eastern mainland Australia. Figures in parentheses indicate erosion index (EI30) at each centre. (After Freebairn, 1982.)

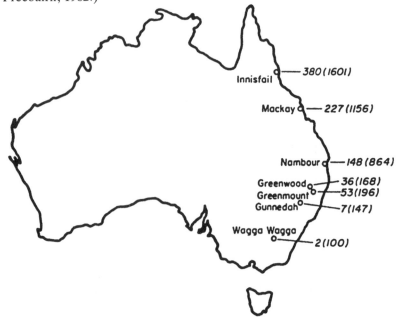

the losses from pasture, crop, and continuous bare-fallow in Table 6.3).

3. Losses from individual events can be many times the mean annual loss and can reach high levels, of the order of 300 to 700 t/ha. Many reports of soil erosion use such events to emphasize the importance of erosion and can give a biased estimate of long-term rates.

4. Losses from forests are likely to be in the range 0 to 1 t/ha/y. Accelerated rates, of the order of 10 to 50 t/ha/y, can occur after bushfires and before regeneration occurs, and during logging operations.

5. Under pasture, mean annual losses are of the order of 0 to 1 t/ha/y.

6. Under cropping, rates are of the order of 1 to 50 t/ha/y but are more likely to be at the lower end of this range. Winter cropping in the southern parts of Australia will yield the lowest values while summer cropping in the north will yield the greatest.

7. The growth of crops such as sugar cane in tropical and subtropical areas can lead to losses of 100 to 500 t/ha/y.

8. Bare-fallow conditions on lower slopes in temperate climatic areas can be expected to yield soil losses of the order of 50 to 100 t/ha/y. Much higher levels of loss would be expected in the tropics and subtropics.

Factors and policies intensifying erosion

Many forms of land use have the ability to lead to accelerated erosion. In addition to agricultural land uses these include forestry, mining, industry, urban and peri-urban development, tourism, and recreation, as well as transport and communications. Although agricultural pursuits are not the only ones causing accelerated soil erosion in Australia, because of the large area involved, their potential effect is great. The history of land use and agriculture in Australia has had a major effect in determining the degree of degradation of its soils. Prior to settlement by Europeans two centuries ago, the Australian Aborigines did not exert great pressure on the environment and the landscape remained in a state of dynamic equilibrium. The introduction of livestock and farming by the Europeans, coupled with their lack of understanding of the climate and soils of Australia, had major deleterious effects on the environment.

In addition to the introduction of grazing animals, particularly sheep and cattle, the rabbit exerted a major effect. In 1859 wild rabbits were introduced to Victoria and, because of suitable environmental conditions and a lack of predators, their numbers increased rapidly. By the end of the century rabbits had infested the southern two-thirds of the continent, all but the desert and tropical areas. The rabbits, combined with sheep and cattle, reduced ground cover dramatically over wide areas and induced instability in much of the landscape. Myxomatosis, a virus disease, was introduced from South America as a biological control and this effectively reduced the rabbit population to acceptable levels in south and east Australia in 1950 and 1951 (Donald, 1967). Populations have increased since that time because of reduced effectiveness of the virus, and control programmes using other techniques are still undertaken.

The advent of cropping, particularly for wheat, had a further adverse effect on soil degradation. Wheat yields per hectare, while never high owing to low initial soil fertility, declined until the end of the 19th century chiefly because of the depletion of soil phosphorus and nitrogen. From about the turn of the century, the introduction of superphosphate fertilizer as a source of phosphorus and the use of bare cultivated fallows led to an increase in wheat production. The use of bare cultivated fallows, of about ten months' duration, was thought to increase soil water storage for the next crop, but its greatest effect was probably to promote the breakdown of soil organic matter, freeing more nitrogen for crop growth (Donald, 1967). Therefore, while positive responses were initially obtained from fallowing, serious loss of soil structure resulted, causing widespread wind and water erosion. The increase in wheat yields had ceased by about the 1930s.

While more conservative practices are now used in cropping, cultivation of crops still presents the greatest potential for erosion. Annual winter crops are grown with a summer fallow period which generally coincides with the period of highest erosive rains, particularly in northern New South Wales and southern Queensland. In these latter areas, where it is possible to grow both summer and winter crops, a long fallow of 12 to 15 months, with its attendant increased erosion hazard, may be used when changing from one type of crop to another. Sometimes, however, it is possible to double crop in one year, thus minimizing the time when the ground surface is bare.

The growth of sugar cane in coastal Queensland is a practice associated with high soil erosion rates. Seedbed preparation can extend over a period of 9 to 12 months that includes the summer wet season. Inter-row cultivation is frequently used during crop growth. The potential for erosion from such a system is great.

Two features of Australian agriculture that are of significance in intensifying the erosion problem are its resource base and its large areas of low and highly variable rainfall. The resource base is characterized by a large area of land per unit of output; low labour input per unit of output; high capital investment on machinery per unit of labour; low input (fuel, fertilizer, and pesticides) per unit area; and, recently, declining terms of trade and profitability (Robertson, 1987).

The first four of these features are the reasons for Australia's success as an exporter of agricultural commodities yet, at the same time, they have meant little effort has been spent on identifying, investigating, or monitoring land degradation problems (Robertson, 1987). The fifth feature (declining terms of trade and profitability; Table 6.5) means increased pressure on the land to maximize production to ensure that agriculture enterprises remain viable. As well, it means that there are few funds available for continuing or expanding soil conservation programmes.

Changes in commodity prices, particularly of grains, have a serious effect on stability of the landscape especially in areas marginally suited for agriculture. As grain prices increase there is pressure to extend farming at the low rainfall margins of the cropping belt. Uncertain rainfall can result in cultivated ground with little or no protective ground cover from poor or non-existent crops. The soil is hence susceptible to both water and wind erosion. When prices subsequently fall, such lands are no longer economic to farm, but, as suitable pasture species and pasture establishment techniques are frequently not available for these areas, low-grade, volunteer pastures result. These are then vulnerable to further degradation.

The climate of Australia is characterized by low rainfall over large areas and droughts of varying duration and intensity. The arid area of Australia,

Table 6.5. *Index of declining terms of trade for agricultural producers in Australia (Robertson, 1987)*

Comparison of quantity of goods that could be purchased by 1 tonne of wheat	1974–75	1984–85
1 tonne of wheat (average return less freight, $/t on farm)	92.50	124.85
Superphosphate (tonnes)	6.29	1.28
Urea (tonnes)	1.06	0.43
Chemical ('Spray Seed', litres)	26.4	17.0
Fuel (diesel, litres)	1592.0	227.0
Freight and handling (per tonne)	7.34	3.96
Tractor power (75 kW tractor, kW)	0.63	0.22
Tractor power (130 kW tractor, kW)	0.48	0.18

where rainfall is too low or unreliable to allow regular cropping or establishment of improved pasture, occupies some 70% of the continent. About two-thirds of the arid zone is used for extensive grazing by livestock.

Periodic droughts are experienced in all parts of the country. They typically result in a reduction in the level of vegetative ground cover through poor crop growth and increased grazing pressure. This in turn increases the erosion hazard when drought-breaking rains subsequently fall.

Some government-funded drought assistance practices can exacerbate erosion problems, particularly those practices that encourage landholders to maintain livestock levels through the use of schemes that subsidize the purchase of fodder for livestock. Strategies that encourage destocking of drought-affected areas would be more beneficial from an erosion point of view. In general, not enough emphasis is given to the soil component of the animal–plant–soil system in terms of drought management. This is a major concern as this component is not replaceable, as are the other two components.

Other factors have had an indirect effect on land degradation, including the existence of subsidies for clearing of land up until 1983, and land settlement policies that frequently encouraged unskilled individuals to take up holdings that were too small to be viable. Such policies were aimed at encouraging 'development' of rural Australia and to provide occupations for ex-soldiers and British migrants after World War I (Shaw, 1967).

A major failing in policies to date has been that economic incentives and regulations typically apply to a few inputs or activities (e.g. banking systems and waterways) in circumstances where many activities may have an impact on soil erosion. With such policies, 'land users will not only select the socially least-cost methods to combat land degradation, but they may also respond by altering levels of other activities (e.g. increase stocking rates) with the result that there is little if any abatement of degradation' (Chisholm, 1987).

As well as factors intensifying the erosion problem, three further factors have been identified as limiting the prevention of land degradation (Burch *et al.*, 1987):

1. Insufficient monitoring of the extent and magnitude of land degradation;
2. inadequate application of existing information; and
3. lack of appreciation of the costs of degradation, especially long-term economic and social welfare costs.

These reflect failings on the part of both private landholders and government agencies with responsibilities for land management.

In addition, perceptions of, and attitudes to, erosion and conservation seem to be a major impediment to the implementation of successful erosion control programmes. Erosion generally has a low level of perception among the general populace until major events occur (such as the major dust storm over Melbourne in 1983). Even landholders frequently fail to recognize that they have an erosion problem (Edwards and Wild, 1987) although they can recognize that a problem exists on a neighbour's property (Rickson *et al.*, 1987). Furthermore, the link between erosion and management is not always clearly seen and its offsite effects are even more difficult to perceive.

These points suggest that soil conservation agencies need to re-evaluate their monitoring, education, and extension programmes. In doing so they should consider the possibility that better information does not necessarily lead to a better recognition of the cause of a problem or to a greater acceptance of responsibility for that problem (Rickson and Stabler, 1985). Other avenues, such as increased regulation, may warrant further investigation. The situation where less than 15% of the land is held under freehold title, with the remainder belonging to the state, has been criticized as in part exacerbating the degradation problem, as a lack of security tends to encourage more exploitative land uses. However, a pattern exists where lands in the more fertile regions tend to be mostly freehold, while the less productive lands are held under Crown lease or licence. Generally, it is on these freehold lands that the worst degradation is occurring (Messer, 1987), and so the lack of security of tenure may not be a real cause of land degradation in Australia.

Factors and policies lessening erosion

As a consequence of soil structural decline and increasing erosion, soil conservation agencies were established. The first was in New South Wales in 1938, and this lead was followed by all the other states. Because of the system of government in Australia, soil conservation matters are the responsibility of the various states, and there is no national soil conservation service. Most activity in the early years of these agencies concentrated on the use of mechanical means of soil conservation (construction of banks and waterways) with less attention being given to land management practices. As time progressed, increasing attention was directed toward integrating soil conservation with production following the realization that sound cropping and grazing practices were the key to soil conservation.

In addition, legislation now exists to limit the rights of individuals to clear timber. This generally only applies to certain areas; for example in New South Wales there are restrictions on clearing on slopes over 18° in prescribed catchment areas, within 20 m of defined streams, and in state-owned land held under lease by individuals.

Another boost to soil conservation came through the widespread adoption of legume-based pastures in the post World War II period. In this system crops were separated by one to four years of grazed legume pastures (principally *Trifolium subterraneum*). Such systems were adapted to suit a range of environments from southwest Western Australia to southern New South Wales. The combined effect of nitrogen from these pastures and the reduced frequency of cultivation resulted in improved soil structure. Crop yields were also improved.

As already noted, the introduction of myxomatosis in the 1950s to control rabbit populations markedly reduced the pressure on the land and reduced the erosion hazard. This was of major significance for soil conservation.

In the 1970s the use of alternative cultivation practices was pursued with greater interest due to an increasing awareness of the soil erosion problem, a strong campaign by chemical companies and, finally, due to increasing fuel costs. The alternatives included stubble mulching as opposed to stubble burning after harvesting, reduced numbers of tillage operations, and the use of less-aggressive tillage equipment. In many areas, the disc plough has been replaced as the standard implement for primary tillage by the scarifier and chisel plough. Scarifiers are commonly used for secondary tillage and the general trend has been to place less emphasis on preparing a fine seedbed. Crop residues are left as a protective mulch rather than being incorporated or burned. No-tillage crop production is used in some cases. In the Darling Downs of Queensland, the effect of alternative management practices on soil loss has been well demonstrated at two sites (Freebairn and Wockner, 1986a,b). On a black earth, mean annual losses over an eight-year period ranged from 2.1 t/ha/y for zero tillage to 61.0 t/ha/y when bare-fallows were maintained prior to sowing of the crop. Stubble mulching resulted in losses of 5.3 t/ha/y, stubble incorporation 17.9 t/ha/y, and summer cropping with a winter fallow 22.3 t/ha/y. On a grey clay soil over a six-year period winter losses were 31.6 t/ha/y (bare-fallow), 19.8 t/ha/y (summer crop), 7.8 t/ha/y (stubble incorporated), 3.9 t/ha/y (stubble mulch), and 1.8 t/ha/y (zero tillage).

Similar reductions in soil loss have been achieved in sugar cane production through the use of different tillage practices. For example, using standard techniques, losses of 70 to 150 t/ha have been recorded compared with < 23 t/ha (no cultivation, burning) and < 5 t/ha (no cultivation, no burning) (Prove, 1984).

This conservation farming approach has not been without its problems. It generally requires specialized machinery not owned by all farmers and also demands a higher degree of management skill than do conventional tillage practices. Reduced yields sometimes result from the introduction of conservation practices and it is now widely recognized that individual approaches to conservation farming will be required for each property based on its physical properties and the ability of its manager.

In Australia, as elsewhere, the last two decades have seen a rise in the conservation movement and this has had some benefit for soil conservation programmes. Land degradation is now recognized as being the single most important conservation issue by both conservation groups and the federal government in Australia, and the soil erosion problem is attaining a higher profile in the broader community (Mosley, 1984; Sinden, 1987). The extent to which this can be translated into more effective action is not clear although Hallsworth (1987a) sees this recognition as being of major significance for the future control of land degradation in Australia.

Encouragingly, the Australian government has instituted a National Soil Conservation Programme (NSCP) to assist in soil conservation prevention, rehabilitation, and research on a national basis. While the funds being committed are modest, this recognition at a national level is important and is forcing a more integrated national approach.

Effect of erosion and soil structure decline on water conservation on agricultural lands and reservoirs

Water is the factor limiting agricultural production in much of Australia. Any decrease in its availability for plant growth is therefore of great consequence. Soil degradation can have a significant effect on soil moisture availability. Through cultivation, soil structure can be degraded, for example through the formation of surface seals and plough layers of low permeability. Rates of infiltration of water can hence be reduced and the water available for plant growth lessened. The decline in infiltration rates can be most marked; for example, after 70 years of cultivation the infiltration of water can be reduced by more than 90% (Table 6.6).

This is reflected in much-increased runoff amounts. At two sites in New South Wales, Edwards (1980) showed that runoff from pasture was 17–25% of that from wheat lands. This extra runoff is a loss to the production system.

Water supply dams and weirs for small towns are often sites of significant sediment deposition. Out of 89 such structures in the state of New South Wales, 15 have sedimentation problems with capacities reduced between 25 and 100% (Outhet and Morse, 1984). Water supply structures for larger towns and cities have not suffered the same fate, although water quality decline due to turbidity is related to erosion in the catchment.

As well, small irrigation dams suffer from the effects of erosion and sedimentation. For example, the Pekina reservoir in South Australia, built to

Table 6.6. *Effects of cropping on some soil physical and chemical properties*

Soil/site	Property*	Virgin	Cropped	Crop/ pasture	Source
Loam, WA	OC (%)	1.33	0.85		Stoneman (1962)
40 years crop	WSA > 1 mm (%)	19.64	18.5		Stoneman (1962)
	BD	1.24	1.68		Stoneman (1962)
Silt loam WA	OC (%)	2.79	1.19		Stoneman (1962)
37 years crop	WSA > 1 mm (%)	40.40	11.29		Stoneman (1962)
	BD	1.05	1.38		Stoneman (1962)
Loamy sand, WA	WSA (%)	54	17		Stoneman (1973)
Fine sandy loam,	Total N (%)	0.14	0.08		McIntyre (1955)
SA	WSA > 0.25 mm (%)	58	9.8		McIntyre (1955)
Sandy loam, NSW 10 years crop, 5 years pasture	WSA > 0.5 mm (%)	35.0	13.4	28.1	Hunt (1980)
Sandy clay loam, NSW	OC (%)	1.25	0.78		Harte (1984)
15 years crop	Infil. (mm)	124	70		Harte (1984)
Sandy clay loam, NSW	OC (%)	1.12	0.82		Harte (1984)
15 years crop	Infil. (mm)	224	82		Harte (1984)
Sandy clay loam, NSW	OC (%)	2.40	1.61		Harte (1984)
15 years crop	Infil. (mm)	295	213		Harte (1984)
Loam, SA 70 years cropping, pasture	Infil. (mm)	230	12	190	French, quoted in Young (1983)
Grey clay, QLD 10 years crop	Total N (%)	0.33	0.23		Russell, quoted in Clarke (1986)

*OC = organic carbon, WSA = water-stable aggregation, BD = bulk density, Total N = total nitrogen, Infil. = infiltration rates

irrigate 140 hectares in 1907, had its capacity reduced by 63% and was useless for irrigation after a 56-year period (Young, 1983).

In contrast, sediment has not been a problem in the large irrigation reservoirs of New South Wales. The numerous sediment sinks, especially farm dams, in their large catchments have proved to be highly efficient at retaining most eroded material (Outhet and Morse, 1984). These farm dams, have had their storage capacity reduced by sediment at the rate of approximately 1% per annum where their catchment areas have been subject to cultivation (Neil and Galloway, 1989). The rate of reduction is 80% less where catchments are under pasture.

Loss of nutrients with erosion

There is a paucity of data concerning losses of nutrients with erosion. Some general comments can be made with respect to phosphorus losses (Cullen and O'Loughlin, 1982): (a) there is considerable consistency in total phosphorus losses from catchments across southern Australia with mean rates of the order of 0.4 kg/ha/y; (b) land-use impacts on phosphorus loss from a catchment are marked, with irrigation areas losing 0.6–1.0 kg/ha/y, forest areas 0.01–0.2 kg/ha/y, and other agricultural areas at levels between these two; (c) most phosphorus is lost during storm events. Other data show that losses of nutrients tend to vary with erosion rates. Pastures experiencing low erosion rates in southern Australia lose nitrogen, phosphorus, potassium, and sulphur in runoff at low rates of 0.62, 0.12, 1.93, and 0.06 kg/ha/y, respectively (Costin, 1980).

At another site with little or no accelerated erosion, 0.03 kg/ha phosphorus and 0.9 kg/ha of nitrogen were lost over a one-year period (Chittleborough, 1983). Where intensive horticulture was practised levels were accelerated to 6.4 kg/ha phosphorus and 20.1 kg/ha nitrogen. Most of the nitrogen and phosphorus were lost in particulate form, not in solution.

Phosphorus is strongly adsorbed on to clay particles and therefore its movement is largely controlled by the movement of soil material. Management practices aimed at minimizing soil erosion therefore are expected to conserve nutrients, especially phosphorus, as well as soil.

Comparison of nutrient loss with fertilizer application

The majority of Australian soils require regular applications of phosphatic fertilizer to maintain levels of production when cropped. Superphosphate is the usual form of application. Although annual rates of application of superphosphate vary considerably with crop types, soil type, and location, rates of 100–150 kg/ha/y superphosphate (approximately 10–15 kg/ha/y phosphorus) are typical for many situations.

Some of the higher fertility soils such as the black earths have phosphorus levels of approximately 1800 kg/ha; soils such as red brown earths, red earths, and grey and brown clays that are used for much of the cropping in southeastern Australia have levels of about 600 kg/ha; while the low fertility of solodics have levels of about 200 kg/ha (McGarity and Storrier, 1986).

Sites experiencing soil erosion at rates of 5 t/ha/y would be losing about 1–2 kg/ha/y of phosphorus; i.e. equivalent to about 10% of the fertilizer application. However, as noted before, much of the sediment, with phosphorus adsorbed, is deposited within the same paddock and is not lost to production except in major erosion events.

The rates of loss given by Costin (1980) above should be compared with fertilizer inputs of 12 and 15 kg/ha/y for phosphorus and sulphur, respectively, and of 100–500 kg/ha/y for legume-generated nitrogen in pastures. On irrigated, or intensively cropped areas, such as sugar cane, general levels of fertilizer application will be considerably higher.

Effect of cultivation and erosion on soil organic matter

Agricultural practices in cropping regions generally aim to modify the soil environment to provide a suitable seedbed and to control weed growth in the prospect of better crop yield. In so doing they also degrade the soil environment through reducing aggregate stability, increasing bulk density, reducing soil organic matter and nutrients, and increasing erosion. The levels of soil organic matter have a major controlling effect on the behaviour of soil aggregates, especially in medium to coarse-textured soils. Traditionally, stubble has been burned after harvest in most cropping lands, resulting in a decline in organic matter levels in soils already low in organic matter. This, combined with relatively low levels of productivity of much of the extensive cropping lands and the physical impact of cultivation machinery, has meant a rapid decline in organic matter content of most cropping soils relatively soon after cultivation. This has generally been reflected in a decrease in the physical attributes of the soil. Aggregate stability can be improved under pasture and under reduced tillage practices of various forms.

A number of comprehensive reviews examining the effects of cultivation on erosion and soil properties have been compiled including Clarke (1986), Conacher and Conacher (1986), and Harte *et al.* (undated). A range of data is available to exemplify some of the effects noted in the preceding paragraph (Table 6.6). The general decrease in water stable aggregation, organic carbon, and infiltration rates is clear. Bulk densities increase with cultivation.

Effect of erosion on rooting depth

Soil erosion reduces the depth of soil material and in so doing reduces the water-storage capacity of the soil. Erosion rates of 10–15 t/ha/y represent a loss in depth of approximately 1 mm/y. As previously noted, water shortages limit production in many of the cropping areas of Australia. Production of winter cereals in summer rainfall areas depends on soil water stored during the summer fallow. In winter rainfall areas the dependence on stored soil water is not so great. If a crop requires the soil water stored in say 90 cm of soil profile to produce a satisfactory yield, then a 10% loss of profile depth and associated water storage may reduce yields below economic levels. Such a reduction in soil depth would occur in less than 100 years of cultivation. Many soils in Australia have been cultivated for less than half this period.

Rates of erosion therefore need to be examined in the context of the depth of the soil profile as well as the rate at which soil is replaced by other processes including deposition from upslope and formation of new soil material. Most soil eroded from one point in the landscape does not move great distances except in major events and is deposited further downslope, probably in the same paddock. Over time, however, the material is moved out of the system under the effect of gravity and of flowing water.

Some new soil material is also being formed, but in Australia the general consensus is that the formation rate is extremely slow. Rates of 3 mm/100y (0.4 t/ha/y) 'are indicated for the development of soils with strongly developed texture profiles in alluvium. These rates are almost certainly higher than for soil development on bedrock' (Walker, 1980).

On this basis it can be concluded that the existing depth of soil is all that will be available to both future and current generations and that all but the most conservative agricultural practices are mining the soil resource (Beckmann and Coventry, 1987; Edwards, 1988). Therefore any reduction in soil depth should be regarded as undesirable.

Combined effects on crop yield

The effect of soil erosion on productivity cannot be isolated from the broader effect of soil degradation on productivity; indeed, on the basis of a review of soil loss/productivity relationships, Aveyard (1983) noted that 'changes in physical condition represent a more important component of yield decline than the actual amount of soil lost'. Such changes can include a decline in soil structure, increases in bulk density, and the formation of pans.

In some cases the effect of erosion is direct and immediate; for example, a crop can be completely destroyed by sand blasting during a wind storm (Birch, 1981) or sheet and rill erosion can remove seed and fertilizer and lead to a total loss of production in the affected area. However, most of the time the decline in productivity is much less marked and, given the major variations in production from year to year as a result of changes in rainfall, the effect of erosion is often unnoticed. The effect of improved technology by way of fertilizers, cultivars, and machinery further reduces the effect of erosion.

Most work in Australia has concentrated on mapping the extent of erosion and estimating the cost of rehabilitating or protecting eroded lands rather than investigating the costs associated with lost production resulting from erosion (Blyth and McCallum, 1987). As a consequence there is a paucity of data on the effect of erosion and soil degradation on crop productivity in the country. Studies have focused on comparing yields from plots with either a known erosion history or with defined amounts of soil removed mechanically (Marsh, 1982; Aveyard, 1983). Yields from the various soil loss levels are then compared with those from zero soil loss plots.

Marked, but variable, reductions in yield have been associated with varying levels of soil loss (Table 6.7). However, most of the levels of soil removal are much higher than would be realistically expected to occur (for example 75 mm and 150 mm), especially in the southern part of the continent. Additionally, it is uncertain how such removal compares with that involved in the selective, more gradual erosion process.

One of the most useful sets of data is that reported by Aveyard (1983) from long-term plot experiments. Subsequent to the termination of the experiments the plots were sown to wheat and yields recorded. Relationships were examined between total prior soil loss per plot and wheat yields (Fig. 6.2) and the decline in yield was major. For example, with soil losses equivalent to the mean total soil loss, the yield was reduced by 17% at Gunnedah and 28% at Wagga. At twice the mean total soil loss the yield decreased by more than 50% at Wagga but by only 21% at Gunnedah. The loss in productivity is thought to be due to other factors in addition to the decline in soil volume as, for example, a significant relationship between organic carbon levels and yield was evident at Wagga. The recovery of productivity of the soil subsequent to erosion takes place at a variable rate and will depend on the extent of erosion and on land management factors including land use and fertilizer applications. The depth and fertility gradient of the soil profile will also be major determinants of the effect of erosion. In a summary of soil erosion effects on production it was reported that in Western Australia depression in yield was still evident some 27 years after the original removal of soil, yet in Victoria yield effects had disappeared following four years of legume pasture after the removal of 75 mm and 150 mm of topsoil (Aveyard, 1984).

Table 6.7. *Effect of soil loss on productivity*

Location	Soil loss	Yield decline %	Source
Cowra, NSW	75 mm	28	Hamilton (1970)
	150 mm	52	Hamilton (1970)
	99.5 t/ha	25	Aveyard (1983)
	2 mm	4	Aveyard (1983)
	5 mm	0	Aveyard (1983)
	10 mm	15	Aveyard (1983)
Gunnedah, NSW	75 mm	10	Hamilton (1970)
	150 mm	29	Hamilton (1970)
	221.1 t/ha	28	Aveyard (1983)
	10 mm	6	Aveyard (1983)
	20 mm	8	Aveyard (1983)
	30 mm	13	Aveyard (1983)
Inverell, NSW	75 mm	6	Hamilton (1970)
	150 mm	19	Hamilton (1970)
	287.5 t/ha	40	Aveyard (1983)
	10 mm	3	Aveyard (1983)
	20 mm	−1	Aveyard (1983)
	30 mm	10	Aveyard (1983)
Wagga, NSW	75 mm	46	Hamilton (1970)
	150 mm	51	Hamilton (1970)
	80.8 t/ha	44	Aveyard (1983)
	2 mm	−8	Aveyard (1983)
	5 mm	−8	Aveyard (1983)
	10 mm	7	Aveyard (1983)
Wellington, NSW	75 mm	21	Hamilton (1970)
	150 mm	31	Hamilton (1970)
	19.9 t/ha	49	Aveyard (1983)
Mallee, VIC	150 mm	6	Hore and Sims (1954)
Walpeup, VIC	150 mm	8	Molnar (1964)
Wheat belt, WA	1 mm	3–8	Marsh (1982)
	8 mm	10–25	Marsh (1982)

As a result of a study collating all available soil loss/productivity data in New South Wales Aveyard (1983) reached the following general conclusions:

'(i) that significant losses in soil volume can result in major losses of soil productivity.

(ii) that similar levels of productivity loss can occur as a result of visually undetectable amounts of soil loss when accompanied by soil deterioration.

(iii) that there are major differences between soil types and climates in productivity response to erosion.

(iv) that the productivity response to erosion can vary markedly from season to season.

(v) that the previous standards regarding tolerable rates of soil loss were far too optimistic.'

Benefits and costs of soil conservation

Soil conservation approaches vary from agency to agency within the country and include the construction of earthworks, pasture improvement, and modification of management practices such as conservation farming and livestock grazing rotations. Different forms are appropriate for different erosion problems.

There have been few comprehensive economic studies of soil conservation projects in Australia. A landmark study was that of Dumsday (1973) where a

Fig. 6.2. Relationship between total soil loss at Gunnedah (26 years) and Wagga (30 years) and mean relative grain yield over two and four years, respectively. (After Aveyard, 1983.)

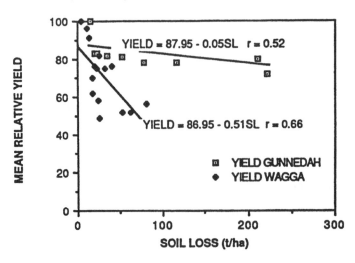

range of management systems were simulated and the economics of each were examined. The lack of soil loss and productivity data meant that extensive testing of any model was not possible. However, the studies indicated that the use of earthworks in the wheat belt of northern New South Wales and southern Queensland was questionable in most situations. The best systems involved shorter periods of summer fallow than were commonly used in the study area.

Case studies of two individual soil conservation projects are available (Department of Environment, Housing, and Community Development, 1978). In both cases soil conservation was achieved through an intensification of agriculture and utilized improved management practices as well as earthworks. The conservation works did not contribute directly to improved productivity but supported improved practices. Benefit/cost ratios for the two projects were 1.5 and 2.0.

The effect of land degradation is reflected in prices paid for properties – with land having minor degradation potential selling for higher prices, if all other factors are held constant (Sinden, 1987). In this restricted study in northern New South Wales it was also found that each dollar invested in soil conservation works resulted in increased land values of A$2.28. The same author estimates that, on all of the non-arid section of New South Wales, at the margin each extra hectare of land that becomes affected by combined sheet and gully erosion results in a reduced output of A$18. In total this represents about 5% of the gross annual production of the area. This figure differs considerably from that in Table 6.1, but reflects a different approach in methods of estimation.

In assessing benefit/cost ratios of soil conservation the offsite effects of erosion also need to be considered. Such costs may exceed the onsite costs in some situations. The database on which to make decisions in Australia is very poor.

Recommended approaches and policies for soil conservation

Day to day management of the land is in the hands of individual land users and it is these people who will determine whether or not degradation of the land continues. Any soil conservation policy must recognize this if it is to be successful. In general, solutions are known for most land degradation problems yet degradation continues. Clearly the options being presented to landholders are unacceptable either economically, socially, or politically.

Therefore effective policies to control land degradation need to take account of a land user's behavioural response; for example, subsidies for structural soil conservation works should be made conditional on the land user adopting acceptable cropping practices and stocking rates (Chisholm, 1987).

The data presented in this report indicate that the base for making decisions on land-use and management options is quite small. Likewise, the implications, both economical and physical, of making alternative decisions are not well documented. As well, techniques for making these decisions are not well developed. Policy development therefore should be based on founding a better database and on improving our means of manipulating those data. To ensure full use of the database, methods of quantifying the effects of contemplated land-use changes should be utilized. Predictive models need to be used in general practice so that 'what if' questions can be answered. Such models exist already and have been widely used outside Australia, but their use in Australia has been limited to date. The simplest of these models, the universal soil loss equation (Wischmeier and Smith, 1978), has been modified for use in New South Wales using locally derived data for validation (Rosewell and Edwards, 1988).

Implicit in this approach is the much greater use of land capability principles. Indirectly the question of the control of land use through regulation is also implied. Greater research effort of an interdisciplinary nature and having greater consideration of the input of management, not simply of modification of some aspect of the biophysical environment, is required. To ensure that desirable management systems are adopted it will be necessary to ensure that these systems not only reduce soil erosion/degradation but also maintain or improve the level of economic efficiency. If not, landholders simply will not be interested in applying such systems unless forced to do so by regulation.

The concept of total catchment management that 'aims to achieve coordinated use and management of land, water, vegetation, and other natural and physical resources to ensure minimal degradation and erosion of soils and minimal impact on water yield and quality and on other features of the environment' (Cunningham, 1987) needs to be fostered. In essence, what is needed is the development of sustainable forms of land use. Any one form of degradation, such as erosion, cannot be treated in isolation. Policies must therefore be integrated rather than directed toward single issues.

The five goals of NSCP largely encapsulate these aims and provide a useful framework for developing strategies to control land degradation and maintain, or increase, the productivity of the land (Kerin, 1987). The five goals are:

1. Sustainable land uses appropriate to land capability;
2. total catchment planning;
3. recognition and acceptance by governments and land users of their responsibilities for land conservation;
4. cooperation and coordination between all involved with land management; and
5. adoption of a land conservation ethic.

7

Soil erosion and conservation in Argentina

J.S. MOLINA BUCK

Severity of soil erosion in Argentina

Soil erosion in Argentina is a very serious problem. Although no official data exist on the entire country's soil losses per hectare, there are some data related to the principal areas with soil erosion problems. My technical experience in Argentina and many trips to the United States during the last 30 years, lead me to believe that estimates that an average of 6–12 t/ha/yr of soil are lost could be realistic. In certain areas, such as the Pampas, the erosion rate is lower than 4–5 t/ha/yr, and in others the rate is higher than 60–150 t/ha/yr. Official data indicate that approximately 24% of Argentina's agricultural land (which totals 34 254 000 ha) shows some degree of erosion (Musto, 1979). Water erosion affects 53.4% of Argentina's agricultural land area (18 300 000 ha) (Table 7.1), and wind erosion affects 47% (16 000 000 ha) (INTA, 1979) (Table 7.2).

No quantitative data have been gathered for the arid region, which accounts for 61% of the total area of Argentina (171 000 000 ha). Regional

Table 7.1. *Water erosion in the more humid areas of Argentina. Data based on a detailed study of 4 600 000 ha conducted by INTA (1957)*

Degree of erosion	Hectares	Percentage	Soil lost (t/ha/yr)
Slight	975 000	21.2	31–60
Moderate	565 000	12.5	61–120
Severe	103 000	2.2	121–360
Very severe	17 800	0.3	360+

data measured with precision at the Tezanos Pintos Experimental Station (INTA) indicated a loss of 20–50 t/ha/yr (2–4 mm of soil). Tables 7.3 and 7.4 indicate losses in the 'terra roxa' area of Sao Paulo, Brazil (Cuba, 1949; Marques, 1949). In the Province of Misiones, which has red tropical soils very similar to the 'terra roxa' soil of Brazil, soil loss could be as high as 150 t/ha/yr, with an average of 5–50 tonnes. Table 7.5 lists soil resistance to water erosion measured by the McCalla method (McCalla, 1942).

As an overall indication of the soil loss by water erosion it has been demonstrated that for each tonne of cereal for export, one tonne of silt must

Table 7.2. *Wind erosion in semi-arid regions (Musto, 1979)*

Degree of erosion	Hectares	Soil lost (t/ha/yr)
Slight	4 500 000	2–10
Moderate	7 000 000	10–15
Severe	4 000 000	15–19
Very severe	500 000	>19

Table 7.3. *Tonnes of 'terra roxa' soil lost in Brazil and percentage of water lost to runoff (Marques, 1949)*

	Tonnes of soil lost (ha/yr)	% water loss by runoff
2-plow operation	5.67	3.10
1-plow operation	3.37	1.42
Straw mulching	1.65	1.36

Table 7.4. *Losses on 'terra roxa' soils in Brazil (Cuba, 1949) on land with 5% slope and 250 mm of rain*

	Soil lost (t/alqueire)*	Water lost (m^3/alqueire)*
Straw burning	60.84	1210
Stubble mulch farming	3.24	338

*1 alqueire = 2.5 ha

be dredged out of the channels of access to the port of Buenos Aires (Kugler, 1983). Argentina's annual cereal export is 20 million tonnes, and it is necessary to dredge 20 million cubic meters of silt from the channel. The cost of dredging the silt is $1.50 per cubic meter, and in 1983, with 21 750 000 tonnes of cereal exported, the total cost of dredging was $61 250 000 (INTA, 1983).

Factors and policies intensifying erosion

The principal factor leading to continued erosion is lack of national awareness of the importance of the problem. Argentine culture is typically urban. Agricultural problems are not very important to the mass of people, including politicians, teachers, journalists, etc. In 1957, after visiting the principal agricultural areas of Argentina, Dr Hugh Hammond Bennett, the founder of the U.S. Soil Conservation Service, indicated that 'Education at all levels, from primary school to special courses for politicians, teachers, journalists, and so on, is fundamental to create a national awareness of the importance of soil conservation for the future of the country' (Bennett, 1957).

During an FAO Conference on Soil Erosion in America held in Buenos Aires in 1969 (FAO, 1969), a complete plan was presented to promote education in soil conservation at all levels, similar to that proposed by Bennett in the United States. The plan was discussed and approved, but was not actually applied and put into practice.

The technical factors that intensify soil erosion in Argentina include the burning of stubble; the lack of rotations in the Pampas; the overgrazing of natural and cultivated pastures and burning and overgrazing of natural pastures in all the arid zones; the shifting cultivation in the subtropical areas that destroys more than 10 000 ha/yr in only one province (Misiones); double culturing of wheat and soybean in one year with the burning of stubble;

Table 7.5. *Soil resistance to water erosion. Data from Molina and Spaini, 1949)*

	Number of drops required to destroy a soil aggregate (media of 10 determinations)
Virgin soil	>300
Eroded soil ('capuera')	7

burning of cotton stubble throughout the Cotton Belt all over the Great Chaco; and the use of bare fallows throughout much of the year; the use of moldboard plows in semi-arid regions instead of one-way discs, heavy harrow discs, 'Graham plows', etc. (Sauberan and Molina, 1956).

Among national policies that intensify soil erosion, the main factor is the increase of crop production instead of cattle production in the Pampas. This prevents the possibility of good grassland farming similar to the ley farming of England, a system that was the common practice in the Pampas until recently (Sauberan and Molina, 1957). The combination of crop production and cattle production even on small farms gives excellent economic results, conserves the soil, and lowers the risks for the farmer. Cattle production in the Pampas is inexpensive because the cattle consume only grass with no grain or other supplement. Cattle do not need to be protected during the winter months because snow is rare, occurring only once every 15–20 years. Crop production on its own under the present Argentine situation is very risky, but combination with cattle production lowers the risk (Molina, 1981a).

Factors and policies improving soil conservation

Since 1957, soil conservation efforts have increased dramatically. One of the biggest efforts was made by ranchers (*estancieros*), producers of meat in the 'Meat Belt' in western Buenos Aires Province, in the semi-arid Pampas. The ranchers created an organization very similar to CETA (Centre d'Études de Technologie Agricole) in France, but with the same philosophy as the Soil Conservation Service of the United States (Cucullu, 1961).

As technical adviser to the first Argentine CREA (Consorcios Regionales de Experimentacion Agricola), which consisted of 42 000 hectares divided into 12 ranches (*estancias*), I directed all our efforts toward controlling wind erosion, which is very serious in this area. By using good 'grassland farming' with a rotation of 4–5 years of alfalfa for grazing with 2–3 years of crops or cultivated pastures such as rye, oats, or barley consumed as grass for cattle; by replacing the moldboards by one-way plows, heavy harrow discs (Goble Massey Harris type) or 'Graham plows'; by leaving 50% of the stubble upon the soil to protect it from rain impact; and by not burning the stubble and by avoiding overgrazing pastures and never burning, it was possible to reduce erosion of soil by wind to near zero in only one year (Molina, 1986). The results of this first CREA group's efforts were so important that in only a few years the number of these groups increased to more than 70, covering a total of more than 3 million hectares.

The same ideas applied by small farmers in the more humid area of the Pampas yielded similar results. In this area, the principal problem was water erosion, which in 40 years had completely ruined the soil. The first CREA group of small farmers covered 2500 ha with more than 40 farms, each farm housing one to four families (Sauberan and Molina, 1962; Aguilar, 1964). By suppressing the burning of stubble and incorporating it superficially with heavy harrow discs, by chiseling the soil with Graham plows made on the farms with old moldboard plows, and by following a good rotation of the grassland farming type, we were able in a very short time (1–3 years) to eliminate the problem of soil erosion and at the same time to control flooding. In practice the official policy has been erratic and not very positive. All efforts are directed toward soil conservation laws and cartographic studies. In the field of practical erosion control, the principal effort has been made by the National Institute of Agricultural Technology (INTA), which has issued many publications on the subject (INTA, 1957, 1958, 1961, 1979, 1983).

All the good results in controlling wind and water erosion in Argentina were obtained by applying a new theory about soil conservation and aerobic decomposition of cellulose (Molina, 1974). A short resume was presented to the Xth International Congress of Soil Science (Moscow). The summary of this paper is as follows: 'After 25 years of laboratory and field experiences a new theory about the relations existing between soil conservation and the aerobic decomposition of cellulose was elaborated.'

In this chapter I describe the application of these theoretical principles to the common farming and range management in the Pampas, which is the main meat and grain producing zone of Argentina.

The utilization of aerobic cellulose bacteria which produce large amounts of polyuronic colloids (up to 16% of the total decomposed cellulose) has reduced soil erosion and increased grain and meat production with a reduction in the unit cost (Molina, 1981b).

Under field conditions it is estimated that wheat and corn stubble have a straw yield of 1–5 t/ha/yr, with 30% cellulose content. As this colloid is active in a 0.1% dose, we can estimate that 100–200 t/ha of soil could be improved each year.

The ecological conditions required for a good growth of this kind of bacteria are (a) good water supply, (b) good aeration of the soil (good structure), (c) the presence of mineral elements (especially Ca), and (d) the presence of nitrogen (which could be provided by aerobic nitrogen-fixing bacteria that use the energy source provided by anaerobic bacteria of cellulose).

Effects of erosion on water conservation on croplands and water storage facilities (reservoirs)

Erosion significantly reduces water retention on agricultural soils. An eroded soil has a much lower water-holding capacity than a non-eroded soil. Exposing the soil to the impact of raindrops diminishes water penetration into the soil and encourages rapid water runoff. At the same time it increases soil erosion and the problem of floods. This type of erosion is particularly serious on very flat soils (Myers, 1950). About 200 liters of water are needed to produce one corn plant and 5000 liters of water are needed to produce 1 kg meat. Each millimeter of water that cannot pass through the soil is at the same time a source of erosion and is lost for agricultural or cattle production.

A second effect of erosion is the hard 'plow pans' that exist in all Argentinian soils without a good rotation with alfalfa or sweet clover. The use of chisel plows ('Graham plows') can provide a quick solution but needs to be combined with a straw mulch on the soil surface. The production of polyuronic colloids during aerobic decomposition of cellulose allows a very high water stability of soil aggregates (Molina and Spaini, 1950; Molina and Sauberan, 1956; Molina, 1968, 1973).

The use of straw mulch in agricultural fields and the dung remains from grazing animals upon the soil surfaces in the natural and cultivated pastures gives very good protection to the soil and allows rainwater to percolate into the soil.

A direct relationship exists between soil organic matter content and its water-holding capacity, especially from the point of view of usable water. When the organic matter content increases from 2 to 5.9%, a 65% increase in available soil water can result. These differences in organic matter content are very common in the Pampas, between the end of the agricultural cycle and the last year of the alfalfa cycle. Alfalfa can produce more than 10 000 kg/ha of soil organic matter per year (Goedewaagen and Schuurman, 1950).

The use of water-storage reservoirs has been successful at the 'Greater Chaco Gualamba' where there is no potable water in the subsoil. The only means of providing drinkable water for humans and animals is to build large reservoirs (10 000 m^3 capacity) (Molina, 1977). On one big ranch the construction of 40 reservoirs completely solved the problem of drinking water. However, if the basins of each reservoir are not protected, in a few years they will be completely filled with silt. But with a good conservation policy there would be no problem (Molina, 1978).

In very large reservoirs, the problem of siltation in an area very similar to Misiones was studied, and it was found that each cubic meter of water that flows into the reservoir contains 1.6 kg of silt (Kugler, 1983). This study gave rise to FAO's 'Umbrella Operation' in the River Plate Basin (Molina, 1976).

Using information from H. H. Bennett, a method was developed in Argentina similar to that in use in the Great Plains of Canada and the United States. Thirty centimeters of available water in the soil depth results in poor yields. Sixty centimeters indicates a medium yield possibility, 90 cm depth of moisture gives the farmer an 80% chance of obtaining a good yield. The determination of soil moistures is made at the seeding time.

In an experiment at the Agricultural Experimental Station at Tezanos Pintos (INTA), the losses of water and soil as listed in Table 7.6 were noted.

Quantity of soil nutrients (fertilizers) lost annually from erosion

Fertilizer use in Argentina is very low, except in areas with irrigation or commercial crops, such as sugar cane, fruit production, or vegetable culture. In the 'grassland farming' system with four to five years of alfalfa and two to three years of cash crops, commercial fertilizers are unnecessary because the alfalfa adds nitrogen to the soil and the other nutrients are conserved.

The virgin soil of the Pampas is rich in phosphorus, but in exhausted or eroded soils this phosphorus is not available for crops. However, with only the incorporation of organic matter (rich in cellulose) such as stubble, this unavailable phosphorus is made available by a reaction similar to that described by Bradley and Sieling (1953) and Molina and Sauberan (1957).

The loess subsoil has adequate calcium and, after four to five years of planting alfalfa with good management, enough calcium is transported from the subsoil to the surface for production. The decrease in the area planted to alfalfa over the last ten years has led to increased acidity of soils throughout

Table 7.6. *Losses of water and soil in 1978–79 with 1009 mm of rain*

Agricultural system	Soil loss (t/ha)	Water loss (mm)
Soybean (without contouring)	19.3	115
Soybean (contour)	10.3	77
Soybean (stubble mulch)	0.2	31
Wheat (without contouring)	0.2	11
Fallow (bare fallow)	22.0	234
Natural pasture	0	9
Cultivated pasture (alfalfa with fescue)	0	7

the Pampas (Molina and Delorenzini, 1978). This phenomenon, however, is not confined to the Pampas. In many of the principal agricultural areas of Argentina, the acidity of the soil has now increased so much that it is difficult to obtain good results with new seedings of leguminous plants, especially alfalfa.

In the last ten years, the policy of the Argentine government has been to increase the use of commercial fertilizers, especially nitrogen compounds such as urea. The cost of fertilizers given to the farmers is paid with wheat in a relation of 2.5 kg of wheat per each kilogram of urea given by the government. The results, in my opinion, are poor especially in the eroded areas where there is no response to the fertilization.

I estimate that the loss of applied fertilizers in Argentina is very similar to that in the United States. According to data from the 'Workshop on Nitrogen Cycling in Ecosystems of Latin American and the Caribbean,' organized by the Royal Academy of Sciences at Centro Internacional de Agricultura Tropical, Cali, Colombia, in 1981, only 25% of the nitrogen applied to the soil is recovered in the crop (Robertson *et al.*, 1982).

On the 7 to 8 million hectares of alfalfa in Argentina, if we could obtain 700 000 to 1 600 000 t of nitrogen fixed from the air (calculating 100–200 kg of N/ha/yr), that would be enough for the needs of most crops. Recent studies on non-symbiotic nitrogen-fixing bacteria (*Azotobacter* and *Beijerinckia*) show that it may be possible to double nitrogen-fixation by using the cellulose from stubble as an energy source (Molina, 1963, 1970, 1981b; Sauberan and Molina, 1964; Molina and Sauberan, 1965). Many practical field experiments strongly indicate this possibility, with a soil nitrogen increase from 113 kg/ha/ yr to 292 kg/ha/yr. (These unexpected results led to good crops of wheat and cotton in completely exhausted soils that normally have no production at all, without using leguminous plants.)

Laboratory research indicates that many common products of cellulose decomposition under anaerobic or semi-anaerobic conditions are very good sources of energy for the aerobic nitrogen-fixing bacteria, such as *Azotobacter* (Molina, 1963, 1970, 1981b; Molina and Delorenzini, 1978). Loss of soil nitrogen due to wind erosion is quite high because of the removal of soil organic matter. A good sandy soil can be transformed into a dune after a big storm. On bare soil, water erosion causes the loss of 90 kg nitrogen and 5 kg of assimilable phosphorus per hectare per year (INTA, 1983). On the same soil with corn planted on the contour the losses are 3.3 kg/ha/yr of N and 0.15 kg/ha/yr of P. Table 7.7 gives details of soil and nutrient losses in soils very similar to the tropical red soils of Misiones.

Comparison of nutrient loss with annual fertilizer application

The application of chemical fertilizers is extremely low in Argentina. In my opinion, fertilizers cannot replace nutrient loss caused by soil erosion. It is necessary first to stop soil erosion and then to attempt to replace the soil nutrients used by crops.

The cost of replacing soil nutrients lost by erosion can be illustrated as follows. In only a few years, wind erosion could extract nearly all the organic matter and the nitrogen and most of the phosphorus from a sandy soil; potassium loss is minimal (Table 7.8, Bennett, 1947). To replace nitrogen lost by wind erosion 15 364 000 kcal/ha (1000 kg of N) and 2 934 800 kcal/ha phosphorus (1000 kg of P) are needed. We estimate that the loss of soil nutrients (N and P) by wind erosion could be more than six to ten times higher than the use of nutrients by the crops. The cost of replacing all the nutrients used by the crops and pastures has been estimated at more than $5 thousand million per year. This figure is based on local production estimated to total $26 thousand million over five years for the installation of 44 factories at $200 million each and the use of 2.2 liters of naphtha for each kilogram of nitrogen produced. We obtained these figures after a symposium sponsored

Table 7.7. *Loss of soils and nutrients in Sao Paulo, Brazil (Marques, 1949)*

	Tonnes of soil lost (ha/yr)	Value of nutrients lost (ha/yr) (in cruzeiro)
Virgin soil	0.05	0.06
Pasture	1.0	14
Coffee plantation	2.0	25
Cotton plantation	38.0	450

Table 7.8.

	Soil before wind erosion	Soil after wind erosion	Difference
Nitrogen (kg/ha)	1380	460	−920
Phosphorus (kg/ha)	920	trace	−920
Potassium (%)	2.5	1.77	−0.73

by 'The Friends of the Land of Argentina' with the participation of represen-
tatives of all sectors involved with the problem based on data in Pimentel
(1976, 1984).

Impact of erosion on organic matter loss

When erosion removes the topsoil, the loss of organic matter is
nearly total. The principal problem in reclaiming eroded soils is to restore the
organic matter content to a minimum compatible with the water stability of
soil aggregates. Virgin soil has about 1.06% organic matter, while eroded soil
has about 0.33% organic matter, a difference of 0.73%. The soil transported
by the wind has 3.35% of organic matter.

The net loss of energy from the eroded soil in relation to virgin soil is 0.42%
of organic carbon, which has an energy value of 76 204 800 kcal/ha. The
76 204 800 kcal lost by the soil per hectare (similar to an acre furrow slice)
indicates the severity of the wind erosion. On the 4 500 000 ha suffering from
severe to very severe wind erosion in Argentina, the resulting energy loss is
tremendous (Table 7.9). It is equivalent to 1 TEP (Tonnes-Equivalent-
Petroleum) per 10 000 kcal. This energy loss explains the difficulty in
reclaiming a soil with severe wind erosion.

If we multiplied the loss of kilocalories per hectare by the 4 500 000 ha
suffering from severe to very severe wind erosion, we would have an
astronomical quantity of 40 725 000 TEP for the entire country. (The total
petroleum production of Argentina is only 30 000 000 tonnes). Water erosion
losses of organic matter obtained by INTA (1983) at the Tezanos Pintos
Agricultural Experiment Station (Entre Rios, Argentina) are 1650 kg/ha/yr
on a bare soil and 56 kg/ha/yr on corn cultivated on the contour.

On some of the best soils in Argentina (Los Surgentes–Provincia de
Cordoba) in the middle of the 'ROSAFE' area, which is the principal wheat

Table 7.9. *Energy needed to re-*
place nitrogen, phosphorus, and
organic matter on an eroded soil

	Energy (kcal/ha)
Nitrogen	15 934 000
Phosphorus	2 934 800
Organic matter	76 204 800
Total	90 503 600

belt, we obtained the data in Table 7.10 for resistance to erosion (Molina and Spaini, 1949). Based on our experience with all types of soils, the McCalla method is the best means of appreciating soil resistance to erosion. It indicates the number of raindrops necessary to destroy one soil aggregate (McCalla, 1942). In relation to the importance of energy content in the soil, Lyon and Buckman (1950) indicate that, according to experiments at Rothamsted, England, the loss of energy from the soil is 2.5 million kcal/ha/yr in a normal soil and 37 million kcal/ha/yr in soil with plenty of manure. This loss of energy is absolutely necessary to maintain the fertility of the soil. A normal soil contains 380 million kcal/ha, and the normal consumption is 25 million kcal/ha/yr. A loss of 90 503 600 kcal, such as that indicated by Lyon and Buckman (1950), represents the energy necessary to maintain the energy balance of the soil for 3.6 years.

Each time the stubble of cereals or the natural pastures in semi-arid countries is burned, large quantities of the energy needed by the soil to maintain its fertility are lost. The first measure to be taken to control soil erosion and to avoid organic matter losses is to stop burning cellulose material.

Effect of soil erosion on soil depth

The principal problem of soil erosion in the more humid Pampas is that the water erosion in its last phase could remove all the arable land, leaving only the 'plow pan' at the surface. The soil depth to the 'plow pan' is the best indicator of water erosion intensity on the flat lands.

Root depth varies according to the degree of soil erosion. In nearly all the Pampas soils, the principal problem is water erosion (Myers, 1950). There are no gullies on the flat lands, but instead the soil is transformed into 'Sunday soils' with 'plow pans' very near to the surface and soil compaction on the surface itself (Molina, 1980, 1981b, 1986). (According to Myers (1950), farmers gave the name 'Sunday soils' to those soils where it is impossible to work on Saturday because they are too humid and on Monday because the soils are too dry. Farmers in certain areas of the Pampas changed

Table 7.10.

	Organic matter (%)	Resistance to erosion (McCalla method)
Virgin soil	3.9	>300 drops
Eroded soil	3.1	13 drops

this name to 'midday soils' because it is impossible to stop plowing to eat lunch.) Rooting depth diminishes in relation to soil erosion intensity until only a 'plow pan' is left on the surface, where seeding is very difficult if not impossible. Instead of growing vertically, the roots of the cotton plants grow horizontally on the 'plow pan'. This problem is very common in the Pampas but is the worst problem in the Cotton Belt at the Great Chaco Gualamba. In the Los Surgentes area (about 150 km west of Rosario City, Cordoba Province), before a soil erosion plan could be implemented on the land, heavy rainfall of 200 mm during one night removed all the topsoil, leaving the 'plow pan' on the surface (Aguilar, 1964). The diminishing soil depth in controlled conditions at an experimental station (INTA) gave results of 2–4 mm/yr of soil loss, with an erosion rate of 40 t/ha/yr.

The loss of one millimeter of soil from the surface may not appear to be important, but according to investigations this surface millimeter is the most important part of the soil. It is similar to animal skin; one of the best Uruguayan farmers calls it the 'skin of the soil.' Also, this 'skin' is important in relation to cellulose decomposition under aerobic conditions to produce polyuronic colloids; the process is very active only in the water–soil–air interface.

Relative to the problem of nitrogen fixation by non-symbiotic bacteria such as *Azotobacter*, working with a 'giant pore of soil' (Molina, 1963), it was possible to demonstrate that in the first millimeter of soil nitrogen fixation is very active, with an increase in nitrogen percentage from 0.198% in the control to 2.053% in the surface millimeter. The difference in percentage nitrogen between control and soil taken at 4–5 mm is insignificant. The nitrogen content of the superficial millimeter is very similar to that of nodules of leguminous plants.

The 'plow pans' of the eroded soils are not only impervious to water but also to diffusion of gases, and on the 'plow pans' the soil conditions are very anaerobic. Under these anaerobic conditions, when we incorporate the stubble with the mold-plow, there is an exaggerated production of ethyl alcohol and other volatile acids such as acetic, propionic, and butyric acid. These are, according to our experiments, very powerful inhibitors of seed germination.

Effects of loss of water, nutrients, and organic matter, and reduced rooting depth on soil productivity and grain yields

All these factors act as part of a system, and it is impossible to differentiate their individual effects. The loss of water capacity is the principal factor affecting yields, but yields depend on rooting depth and also on soil nutrients and organic matter. The organic matter itself regulates the water stability of soil aggregates and the soil's resistance against the impact of raindrops.

In an eroded soil with little resistance to water impact, the soil surface is impervious. There is a 'plow pan' at a low depth, and a low capacity for water retention. Soil nutrients are not only reduced, but many are not even available. These factors combine to reduce soil productivity. Yields are extremely low, and after a few attempts to grow crops, the land is often dedicated to cattle production with very low yield in meat or milk.

The surface of virgin or reclaimed soils is very resistant to the impact of raindrops and the water can soak into the soil without problems. In addition, there is no 'plow pan' and the roots can develop quite well, extracting water and nutrients from the lower levels of the soil, down to 1–2 m in grain crops and 7–8 m in alfalfa. These soils have a large capacity for water retention, and the surface soil has ample available phosphorus due to the reaction described by Bradley and Sieling (1953). The polyuronic colloids produced during the aerobic decomposition of cellulose in combination with the non-available phosphorus (compounds of iron and aluminum) react to form available phosphorus, and there is a rapid increase in organic matter. The same applies to nitrogen and other nutrients. Yields have increased from zero to more than 20 bushels of wheat in only 1–2 years, with a normal production of 30 bushels/acre (1 bushel/acre = 90 kg/ha) for the last 25 years. Indicated as quintals per hectare, the results are 12 and 18 qq/ha, respectively. These yields are obtained without fertilizers but employing an alfalfa rotation. In contrast, on a farm whose soil was exhausted and eroded after more than 50 years of wheat and corn crops, with no rotation or fertilizers, and burning the stubble, the yields were low and unprofitable. As a general statement, we can estimate that wheat yields are reduced by about 3 bushels/acre for each inch (2.5 cm) of soil lost. These data are quite similar to those of the United States.

With three to five years of good management, i.e., alfalfa rotations; no burning of stubble, incorporating the stubble with a heavy harrow disc into the superficial soil, leaving 50% on the soil surface, etc.) the situation could be reversed (Table 7.11).

Economics of increased grain production compared with the low cost of soil conservation practices

The economic effects of erosion are so great that they account for the difference between a very expensive agriculture and cattle production with low yields and a low-cost agriculture with very good yields. With high yields, the CREA group was able to buy new ranches in a few years. Relative

to meat production with cattle raised on pastures year-round, local production is 90–100 kg/ha/yr of meat (living weight), but within the CREA
group, production increased to 200–240 kg/ha/yr in only 2–3 years (Molina,
1986). In the subtropical areas of 'terra roxa' (tropical red soils) of the
Province of Misiones the production of 'Yerba Mate' ('Paraguayan tea') is
700 kg/ha/yr, but only for 12 years, after which there is no more production.
In contrast, in virgin or reclaimed soils a production of 2000 kg/ha/yr has
been maintained for more than 25 years. On farms with annual crops, all
topsoil is lost to erosion after three years on the tobacco or corn farms in
Misiones. In an estimate of the cost of production, a tobacco cooperative
stated that the soil must be changed every three years. The soil lasts less time
than a tractor! On a large-scale project of 90 000 ha in the heart of Buenos
Aires Province, we estimated that at a cost of 80 000 pesos (1969 value) over
three years we could obtain an increase in production of 67 million pesos
(Molina *et al.*, 1974).

In relation to wheat yields, we found that in the drier region of the western
Pampas near Eduardo Castex, water retention for good management practices ('stubble mulch farming') could increase production in a very similar
way to the United States (Table 7.12; Arregui and Bautista, 1965). That is,
yield increases approximately ½ bushel/acre for each 10 cm of accumulated
water.

The cost of plowing in the well-managed soils diminished by 30–50%. In
the eroded land of 'Los Surgentes,' varying labor inputs were needed

Table 7.11. *Production of wheat (without fertilizers)*

Variety	Area (ha)	Yield (qq/ha)	Protein content (%)
1	8.5	39.2	13.06
2	40.0	33.3	14.5
3	27.5	34.3	12.0
4	12.2	29.0	15.12
5	20.5	29.0	13.0

Table 7.12.

Water use in Argentina (m of depth)	Yields (bushels/acre)
0.50	16.69
1.00	19.14
1.50	21.58
2.00	24.03

according to the soil conditions (Table 7.13). A soil with a good organic matter content weighs 700 t/ha less than an eroded soil (arable layer). In this area of small farms, recommendations include total suppression of the burning of stubble, and instead the subsuperficial incorporation of it, rotation with alfalfa, the use of 'Graham plows' made on the farm with old moldboard plows, the use of straw spreaders, etc. When all these measures are taken simultaneously the results are very good – no more erosion, no more floods in the lowlands, no more soil compaction, no more 'Sunday soils,' no more 'plow pans,' a very good growth of *Azotobacter* and *Cytophaga*, a large increase in organic matter content, the appearance of available phosphorus, etc. without using chemical fertilizers.

Recommended soil conservation programs based on the ecology, culture, and economy of Argentina

In Argentina, the Pampas land is the principal producer of wheat, corn, soybean, meat, and milk. The soils are very deep with loessial subsoils very rich in calcium carbonate and with a good phosphorus (available P) content in virgin or 'reclaimed' soils. The growth of alfalfa is very good on the heavier soils from the east to the sandy soils of the west. The principal problems are water and soil structure, not mineral fertilizers.

In this region, the first technical measure to be taken would be to suppress the burning of stubble, which contains 50% of the energy fixed from the sun by the plant. If we take all the energy needed to avoid erosion, increase nitrogen fixation, transform non-available phosphorus into available phosphorus, etc., and transform it into smoke, it is very difficult to avoid erosion (Sauberan and Molina, 1956).

The use of heavy disc harrows to incorporate 50% of the stubble on the soil surface and the other 50% into the first 5–10 cm of depth affords very good

Table 7.13. *Labor input relative to soil condition*

Soil condition	Number of plowings	Yield of wheat (quintals/ha)*
After 50 years of poor management	5–6	14
After 40 years of poor management	3–4	18
After 30 years of poor management	2–3	22
'Reclaimed soil'	2–3	23

* 1 quintal = 100 kg.

physical protection against impact of raindrops. In addition, the bacterial decomposition by *Cytophaga* and other aerobic cellulose bacteria with the production of great quantities of polyuronic colloids very active in the stability of soil aggregates in water; the big increase in nitrogen fixation by aerobic bacteria (*Azotobacter* and *Beijerinckia*); the anaerobic decomposition of cellulose that provides energy for the aerobic nitrogen fixation, etc. permits in a very short time using rotations with alfalfa to diminish costs and to increase yields, avoiding completely the problem of erosion, flooding, soil exhaustion, lack of nitrogen, etc. This is the same system used by nature to dispose of cellulose production.

The practical application of a new theory about soil conservation and aerobic decomposition of cellulose in the Pampas was presented at the Xth International Soil Science Congress held at Moscow (1974). By working with an expert in farming and rangeland administration, it was possible to take the results obtained in the laboratory to the field for practical application as listed below (Sauberan and Molina, 1958). The basic principles were very similar to those of stubble mulch farming (Duley and Kelly, 1939).

1. *Cellulose materials:* The basic factor needed to achieve a plentiful production of polyuronic colloids was the availability of materials rich in cellulose. This was possible by using the stubble of grain crops and the remains of natural or cultivated pasture. The straw yield of a wheat harvest can be roughly estimated as 2–4 t/ha and the remains of a well-managed pasture as between 1 and 5 t/ha. The composition of straw is approximately 30% cellulose. In laboratory experiments, 17–18 g of polyuronic colloids are obtained from each 100 g of decomposed cellulose. As this colloid is active in a 0.1% dose, we estimate that 100–200 t/ha of soil could be improved each year. This shows the possible importance of these colloids in field practice. The decomposition of these colloids by soil bacteria is very slow, and its effects remain for one to three years, which is exactly the time that a soil would have a good structure after cultivation in tropical conditions.

2. *Good water supply:* Water was obtained through an adequate plan of seasonal fallows complemented with the use of 'fallow plants' such as grazing corn (Molina, 1980).

3. *Good aeration:* These bacteria need plenty of water and air; to be sure that in both the rainy season and the dry season aeration would be suitable for their development, it was decided to leave 50% of the plant residues on the soil surface and the other 50%

buried in the first 5–10 cm of the soil (Molina and Spaini, 1950). One-way plows and heavy harrow discs are good implements for this kind of work (Sauberan and Molina, 1955).

4. *Good supply of nitrogen:* The nitrogen supply was obtained by means of rotation with leguminous plants. The use of a three-year rotation of 'cash crops' and five years of alfalfa mixed with grasses enables us to obtain enough nitrogen for a good development of the cellulose bacteria. The fixation of nitrogen by *Azotobacter* in the few superficial millimeters of the soil ('the soil skin') also aids these processes (Sauberan and Molina, 1957; Molina, 1981b).

5. *Mineral elements:* Mineral salts and the neutral reaction needed by these bacteria are not considered to be a limiting factor in the Pampas. The alfalfa roots enable the extraction of calcium from the calcareous subsoil and transport it to the soil surface. In very rare cases where the soil does not have calcareous subsoils, it may be necessary to add lime to the straw for easy decomposition.

By using these methods, the erosion problems, which are very severe in many regions of the Pampas, especially wind erosion in the west and water erosion in the east, were controlled or eliminated in a very short time (1–3 years). The yields increased substantially, but what was really important and struck our attention was the decrease in cost. A better soil structure and a higher percentage of organic matter permitted a reduction in the number of tillage operations. In addition, due to the lower weight of the soil, tillage became gradually easier and more economical (Sauberan and Molina, 1965).

The biological improvement of the soil leads to a notable increase in yields at a reduced cost (30–50% in only one year). I believe that the low cost of the biological methods suggested in this study permits their easy application under the conditions existing in most developing countries. Especially important could be their use in tropical and subtropical regions on oxisol soils, since, under these climatic conditions, biological improvement of the soil is achieved by the action of the iron and polyuronic colloids (Molina and Spaini, 1948).

The average difference between well-managed farms and ranches and the common management (stubble burning, bare fallows, no rotation with legumes, etc.) is summarized in Table 7.14.

Reclamation of non-saline alkali soils

Non-saline alkali soils are the principal soils with special problems in the Pampas. 'The Reclamation of Sodic Soils by Means of Methods Founded on the Ecology of Microbes' is the title of a special study that has been in progress for more than 30 years. The problem of reclamation of sodic soils is very important in Argentina and in many other countries around the world.

In the semi-arid zone of the Pampas, and also in the Greater Chaco Gualamba, there are huge areas of sodic soils, especially in the lowlands without outlet. Many of these soils are very near to Buenos Aires and other big Argentine cities. The problem concerning the improvement of these soils, with neither irrigation nor drainage and in zones where the land is quite cheap, does not allow the use of common methods with chemicals in irrigation zones.

In vast regions of Argentina the use of a new method founded on the utilization of ecological principles concerning the soil microbes has changed barren alkali soils into good grazing pastures, with a high production of meat and milk.

Some preliminary observations made in several Latin American countries particularly Mexico, and in Romania in Europe make us believe that similar results might be obtained in other countries. Quite recently we have learned that, using our methods, it was possible to reclaim 5000 ha of alkaline soils in Pakistan, near Lahore (Malik, pers. comm.). The reclamation of sodic soils using low-cost ecological methods is a real possibility. The principal problem of sodic soils is a very high pH, a very poor structure, and low phosphate availability. The high pH can be lowered successfully by cheap biological methods using carbon dioxide as an acidifying factor (Sauberan and Molina, 1964). The carbon dioxide produced during the decomposition of vegetal residues and exhaled through the roots of very resistant plants is so effective that the excessive pH of sodic soils is rapidly lowered, permitting the development of highly productive pastures at a very low cost.

Our experiences concern soils of many different textures generally with

Table 7.14.

	Yields with common management (kg/ha)	Yields with good management (kg/ha)
Meat production (living weight in the field)	90–100	200–240
Milk production (fat)	25–30	90–100
Wheat	1500	2500
Corn	2500	4000

little or no structure and offering in all cases an impermeable layer in one or another horizon. This gives them the general characteristics of lands that are very easily flooded, especially during periods of heavy rainfall. In the wet regions these soils are found in the low areas with little or no drainage.

The presence of salts such as chlorides and sulfates is generally low and they are very often found only as vestiges. However, the pH is very high. In the superficial layers of soil during the period of drought in summer, we can find values of pH 11.6. Between 20 and 50 cm there is generally a hard layer very rich in calcium carbonate.

The process of recovery begins with the sowing of a producer of a large quantity of cellulosic materials, such as 'broom corn' (*Sorghum technicum*). Generally the sowing results are unequal. In areas with very high pH, crops fail or remain very poor. In areas with better soils, the crops often grow to higher than 2 meters. Afterwards the 'broom corn' is submitted to an intensive grazing with 8–10 cows per hectare, for a suitable time to permit the cattle to consume 60–70% of the pasture. The remaining 30–40% is incorporated into the soil via a one-way plow or a heavy harrow disc. This method leaves a thick stubble on the ground.

Besides lowering the cost of the recovery process, intensive grazing helps to equalize the soil, eliminating the 'spots' of sterile soil. The cattle act as removers of fertility from the good parts of the soil to the poor ones. A fact that contributes to this is the cattle's preference for sleeping in places without vegetation, which are precisely the 'alkali spots'. Another important factor is that cattle dung contains microflora with a great quantity of anaerobic bacteria that decompose cellulose with production of ethyl alcohol, and acetic, propionic, and butyric acids, etc. These organic acids are also very important in lowering the pH of these soils (Molina, 1970).

The problem of low phosphorus availability was resolved by the increase in carbon dioxide concentration and also by the organic acids formed in the anaerobic decomposition of cellulose. In laboratory tests using the soil plaque method of Winogradsky (1949), modified by Molina (1963) for testing phosphorus availability, the results were very positive.

Tests made in the laboratory seem to prove the lack of anaerobic bacteria in alkaline soils; that is why the addition of dung acts as a 'starter' that accelerates the decomposition of cellulose. The importance of anaerobic decomposition of cellulose in the production of organic acids that could be used by *Azotobacter* as an energy source to fix nitrogen from the air (Molina, 1970) lends special interest to this aspect of the problem.

The very poor structure of sodic soils can be greatly improved by means of the action of polyuronic colloid produced in the aerobic decomposition of cellulose that remains upon the soil, by bacteria belonging especially to the genera *Cytophaga* and *Sporocytophaga*.

The increased phosphorus availability and the presence of abundant carbon sources, such as acetic, propionic, and butyric acids, permits a great increase in the number of *Azotobacter*. In many soils its numbers can be increased from 10 to 20 colonies per gram of soil to more than 100 million. The use of field methods founded on these ecological principles can transform a barren, unproductive alkaline soil into a good grazing pasture, which might produce more than 200 kg of meat per hectare (living weight in the field).

In the first year of using these methods we reclaimed 400 ha of alkaline soil, 14 years later the number of hectares transformed was more or less 150 000, and now its number is 500 000. The 5000 ha of alkaline soils reclaimed by Dr Malik at Lahore, Pakistan, by means of the same basic theoretical principles but using different plants indicates the great potential of this method.

Because these basic principles are possibly the same in all sodic soils, their application on a worldwide scale could be an important contribution to today's protein shortage. The low cost of the field application of these ecological methods enables their use by developing countries.

Soil conservation, however, cannot be accomplished by laws or governmental decrees. It is necessary to develop an intense extension program by means of special groups of well-trained technicians with the aid of radio, television, and journals. The basic factor – as explained by Dr Bennett – is 'Education at all levels, but specially in the primary schools.' It is a very good policy to give special prizes to the farmers who make efforts to control erosion on their ranches or farms. It is also very important to instruct political leaders, teachers, and priests about the real importance of soil erosion.

It is necessary to take action very similar to that of the U.S. Soil Conservation Service. But it is imperative that the work is not just done by the government. The number of well-trained technicians at the official level is very low; a change is needed in the official attitude towards economic measures to be taken to solve the problem of soil erosion. Erosion needs to be controlled on the farms and not only at the experimental stations.

For many years, especially under military governments, official action was directed to combat the non-governmental organizations (NGOs) that fight against soil erosion, because it was thought that only official action would be effective. Now the situation is changing, but not quickly enough to obtain good results in the field practice in a short time.

Many of the principal results obtained in the field of practical control of soil erosion were made by NGO groups such as 'Friends of the Land' and

'CREA' groups, and by university groups such as the 'Agrotechnical Institute of the North Eastern University,' and the cathedra of 'General Agriculture' and the 'Agrotechnical Institute', both at Buenos Aires University. Unfortunately, these last two centers of work were closed during the last military government in 1980 and all their technical personnel were fired.

The universities face the major job of preparing good technicians knowledgeable about soil erosion and conservation. We need to change the policy of no cooperation between official and non-official organizations. After more than 50 years of political instability, the government structure is so damaged that without the non-official organizations it would be very difficult to do a good job in a short time. At present, due to the critical financial situation in Argentina, many technicians have no budgeted fuel for their cars during many months of the year.

During the last few years several scientific conferences on soil erosion have been held, but it is not possible to stop erosion by lectures alone. We need an effective task force with executive power capacity formed by government technicians, NGOs, farmers, and ranchers. It must be composed of people with many years of experience in the fight against soil erosion. This task force will have to be provided with funds, which are very difficult to obtain at this time.

This proposal is not unrealistic; during one of the short periods of democratic government in Argentina, a very important meeting was held with official agencies, NGOs, technical associations, ranchers, and farmers, with the aid of the United Nations. This meeting published a very good report (INTA, 1957), but no action has been taken to bring these conclusions into practice.

Hugh Hammond Bennett, in his first letter after we invited him to visit Argentina, indicated a three-year plan with 160 technicians at work in the field. At the time of writing we have only 30–40 technicians with very low funds to make a real effort to control soil erosion. The NGOs have only the money provided by their associates and a few contributions from banks such as the National Bank of Argentina, but all the contributions are very low, no more than a few hundred dollars per year.

The current soil conservation effort in Argentina is not good. It is necessary to show the farmer and the rancher that 'soil conservation is a good business.' Fortunately, it has been possible to obtain good results all over the country from big ranches to small farms.

This report represents my opinion after more than 40 years of work all over Argentina and in many Latin American countries, including Brazil, Venezuela, Paraguay, Chile, Uruguay, and, quite recently, Bolivia (Santa Cruz de la Sierra). It is also the result of the work of many technicians who worked in NGOs such as the 'Friends of the Land' and in many university centers of investigation in the North East University and Buenos Aires University and of technical advisors of 'CREA' groups all over the country.

This team includes Argentinian, Peruvian, Colombian, Brazilian, Uruguayan, Chilean, Bolivian, and Venezuelan technicians. Many of them have studied in the universities of the North East and Buenos Aires and in the Research Center on Biotechnology and Microbial Ecology at Buenos Aires.

Although the work described was conducted by many technicians, the conclusions are mine alone. It is quite possible to take a system as complex as soil erosion and implement sound soil conservation practices. This will benefit individual farms as well as the nation.

I owe much to H.H. Bennett and his contributions to solve soil problems in Argentina. The National Day (7th July) of Soil Conservation is dedicated to H.H. Bennett's death on that day.

8

Soil erosion and conservation in the United Kingdom

C. ARDEN-CLARKE AND R. EVANS

The presence of accelerated soil erosion in the United Kingdom (UK) has been documented since the latter half of the 19th century. For example, Fisher (1868, quoted in Douglas, 1970) records that in Norfolk:

> Upon the land surface, a certain amount of the finer material is being carried into the rivers, and by them deposited at the head of the Broads, or, where such do not exist, in the sea. This denudation by pluvial action is undoubtedly greater where the land is under the plough than it would be otherwise.

Large-scale erosion events have been despoiled quite frequently in the literature, reviewed by Douglas (1970), Evans (1971), Reed (1979), Morgan (1980), Fullen (1985a), Arden-Clarke and Hodges (1987), and Speirs and Frost (1987). However, it is only within the last two decades that information has been systematically collected on erosion (Reed, 1979; Evans, 1980), after a plea in 1971 (Evans 1971) that the conservation of soil in the UK should not be ignored. The plea ultimately led to the establishment of a project to monitor water erosion in a number of localities spread throughout England and Wales (Evans and Cook, 1987; Evans and Skinner, 1987; Evans, 1988a). This project was carried out under the auspices of the soil science arm of the Agricultural Development and Advisory Service of the Ministry of Agriculture, and what was then the Soil Survey of England and Wales. This survey largely fulfils Stocking's recently stated (1987) requirements for a soundly based monitoring scheme, and is probably the first in the world to do so.

With the exceptions recorded above and a noteworthy paper on wind erosion in the East Midlands (Wilkinson *et al.*, 1969), prior to the mid–late 1970s, soil erosion was recognized as a problem only in some upland areas of the UK (reviewed in Evans, 1971; Evans and Cook, 1987). Overgrazing, deliberate burning, and air pollution, all of which increase the exposure of soil and peat to high rainfalls in these uplands areas, are the causal factors, and it appears that some 4.4% of the land area of England and Wales is at risk from moorland erosion (Boardman, 1987b; Evans and Cook, 1987). Evans (1977) cites overgrazing by sheep as the particular cause of soil erosion (by water) at rates of up to 34 tonnes/hectare/year (t/ha/yr) in a small drainage basin in the Peak District. Erosion was initiated where the stocking density of sheep exceeded two per hectare. These areas are of minor significance in agricultural terms (though of considerably greater significance in wildlife conservation terms), compared to lowland arable areas where erosion now appears to be increasing dramatically. With the exception of three systematic surveys carried out in England and Wales (Reed, 1979; Evans and Cook, 1987; Boardman, 1982–87, unpublished), monitoring of UK erosion is patchy, especially in Scotland (Speirs and Frost, 1985; Watson, 1987), and very few data exist for Northern Ireland. The scarcity of published data requires the present discussion to focus on erosion mostly occurring in England. It is only in the last decade that soil erosion, largely water-induced, has come to be recognized as having an important agricultural impact in lowland areas of the UK. Attitudes towards this phenomenon are changing (Boardman, 1988), even among farmers (Evans and Skinner, 1987). The Royal Commission on Environmental Pollution noted, with regard to eutrophication of reservoirs in agricultural catchments, 'the desirability of adopting farming techniques to reduce the nutrient losses due to erosion' (RCEP, 1979). Detectable erosion of lowland agricultural soils was recorded well prior to this (e.g. Stamp, 1948; Jacks, 1954), but apparently not at rates or frequencies to excite the attention of agricultural scientists. This situation began to change in the mid-1970s, as individual soil scientists accumulated records of erosion incidents (Reed, 1979; Evans, 1980). By 1981, Evans (1981a) knew of 1300 eroded fields in England and Wales, this total building up from less than 100 in 1977 when he began deliberately searching for evidence of erosion. Only three systematic surveys have been carried out in Britain, one in the West Midlands of England (Reed, 1979, 1983), one at 17 localities throughout England and Wales (Evans and Cook, 1987), and one on the South Downs around Lewes (Boardman, unpublished). None of these have been running for enough time to demonstrate variations in erosion rates over the long term. However, there are now sufficient data to suggest that the problem is on the increase, especially where winter cereals are grown (e.g. Evans and Cook, 1987; Speirs and Frost, 1987). Water erosion of lowland agricultural soils is now regarded as a major cause for concern in terms of areal extent, frequency of occurrence, and local impact (Boardman, 1987a).

One estimate suggests that a combination of water and, to a lesser extent, wind erosion is an appreciable risk on some 37% (*c.* 25 500 km^2) of the total arable area of England and Wales (Morgan, 1985). Another (possibly more accurate) estimate puts the figure at 44% of the arable land of England and Wales (ENDS, 1984).

Wind erosion is largely restricted to a few well-defined areas of sandy soils in eastern England and parts of eastern Scotland, together with some peat soils in the Fens, Lancashire, and Humberside regions of England (Davies, 1983). The majority of the following discussion relates to water erosion of lowland soils, about which more is known than wind erosion, and which is of greater significance in economic, agricultural, and environmental terms.

Rates, occurrence, and frequency of soil erosion in the UK

Recorded rates of water-induced soil erosion in lowland Britain vary widely, from levels of less than 1 t/ha/yr, to very destructive rates approaching 20 t/ha/yr. The latter are rare and localized events (e.g. Evans and Nortcliff, 1978; Reed, 1979; Boardman, 1983), but clearly have serious implications for the productive capacity of the fields involved. Most rates of erosion are less than 1–2 m^3/ha/field/yr (Evans and Skinner, 1987), but erosion is often confined to small parts of the field (usually between 1 and 5% of the area), and hence erosion rates are much higher in these limited areas. Severe rates of, say, more than 10 m^3/ha/field/yr occur in less than 10% of erosion events.

Measurements of wind erosion rates are few and far between, perhaps because this form of erosion is regarded as of economic importance largely for its abrading effects on crops, rather than for any long-term yield losses due to reductions in soil depth (Davies, 1983). It is likely that fields at risk from wind erosion erode less frequently than those at risk from water erosion, but at slightly higher erosion rates (e.g. 6–10 m^3/ha in serious single events, Fullen, 1985c). Wilson and Cooke (1980) provide higher estimates for affected sites in the Vale of York, Nottinghamshire, and north Norfolk, ranging from 21 to 44 t/ha/yr, but these apply only to the most vulnerable parts of the field.

The five-year survey (1982–86), carried out by the Soil Survey of England and Wales (SSEW), provides a guide to the frequency of occurrence and generally encountered rates of water erosion on lowland soils in Britain. In 1982, 297 fields at 15 selected localities were visited by soil scientists, 148 of which were found to be eroding at rates varying from 0.1 to 36.8 m^3/ha/yr, or approximately 0.1 to 47.8 t/ha/yr (SSEW, 1983). (All erosion rates in the SSEW survey are recorded as m^3/ha. These are converted in the present account to t/ha by multiplying the given value by 1.3, on the assumption that the mean bulk density of agricultural soils in the UK is 1.3 g/cm^3.) In 1983, 16 localities were surveyed and, on average, over twice as many fields within the sample localities were found to have eroded in that year compared to the previous one (SSEW, 1984; Evans and Cook, 1987). In 1984, the number of eroded fields was approximately one-third of that in 1983, and two-thirds of that in 1982, the wide annual variation in the incidence of erosion being ascribed to variations in the number of erosive rainfall events in winter and spring (SSEW, 1985; Evans and Cook, 1987).

By 1986, an SSEW *ad hoc* monitoring programme had located 1769 fields suffering erosion in England and Wales, the distribution of which (Fig. 8.1) reflects that of arable land use (Evans and Cook, 1987). However, these results were known to under-represent the frequency of erosion revealed in the West Midlands by Reed's (1979, 1983) survey of this region, by an order of magnitude. By 1983, Reed had located over 1000 eroding fields in this region alone, and by 1987 this number had risen to over 2000 (Evans and Skinner, 1987). Estimates based on aerial surveys and fieldwork may lead to an underevaluation of erosion by as much as 40% (Reed, 1983). Estimates

Fig. 8.1. Number of eroded fields in 100 km National Grid Squares. (From Evans and Cook, 1987.)

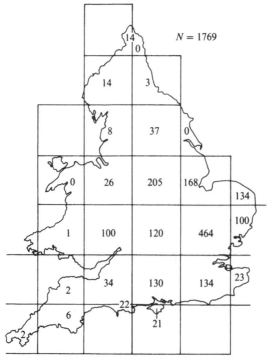

based on aerial surveys and fieldwork may lead to an underevaluation of erosion, but not by a lot, and generally only small-scale events will be missed. All schemes will miss some erosion events unless they are monitored at very frequent intervals, which is often not practicable, given the necessary large fieldwork input.

Less information on soil erosion rates exists for Scotland, where the majority of the important survey work has been undertaken by three workers. Watson (1987) reports on erosion occurring in the northeast of Scotland, but gives details only in terms of soil loss per field. The author noted that several fields in this study area lost over 100 m^3 in a single year, and one lost 420 m^3. Prior to 1969, Speirs and Frost (1985) were aware of only one soil erosion event in the Borders, Lothian, Fife, and Tayside regions of southeast Scotland. In the decade 1969–79, they noted 30 erosion events, and between 1980 and 1984, they recorded a further 300 events. In the winter and spring of 1984/85 alone, Speirs and Frost (1985) recorded another 120 erosion events, bringing the total for Scotland to 450. On 42% of these fields erosion involved the loss of between 1 and 10 m^3 of soil (approx. 1.3–13 tonnes), and on 14% soil losses exceeded 100 m^3 (130 tonnes). Notable individual events involved losses of 900, 800, 650, and 350 m^3 of soil (respectively some 1170, 1040, 845, and 455 tonnes). Speirs and Frost (1985) do not record erosion rates for any of these events on a farm in the Scottish Borders, resulting in soil loss rates of 79 t/ha (Frost and Speirs, 1984). These events are not representative of erosion rates in the study area as a whole, as the median soil loss value per *field* for the entire survey was only 10 m^3, or approximately 13 tonnes (Speirs and Frost, 1985). It is likely that, in general, erosion rates are higher in Scotland than England because soils are often of lighter texture and rainfall is higher.

Data on occurrence and frequencies of erosion events are available only for England and originate from the SSEW survey. In the two years of the survey when erosion was widespread – 1983 and 1986, it occurred in fields covering about 4% of the landscapes photographed. However, the figure for different localities ranged from 1 to 20% (Evans and Skinner, 1987). In years when erosion was less common, some localities suffered none, whereas others suffered a maximum of 5%. During the five years of the survey, it is likely that in some localities up to 40% of the fields have suffered erosion, about a third of them being affected on two or more separate occasions. These are areas with a high proportion of sands or light loams, such as Shropshire, Staffordshire, and Nottinghamshire. On chalk downland, erosion in a bad year affects about 20% of the fields, whereas on heavier textured soils erosion will affect only about 5% of the fields in the same year. Evans and Skinner (1987) provide figures showing that between 1968 and 1977 winds damaging crops in susceptible areas in Cambridgeshire and Nottinghamshire occurred about one year in two, for each month from March to May.

All the foregoing estimates of erosion are likely to understate the problem (Boardman, 1987a), and this should be borne in mind when evaluating the available data. It is safe to assume that rates of erosion in the UK are currently underestimated, but not by a great deal.

Factors and policies contributing to soil erosion in the UK

Physical factors beyond the influence of the farmer, and agricultural management factors which are under the farmer's control, both affect erosion rates. The physical factors of importance are: (a) rainfall characteristics, (b) soil type, and (c) landform. Of these, only the first, rainfall characteristics, is likely to change enough on a human time-scale to affect rates of erosion. Since the onset of the post World War II intensification of agricultural methods in the UK, it would appear that there has been no major change in the climate sufficient to affect erosion rates (Arden-Clarke and Hodges, 1987). With regard to soil type, it is generally those lighter textured soils high in silt or fine sand which are most at risk from erosion (Reed, 1979; Evans, 1980; Evans and Cook, 1987; Speirs and Frost, 1987). In the UK, soils with only 10–25% clay content are particularly vulnerable to erosion (Morgan, 1974). Of 88 sites at which Evans and Cook (1987) measured erosion rates, 25 were clayey soils and had mean erosion rates of 1.9 m^3/ha/yr (2.5 t/ha/yr), 33 were fine silts with a mean erosion rate of 4.8 m^3/ha/yr (6.2 t/ha/yr), and 29 were sandy and coarse silts with a mean erosion rate of 13.6 m^3/ha/yr (17.7 t/ha/yr). The erosion rates quoted here are for drainage basins within fields, not mean rates per field, as are given earlier, so that the rates given are higher than those quoted earlier.

Slopes which are convex in profile appear to be the landform most susceptible to water erosion (e.g. Reed, 1979; Evans, 1980; Boardman, 1984; Boardman and Robinson, 1985; Colborne and Staines, 1985; Evans and Cook, 1987; Jackson, 1987), and valley floors are susceptible in some situations where runoff from the valley sides merges on the floor attaining sufficient volume and velocity to cause erosion (Evans and Cook, 1987). Rills generally occur only in valley floors on heavy textured soils, but occur much more frequently on slopes on light textured soils.

The agricultural management factors directly influencing occurrence of water erosion have been reviewed in some detail by Evans (1983) and Arden-Clarke and Hodges (1987). The following techniques all increase the susceptibility of agricultural soils to water erosion: (a) having land under continuous arable cultivation; (b) converting grassland to arable cropland; (c) increasing the area of land under cereals, and in particular autumn-sown seeds; (d) using heavy machinery which compacts and damages the soil structure; (e) 'tramlining' – the use of the same tractor wheelings for successive agricultural operations; (f) working the land up and down slope; (g) removal of field boundaries; (h) untimeliness of cultivation; and (i) preparation of fine, level seedbeds. The way in which these techniques influence erosion rates is described below, and current trends in their application are explained in terms of the recent intensification of farming practices and the incentives provided by the Common Agricultural Policy (CAP) of the European Community (EC).

Techniques (a)–(c) mediate their effects through reductions in soil organic matter levels and increases in the exposure of the soil surface to raindrop impact. The abandonment of ley–arable rotations in favour of continuous arable cropping, the ploughing up of old pastures for conversion to arable land (e.g. on the South Downs of England), and the growing, and often exclusive, dependence on inorganic fertilization techniques which have been substituted for organic and green-manuring practices, all cause reductions in soil organic matter levels (Arden-Clarke and Hodges, 1987). Falling organic matter levels reduce the structural stability of the soil affected (e.g. Greenland *et al.*, 1975; Chaney and Swift, 1984), and reduce the 'infiltration capacity', thereby increasing surface runoff, which is the major cause of water erosion (Morgan, 1980). Reduction in soil organic matter levels also increases the disaggregation of soil particles under raindrop impact ('slaking'), enhancing the 'capping' of the soil surface by an impermeable crust, thus increasing runoff (e.g. Boardman, 1983). The pasture crops replaced by arable crops provided a more complete cover and therefore better protection of the soil from raindrop impact (Morgan, 1979). On soils with lowered or inherently low organic matter levels, the only method of preventing water erosion on slopes may be the re-establishment of a grass crop (e.g. Evans and Nortcliff, 1978; Boardman, 1983; MAFF, 1984a).

The increased erodibility of land sown to cereals is demonstrated by the data in Table 8.1, modified from Evans and Cooke (1987). The table shows the percentage occurrence of different crops in eroded fields noted in 1978 and 1984. In 1978, of the eroded fields sown to cereals where it was possible to determine sowing date, 87.1% were sown in the preceding autumn and only 12.9% in the spring. In 1984, the respective percentages were 86.2 and 13.8%. While it is clear that erosion occurs most frequently in winter cereals, these data probably underestimate the prevalence of erosion in spring cereals, sugar beet, potatoes, and vegetable crops, as it is often impossible to ascertain the crop sown in a field of bare soil, especially on aerial photographs (Evans and Cook, 1987). Preliminary data suggest that erosion is somewhat more common in sugar beet than in spring cereals, in which it is about 50% more frequent than potatoes, a crop which in turn is about twice as likely to erode as a grass ley.

Evans and Cook (1987) draw attention to the changes in the national pattern of cereals sowing that has taken place between 1969 and 1983. Whereas the proportion of arable land under cereals has shown no major change over this period (74.5% in 1969 versus 74.6% in 1983), the area under autumn-sown (i.e. winter) cereals has more than trebled. Drawing on MAFF statistics (MAFF, 1973, 1984b) Evans and Cook estimate that winter cereals covered some 800 000 ha in 1969, compared to almost 2.5 million hectares in 1983, an increase of 3.1 times. This trend has continued since then, and by 1985 the area under winter cereals was close to 2.9 million hectares (MAFF, 1986). Given that some 87% of eroding cereal fields where it was possible to identify sowing date were down to autumn-sown varieties in both years, the implications of increases in the acreage of winter cereals for erosion rates in the UK, are clear. The switch from spring to autumn sowing increases the exposure of soils to potentially erosive precipitation throughout most of the winter (e.g. Boardman, 1984; Frost and Speirs, 1984; Boardman and Robinson 1985; Speirs and Frost, 1985, 1987; Reed, 1986; Evans and Cook, 1987). Formerly, the sowing of spring cereals meant that farmers could leave the stubble of the last crop down over the winter, or the land was left rough after ploughing.

The switch from spring to autumn-sown cereal varieties is essentially a response to the subsidy structure of the EC's CAP. In 1986, the EC provided

Table 8.1. *Land use and erosion in 1978 and 1984*

| | % of eroded fields sown to each crop* | |
Crop	1978	1984
Cereals	40.2	45.4
Bare soil	40.0	43.1
Potatoes	2.0	2.8
Roots (incl. sugar beet)	2.2	1.1
Ley	1.3	1.1
Horticulture	0.9	1.2

* 1978 sample of eroded fields, $n = 477$; 1984 sample of eroded fields, $n = 1521$

UK farmers with some £358 million in cereal price support (Boardman, 1987b). Subsidies in the form of price support encourage farmers to maximize yields, and yields of autumn-sown cereals are generally substantially higher than those of spring-sown varieties, hence the switch to autumn sowing dates. Average yields of winter wheat quoted by Nix (1985) are 6.75 t/ha and of winter barley are 5.65 t/ha, compared to average spring barley yields of 4.5 t/ha. The size of the EC cereal subsidy is sufficient to encourage farmers in many areas to switch from grassland production to cereal production, even where this involves ploughing up potentially erodible land (Boardman and Robinson, 1985; Boardman, 1987a,b). Winter cereal production is now possible even in the wetter west of the country, because of the development of effective herbicides which control the weeds which formerly choked this crop under wet conditions. Thus, while technical innovations have had a hand in making the spread of winter cereal production possible, it is the EC cereal price support which has encouraged its spread. It is therefore the CAP subsidy which is most directly responsible for increases in soil erosion observed in the UK in recent years.

The problem of soil compaction by agricultural machinery was first highlighted by the Strutt Report (AAC, 1970). Since then risks posed to UK soils by compaction have increased as the weight of agricultural machinery has risen in tandem with the frequency of cultivations necessary in more intensive arable systems (Greenland, 1977a; Reed, 1979). Ten or more passes may now be required to establish and grow a crop in a conventional tillage system (Stanley and Hardy, 1984). Reed (1983) notes that 'widespread effects of soil compaction are being experienced in the intensive arable areas (of the UK)'. Compacted soil surfaces are less permeable to water (Fullen 1985b), so infiltration during rainfall events is decreased, surface runoff is augmented, and, given sufficient continuous rain, erosion is initiated. Tractor wheelings running in a down-slope direction can be converted into gullies over 1 m deep in the course of a single rainstorm, and soil loss may exceed 100 t/ha in such events (Reed, 1983; Boardman, 1987b).

Tramlining is a more localized and severe form of soil compaction, consequent upon the repeated use of a single tractor wheeling for subsequent field operations. In tramlines created by several passes, runoff approaches 100% (Reed, 1983). Repeated applications of pesticides by tractors to the crop throughout the growing season, result in a compacted soil being exposed to raindrop impact and runoff throughout the year (Arden-Clarke and Hodges, 1987). Resultant erosion will generally only be severe when tramlines (and single wheelings) are aligned up and down slope. Down-slope alignment has been shown to be a crucial factor in the initiation of erosion events observed in Norfolk (Evans and Nortcliff, 1978), the West Midlands (Reed, 1979, 1983, 1986) the South Downs (Boardman, 1984) and Somerset (Colborne and Staines, 1985). Evans (1980) noted that only 49 of 302 (16.2%) instances of water erosion that he had often then observed in England and Wales, were in fields worked across the slope. However, many farmers in the UK seem reluctant to adopt across slope techniques (Evans, 1985; Evans and Skinner, 1987). Most tractors in the UK do not have a self-levelling chassis (an expensive feature), are difficult to drive safely and accurately across slopes, and are therefore effectively designed to work up and down slope (Reed, 1986).

Removal of field boundaries can increase the magnitude of both wind and water erosion, but it is generally a more critical factor in instances of wind erosion. Tree shelter-belts with a width of 9 m can give protection over a distance of about 12 times their height to leeward, and a 3-m wide hedge gives protection to about 30 times its height to leeward (Morgan, 1979). Field boundaries reduce the distance of uninterrupted blow (the wind fetch) and by trapping wind-borne particles prevent isolated areas of erosion from becoming contiguous during major sandstorms. Hedgerows on slopes have been shown to be an important factor in the prevention of water erosion as well. Evans and Nortcliff (1978) noted that the removal of old hedges and ditches on valley sides results in less restricted water flow and consequently high surface runoff velocities at which soil particles can be eroded and transported. Evans and Nortcliff (1978) and Reed (1983) recommend retaining such 'key' hedges on long slopes or at points where slope angle increases and is backed by a large potential catchment area. Boardman (1984) cites major erosion events on the South Downs which have been initiated where hedges have been removed on, or just below, the crests of the Downs, which form extensive catchments.

The intensification of arable practices that has occurred since the last war has been characterized by increasing field size and the removal of field boundaries. While the rate of hedgerow removal in Britain peaked at around 9000 miles per annum in the late 1960s and early 1970s, it continued at a rate of 2–3000 miles per annum in the 1980s (Hooper, 1987). As a result, instances of water (and wind) erosion exacerbated by hedgerow removal can be expected to continue increasing. Untimely cultivation of a soil enhances subsurface compaction and the development of a relatively impermeable plough-pan (AAC, 1970). Untimely cultivations are generally those which are undertaken during periods of high soil moisture content, when smearing and realignment of soil particles result in increases in bulk density, reduction in the size and number of pores and channels in the soil, and, consequently, reduction in soil permeability (Reed, 1979). Surface compaction can also be exacerbated in wet conditions (AAC, 1970). The intensification of arable practices and the switch to arable rotations (particularly of cereals) in the UK have led to less flexibility in the timing of field operations. Where a whole farm is under only one or two crops, cultivations have to be completed within a very restricted time frame, and as a result will often have to be conducted when soil conditions are unsuitable and smearing and compaction of the soil are enhanced (AAC, 1970; Evans 1985). The general increase in farm size associated with intensification has also contributed to this problem (AAC, 1970).

Fine, level seedbeds are often prepared for arable crops to enhance the effectiveness of soil-applied herbicides used to control weeds. The over-winter exposure of these seedbeds (and particularly those prepared for autumn-sown cereals) is regarded as an important factor in observed increases in erosion (Boardman and Robinson, 1985; Evans, 1985; Speirs and Frost, 1985, 1987). The increases in soil erosion in the UK resultant upon the switch from spring to autumn-sown cereals was detailed above, and the preference for fine, level seedbeds serves to exacerbate the effect of this switch.

The present review focuses on the erosion occurring in lowland soils in the UK, but it should be pointed out that there have been increases in erosion rates in upland areas since World War II. On certain vulnerable grassy slopes, often formerly covered by heather which has been grazed-out by sheep, the turf mat is being broken down by sheep's hooves, initiating erosion. Where this has happened on the upper slopes of the Peak District, bare soil is expanded in area by about 4% per annum, and has been expanding at this rate for over a decade. Erosion was initiated after the war with the tripling of the number of sheep on the moors. Again, it is the farmer's response to government (and EC) policies (Evans, 1990). Erosion due to over-grazing by sheep is also occurring in Scotland (McVean and Lockie, 1969), Wales (Smith, 1986) and in the Lake District (Pearsall and Pennington, 1973).

Erosion has also taken place in the uplands in England and Wales where coniferous afforestation has not been carried out with sufficient care, or where the moorland has been drained with open ditches (Newson, 1980), and the runoff intercepted by the ditches incises into the unstable channel. Graesser (1979) shows similar effects of land 'improvements' in Scotland. These 'improvements' are usually funded from government grants or, in the case of afforestation, by tax concessions for the wealthy.

The foregoing discussion makes it clear that the recent intensification of arable farming practices in the UK is the major cause of the observed increases in soil erosion. EC subsidies granted under the Common Agricultural Policy have been a major factor in this intensification process leading to increased erosion rates, but mechanical, technical, and chemical developments (e.g. the introduction of heavier, more powerful agricultural machinery and the increased use of pesticides), have also contributed to the increases in soil erosion in the UK. Current rates of erosion of agricultural soils may therefore be seen as a product of both EC policy and specific management techniques which characterize intensive arable farming systems.

Factors and policies improving soil conservation in the UK

Whereas there are certain physical and climatic factors which tend to enhance soil conservation in the UK, there are virtually no policies which encourage it. Indeed, the existing agricultural support system in the form of government policies and the CAP is effectively subsidizing erosion in the UK. Having said that, there are currently some signs of an impending shift in UK government policy, which would help to create a greater awareness among farmers of soil erosion and appropriate preventative measures.

The UK has a temperate, maritime climate, and as a such is likely to have among the lowest rates of natural erosion in the world (Saunders and Young, 1983). Rodda (1970) pointed out that the climatic conditions and diversity of the British agricultural landscape had (until then) allowed farmers in the UK to follow practices which would have proved disastrous anywhere else in the world. Furthermore, of the agricultural land in the UK, 62.8% remains under grass (MAFF 1986), and consequently suffers very low erosion rates (e.g. 0.1–3.0 t/ha/yr on sandy loams, the most erodible soil type, Morgan, 1986). It is generally only arable land, and by and large inappropriately managed arable land at that, which is at risk from soil erosion in the UK. Morgan (1985) has estimated that only 16% of the arable land in England and Wales is at risk from soil erosion when it is used in accordance with its capability rating. Nevertheless, erosion rates on arable land in the UK are increasing, and the application of recognized soil conservation techniques and policies in response to this has yet to be undertaken on a significant scale.

While there is currently no recognizable policy on soil conservation, the modifications to current agricultural practices necessary to reduce the incidence of soil erosion in the UK are well established. Morgan (1980) recommended the use of organic manures to raise soil organic matter levels and improve soil structural stability, as did the Strutt Report (AAC, 1970) ten years earlier. Speirs and Frost (1985) recommended a switch from autumn to spring sowing of cereals where erosion is a continuing problem with the latter. The sowing of grass on steeply sloping or otherwise inherently erodible land (e.g. sandy loams or any soil with low clay and organic matter levels) has been suggested by a number of authors (e.g. Evans and Nortcliff, 1978; Boardman, 1983, 1987b; MAFF, 1984a; Evans, 1985; Morgan 1986). More specifically, Boardman (1987a) suggests that any rational soil conservation policy for the UK would seek to exclude arable cultivation from steeply sloping land (>11°), wherever soil thickness was less than 20 cm (e.g. on the South Downs). Reed (1979, 1983, 1986) and Evans (1985) recommend the mounting of winged tines behind tractor wheels to break up the lug pattern, reduce soil compaction, and increase the infiltration of water into tractor wheelings.

Conducting field operations along the contour rather than up and down the slope has been suggested by Evans and Nortcliff (1978), Morgan (1979), Reed (1979, 1986), MAFF (1984a), and Evans (1985). The maintenance, and even replanting of field boundaries on long slopes, or at key points on those slopes, has been recommended by Evans and Nortcliff (1978), Morgan (1979), Reed (1979, 1986), and Boardman (1983, 1984). The retention of crop residues on the soil surface to reduce its exposure to raindrop impact and increase infiltration rate has been suggested by Reed (1983), MAFF (1984a), Evans (1985), Speirs and Frost (1987), and Robinson and Boardman (1988). Evans (1985) and Speirs and Frost (1987) have recommended that seedbeds be prepared as coarsely as possible to increase water infiltration rates.

The effectiveness of these recommended soil conservation measures in the UK will depend on the extent to which they are adopted by UK farmers. This in turn will depend on the farmers' perception of the erosion problem, and the extent to which they are encouraged to pursue soil conservation practices by the Ministry of Agriculture, Fisheries and Food (MAFF) and other statutory bodies. With regard to the farmers' perception of the problem, as annual crop losses due to erosion are generally small, affecting less than 1% of the field area (Evans and Skinner, 1987), farmers do not consider this a problem. Where erosion rates are higher, leading to significant crop losses or gullying which interferes with how the land is worked, they do not tend to affect the same field at frequent intervals, so again the farmer does not think it important. During their lifetime, most farmers will not see their yields decline because of thinning of the topsoil, indeed they will see an increase in yield owing to technological advances (Silvey, 1981). Even where the yield potential is diminished in the short term, it will be difficult to distinguish this because the variation in yield from year to year attributable to the weather is much greater (Catt, 1986; Evans and Catt, 1987).

With regard to policies directly affecting soil conservation, there are, unfortunately, financial subsidies available as part of the CAP which currently reward farming practices which enhance the susceptibility of agricultural soils to erosion (see above). This implies that in some agricultural situations the best available soil conservation measures will be incompatible with these financial incentives. The most positive soil conservation measures taken by the MAFF to date are the production of advisory leaflets on the prevention of water and wind erosion (MAFF, 1984a, 1985). The leaflets outline areas and soils most at risk, factors encouraging erosion, and the time of year when erosion is most likely to occur, and provide advice on a range of control measures. The leaflets are an important policy development not only in terms of the role that they should play in alerting farmers to erosion, but also in that these publications are a tacit acknowledgement by the MAFF that soil erosion is now a problem of sufficient economic importance to warrant preventative measures. Indeed, the Soil Science arm of MAFF's Agricultural Development and Advisory service (ADAS) is now actively researching techniques to combat erosion, especially in sugar beet. With regard to broader aspects of government agricultural policy, the fact that erosion threatens the productivity of some UK soils in the medium to long term does not yet seem to have been acknowledged. MAFF's chief scientist maintained that there was no evidence that the current farming systems were unsustainable (House of Lords Select Committee on Science and Technology, 1984). Answers to parliamentary questions posed to the Minister of Agriculture in January 1988, indicate that even restrictions on the cultivation of winter cereals are not justified by current erosion problems. The minister maintained that the onus was on farmers to prevent soil erosion by following the code outlined in the aforementioned MAFF pamphlets. These recent statements of the government's position were made notwithstanding the assertion by soil scientists such as Morgan (1986, 1987) that some eroding soils in the UK will not support current cropping regimes beyond the first quarter of the next century. This official attitude will have to change, and appropriate modifications to EC subsidy structures will have to be made, if a suitable political and fiscal climate is to be created for the adoption of soil conservation measures in the UK.

Erosion and water resources

There are few published data on the impact of soil erosion on surface waters (streams, rivers, lakes, and reservoirs) in the UK. This is not surprising given that even on-farm effects of erosion have only recently (since the late 1970s) begun to receive much attention. The data which do exist are generally qualitative rather than quantitative, and often do not categorically identify erosion of agricultural land as the source of sediment pollution. Thus, for example, in 1984 investigations were being conducted by water authorities into the fisheries decline and pollution in three lowland rivers in southern England (the Avon, Itchen, and Test), to establish whether or not erosion was the cause (ENDS, 1984). However, Boardman (1986) explicitly links freshwater fishery declines to erosion-derived sediment pollution. In particular, this author cites the Hampshire Avon in England, as a river where silt from erosion events has increased the turbidity of the water and clogged gravel beds required for spawning by some fish species. Evans (1988b) also cites silting of gravel stream beds on chalk, as an important off-farm impact of erosion in Britain. In the same paper, this author notes that erosion of exposed arable soils may also be the source of a sediment plume photographed by satellite in the southern North Sea in May 1986. Unfortunately, none of these observations are in any way quantifiable. Local water authorities have only sporadic measurements of sediment loads in rivers, which are not related to river flows (G. Petts, pers. comm.), and these observations are insufficient to provide any reliable assessment of long-term trends in sediment pollution.

The majority of existing data on sedimentation rates in water storage facilities are for reservoirs in upland areas, and there appear to be figures for only one reservoir with a lowland catchment including a significant portion of arable land. Cummins and Potter (1967), cited in Douglas (1970), estimate that Cropston Reservoir in Leicestershire receives 12.2 $m^3/km^2/yr$ (0.12 $m^3/ha/yr$) of sediment from its catchment. Walling and Webb (1981) cite a more recent estimate of sediment yield for the same reservoir, which at 45.6 $t/km^2/yr$ (0.46 t/ha/yr) is at least three times greater. The lower estimate is approximately one tenth of the sediment yield of the catchments of two upland reservoirs cited in Douglas (1970), but both these records pre-date the recent increase in erosion rates of lowland soils in the UK. The only recent measurement of sediment yields of reservoir catchments are for upland areas in England, and most of these are for a single region, the Peak District. Butcher *et al.* (1990) quote sedimentation rates for reservoirs in this area of 0.6–2.8 t/ha/yr, suggesting that some reservoirs in this area will lose 75% of their capacity over 100 years. Burt and Oldham (1986) present sedimentation rates for reservoirs in the Peak District which range from 0.2 to 1.5 t/ha/yr, and which imply capacity losses for some reservoirs of up to 75%, 100 years after their construction. The erosion causing this sedimentation is linked to commercial afforestation of these upland areas, which involves deep ploughing of highly erodible peats (Burt and Oldman, 1986). The increase in the sedimentation of the Cray Reservoir in South Wales, attributable to upland afforestation, was so severe as to have made it necessary to switch water supplies, raising the supply costs by a factor of 160 (House of Lords Select Committee on the European Communities, 1984). Upland rivers have also been adversely affected by the afforestation of these areas, and Graesser (1979) attributes the decline in some salmon fisheries in Scotland to increased erosion from the 'improved' lands.

In addition to straightforward sediment pollution of water bodies, there is the problem of the nutrients and agricultural chemicals which bind to the finer, more erodible, soil fractions. The higher the proportion of fines (largely clays and organic matter) in a soil, the greater is the percentage of these fines removed by an erosion event (Evans, 1988a). Some of the pesticides and phosphates which bind preferentially to these fines are, respectively, toxic to aquatic organisms and the cause of eutrophication problems. There are no quantitative data available on this form of pollution in the UK.

Effects of erosion on soil organic matter, soil water availability, and nutrients

Soil erosion has only recently emerged as a problem on lowland soils in the UK, its significance has only recently been appreciated, and, largely as a consequence of this, its effects on soil constituents and properties are poorly monitored and understood. It is clear, however, that water and wind erosion processes selectively remove the finer fractions of the soil. Organic matter, silts, and clays (the 'fines') are the soil fractions most easily stripped by flowing water (e.g. Frost and Speirs, 1984; Boardman and Robinson, 1985; Boardman, 1986; Evans and Skinner, 1987), and tend to be transported further than the other fractions. This means that whereas the coarser particles often remain within an eroding field, the fines are frequently moved beyond the field boundary and are lost altogether. The removal of fines by water erosion is roughly in proportion with their percentage occurrence in the soil (Evans, 1988a).

As these fines are the repository of a large proportion of the nutrients and water-holding capacity of any soil, it is currently impossible to accurately determine the effect of erosion on soil constituents and characteristics. To date, there only exists a theoretical treatment of the problem of rates of organic matter loss caused by water erosion, compiled by Frost and Speirs (1984). These authors calculate that a sandy loam eroding at a rate of 25 t/ha/yr, with an initial organic matter content of 2.5%, will eventually equilibrate at a level of 1.6%. Measurements of organic matter levels on eroding and non-eroding parts of a field covered by the soil in question, appeared to confirm these calculations. Frost and Speirs (1984) made no attempt to calculate the effect of these organic matter losses on nutrient losses or water-holding capacity. The effects of water erosion on these characteristics of other UK soils are currently unknown.

Erosion, soil rooting depth, and cereal yields

The permanent effects of erosion on crop yields in the UK have been assessed with regard to erosion rate, the initial depth of the topsoil, and the nature of the subsoil or bedrock immediately underlying the topsoil. Where topsoils are thick, and the underlying subsoil consists of unconsolidated and therefore workable material, which can be incorporated into the topsoil, effects of erosion on yields will be negligible in the short term (e.g. Evans and Nortcliff, 1978; Frost and Speirs, 1984; Evans, 1985). In general, detectable reductions in cereal yield will only occur where total soil depth is less than 1.2 m (Evans, 1981b; Frost and Speirs, 1984). Frost and Speirs (1984) estimated that fields with a sandy loam topsoil and subsoil with a total depth of 2.0 m, eroding at 25 t/ha/yr, would not suffer significant cereal losses for at least 200 years. Evans and Nortcliff (1978) and Evans (1981b) make similar calculations relating soil loss to yield loss for sites in Norfolk (Hempstead) and Cambridgeshire (Maxey), with soils respectively 0.9 and 1.2 m deep. These calculations show that an erosion rate of 3 t/ha/yr for 100 years will reduce cereal yields in the two soils by only 1.0–3.6% (Table 8.2).

Table 8.2. *Estimated percentage reduction in yield of cereals over 100 years at various rates of erosion. (After Evans, 1981b)*

| Rates of erosion (t/ha/yr) | Reduction in yield (%) | | | |
| | Spring barley | | Winter wheat | |
	Hempstead*	Maxey*	Hempstead*	Mean
1.0	0.3	0.5	1.4	0.7
3.0	1.0	1.8	3.6	2.1
12.0	5.3	7.9	13.1	8.8

* Initial soil depth at Hempstead = 0.9 m. Initial soil depth at Maxey = 1.2 m

A higher rate of 12 t/ha/yr will reduce cereal yields by between 5.3 and 13.1% after 100 years, a yield loss that should not be regarded as tolerable (Evans, 1981b). However, on a regional basis, where such rates of erosion are confined to a small proportion of the agricultural land, long-term yield effects will be small.

Where topsoils are thin and subsoils are rapidly exposed by erosion, or where their incorporation with the topsoil results in a less workable seedbed, yield losses from erosion will become evident much sooner. Evans and Catt (1987) showed that where chalky boulder clay had been brought to the surface on eroding valley sides in Cambridgeshire, winter wheat yields were only 59% of those on the valley floor where the topsoil was intact. Similarly, when a clayey subsoil was incorporated with clay loam topsoil, seedbed conditions deteriorated to the extent that yields fell by over 50%. On soils less than 0.3 m deep, the effects of erosion on yields are likely to be dramatic (Evans, 1981b; Boardman, 1987a). Yields are already low on such soils, and will quickly become lower if erosion occurs. Morgan (1987) calculated that a sandy loam soil with an initial depth of 0.25 m, eroding at 20 t/ha/yr, will reach ultimate physical degradation in 37 years, while a clay loam of similar depth, eroding at the same rate, will have a life-span of only 68 years. Ultimate physical degradation of a soil is taken as the point at which yields fall to 75% below the maximum possible.

Higher rates of erosion on similarly thin soils can destroy its productivity capacity within a very short period of time. Boardman and Robinson (1985) record a single erosion event that removed 100 t/ha of a redzina soil overlying chalk on the South Downs in Sussex. The soil depth at this site varied from

Table 8.3. *Calculated life-span of soils in years*

	Sandy loam			Clay loam		
Bulk density (g/cm^3)	1.4			1.3		
Minimum soil depth (m)	0.20			0.15		
Available soil depth (m)	0.5	0.3	0.25	0.3	0.25	0.2
Life-span at mean annual erosion rates of (t/ha):						
2	7800	1400	700	1950	1300	650
5	700	350	175	488	325	163
10	311	156	78	217	144	72
20	147	74	37	103	68	34
50	57	29	14	40	27	13
100	29	14	7	20	13	7

0.1 to 0.3 m and Boardman and Robinson (1985) estimated that it would only take between 13 and 20 such events of this magnitude to remove all the topsoil from the field.

The relationship between soil depth, erosion rates, and productive life-span of sandy and clay loams is illustrated in Table 8.3, reproduced from Morgan (1987).

The costs of erosion-related yield losses and soil conservation measures

Assessments of the costs of erosion-related yield losses of cereals and soil conservation measures are currently distorted by EC subsidies which effectively provide a major incentive for the continuous cultivation of winter cereals. Whereas the planting of spring barley as opposed to winter wheat or winter barley can be a very effective means of reducing erosion, the gross margin on a winter wheat crop exceeds that on spring barley by around £100/ha (Nix, 1985; Speirs and Frost, 1987). On a thick soil, where erosion is not bringing about any long-term decreases in soil fertility and yields, this difference in gross margins is likely to substantially exceed annual crop losses (due to removal or burial of crop plants), and any costs incurred by the farmer having to clear eroded soil (Speirs and Frost, 1984, 1987). Other erosion control options, such as the sowing of inherently erodible land to grass, are even more expensive for farmers in the current economic climate (Speirs and Frost, 1987). This explains why even those farmers with thin soils on steeply sloping land persist in sowing it to winter cereals, suffering high erosion rates as a consequence (Boardman, 1987b). The resultant erosion rates of up to 100 t/ha/yr will be severely affecting the long-term productivity of affected soils.

The economic pressure causing such mismanagement of soil resources can be illustrated by reference to current cereal prices, long-term yield losses attributable to erosion, and the cost of some remedial measures. The calculations will be based on the model relating soil depth to cereal yields, developed for the site studied by Evans (1981b) at Hempstead in Norfolk. Assuming that a farmer plants winter wheat continuously for the next 100 years and suffers the rather high erosion rate of 12 t/ha/yr as a result, annual yields from this soil (which is initially 0.9 m deep) will be reduced by 13.1% after 100 years. Assuming a constant rate of erosion and correspondingly constant increase in yield losses, the farmer loses 6.5% of his potential yield in the first 100-year period. Assuming a winter wheat yield of 6.5 t/ha/yr (Nix, 1985), at an average price of £110/t (Nix, 1985), the farmer's fiscal loss over 100 years will be, all other factors being equal, £4826. That is 6.5% of 6.75

tonnes, approximately £110/tonne for 100 years. Conversely, if for those 100 years, the farmer plants spring barley, thus reducing erosion to the point where it has no effect on yields (a somewhat optimistic assumption if it is grown continuously), then given that the current gross margin per tonne for spring barley is approximately £100 less than that for winter wheat, over 100 years the farmer will lose £10 000/ha. On the basis of these assumptions it clearly pays the farmer to let his soil erode, even over a period of 100 years. At the end of that time, when the annual winter wheat yield has been reduced by about 13%, the gross margin for winter wheat on the eroded soil will be approximately equal to that on spring barley on the uneroded soil at current prices. Given the relatively short planning horizons of most farmers (and indeed most economists), the fact that permanent yield losses from erosion in the first couple of decades would only be a fraction of natural annual variation in cereal yields (Evans, 1981b), that annual losses due to erosion are also negligible in comparison to natural variation (e.g. Catt, 1986; Evans and Catt, 1987), and that the off-farm costs of erosion rarely accrue to the farmer (e.g. Evans, 1981a; Boardman 1987b), it clearly pays the farmer to let his field erode. For a farmer on steeply sloping land, where the only effective soil conservation measure may be to put the land down to grass, the economic incentive to continue planting winter wheat will be even greater, given similar soil depths and erosion rates. The difference in the annual gross margin between ewes and lambs grazed at 10 animals/ha and winter wheat yielding 6.75 t/ha, is on average £166/ha in favour of winter wheat (Nix, 1985).

Where less drastic erosion control measures such as cultivation along the contour, avoiding unnecessarily fine seedbeds, retention of crop residues, and planting of catch crops are effective, the costs of soil conservation are not so high. Thus, for example, Robinson and Boardman (1988) showed on a site on the South Downs, that direct drilling of winter wheat through the unburnt stubble of last year's crop not only reduces erosion, but also gives a yield some 10% greater than a conventionally tilled (ploughed and harrowed) crop. A direct-drilled plot of winter wheat eroded at a rate of 32.4 m^3/ha and yielded only 6.05 t/ha of grain. These experiments, conducted on a 7° slope, indicate that direct drilling without burning of the crop residue not only reduces erosion but also increases yields, and consequently offers an immediate economic gain to the farmer, in addition to the longer-term benefit of reduced erosion rates. The cost-effectiveness of these soil conservation measures needs to be more thoroughly evaluated for different soil types, slope angle, and other site-specific characteristics, as do any conservation measures which have major implications for farming systems design. In some situations, a return to ley–arable rotations may represent the only effective solution to soil structural and erosion problems (e.g. AAC, 1970; Reed, 1979, 1983; Morgan, 1986), and yet under the current subsidy regime a farmer adopting such a rotation may be heavily penalized for doing so. This is particularly the case where farmers have responded to the incentives provided by specializing exclusively on arable cropping, selling their livestock and dismantling the infrastructure necessary to run a mix of crop and livestock enterprises.

Research and policy development for soil conservation in the UK

The current priorities for research into soil erosion in the UK are to establish where it is happening, at what rates it is occurring, what the causal factors are, and how to estimate the scale of the on- and off-farm costs in agricultural and environmental terms. These subjects have been addressed in the present review, which has made it clear that the research conducted so far provides partial answers for only some of these questions. Yet the research effort being expended on soil erosion and conservation in the UK is currently dwindling. Since the five-year SSEW monitoring programme ended in 1986, field studies on erosion conducted by this body have fallen away. In 1987 only seven flight lines of the original 17 were photographed, none were photographed in 1988 nor in 1989.

The intensive monitoring of soil erosion conducted by Reed in the West Midlands has ceased since the author's retirement, and the researchers in Scotland who conducted the only extensive survey of soil erosion there, currently have limited time and funds available for this work (Speirs and Frost, pers. comm; Watson, pers. comm.). This trend must be reversed, and the coverage of erosion monitoring must be extended to all areas considered at risk, and even to some areas not considered at risk where factors known to exacerbate erosion are operating. The most appropriate methodology for these surveys is aerial photographic transects, followed up by ground surveys to evaluate the aerial photography data and extend the type and quality of information collected. Data on rainfall, soil type, land form, and agricultural management factors should be collected at each site to enable identification of the prime causal factors in the observed erosion, with a view to establishing appropriate soil conservation measures.

To date, no comprehensive attempt has been made to evaluate the on- and off-farm costs of soil erosion. Without such an estimate it is not possible to construct a rational policy on soil conservation. The losses of organic matter, soil nutrients, and water-holding capacity consequent upon erosion must be quantified, as they have major implications for short- and long-term crop yields. The work conducted by Morgan (1987) relating erosion rates and soil thicknesses to the expected productive life-span of that soil should be refined and extended. This information will make it possible to identify the most vulnerable soils in the UK, and, together with the survey work recommended above, will allow for the establishment of appropriate conservation measures. The costings of both short and long-term yield losses consequent upon erosion should be investigated further to improve the accuracy of current assessments. The off-farm costs of road and drain blockage and clearance, property damage, and nutrient and sediment pollution of natural water bodies and water storage facilities will have to be estimated, and given appropriate weighting in the re-formulation of policy. Society rather than the farmer currently bears most of the off-farm costs of soil erosion (e.g. Evans, 1981a; Stammers and Boardman, 1984; Boardman, 1987b), and the farmer's ability to 'externalize' these costs reduces the incentive to adopt income-limiting soil conservation measures.

There is an urgent need to relate the sum of the present research findings on the incidence and cost of soil erosion, to the agricultural policy pursued by the UK government and the EC. This is well illustrated by the fact that, of the current price received by UK farmers for wheat grain, namely £110/t, some £75 (approximately 68%) is effectively subsidies from the government and the EC. In some parts of the UK, this amounts to a subsidy for erosion (Boardman 1987b), as well as the over-production of cereals. Other trends enhancing soil erosion in the UK, such as the specialization in continuous arable cropping, increases in field and machinery size, and the decline in organic and green-manuring practices, are at least in part consequences of the current agricultural subsidy structure which favours high-input, high-output farming systems. Integrated farming systems requiring lower inputs, such as organic farming systems, are inherently conservative of the soil resource (e.g. USDA, 1980; Arden-Clarke and Hodges, 1987; Reganold *et al.*, 1987). Such systems maintain soil cover and soil organic matter levels, retain crop rotations (particularly ley–arable systems), and are more flexible with regard to the timing of cultivations. Furthermore, as the economic margin to be gained from these systems is dependent on minimizing inputs, farmers adopting them will not be rewarded by attempts to intensively crop marginal land or thin soils, which inevitably exacerbate erosion problems. A much larger proportion of the UK's agricultural research and development budget should be diverted to the investigation of integrated and organic farming systems, than is currently the case.

The size of the current EC subsidies in relation to the price of farm produce gives tremendous leverage to the policy-makers who determine and allocate these grants. These policy-makers therefore have the power to bring about dramatic changes in agricultural policy when, as now, it is seen to have both economically and environmentally damaging effects. What is required is the political will, which at present is notable by its absence. This is likely to remain the case for as long as irrefutable scientific evidence of the real resource and environmental costs of current agricultural policies is lacking. Given adequate funding to conduct the necessary research, soil (and other environmental) scientists can provide this evidence and these costings. The costs of this research will be small by comparison with the benefits to be gained from the development of a more resource-efficient and environmentally friendly agricultural system.

9

Soil erosion and conservation in Poland

L. RYSZKOWSKI

Historical background

Agricultural activities on the present area of Poland date back to
the warm, moist Atlantic climatic phase (5500–3000 BC) of the post-glacial
epoch (Hensel and Tabaczynski, 1978). At the time, human impact on the
environment was too insignificant to stimulate erosion processes and alter the
natural cycles of matter. With rising population density and the invention of
more powerful tools, considerable portions of the upland areas of southern
Poland were cleared of forests to take advantage of their loess soil which was
good for cultivation and rather easy to till. The probable decline of soil
fertility in loess upland during the dry subboreal phase of the post-glacial
period (3000–800 BC) may explain an expansion of the cultivation in other
areas (Hensel and Tabaczynski, 1978). In the Iron Age, fuel demand for
primitive smelting furnaces was another cause of deforestation. In Poland,
iron metallurgy began its intense development in the 3rd and 4th centuries
AD in the Swietokrzyskie Mountains, and by the end of the 4th century AD
the population density was estimated to be three per square kilometer (Zak,
1978). At that time, loess areas of the Lublin uplands, and Malopolska
Region, were almost totally deforested. This resulted in intensification of
erosion processes, which is manifested by significant deposits of eroded soil
at slope bases, formed in the period AD 500–1000 (Strzemski, 1964).

By the end of the 10th century, arable land constituted as much as 16%
(Zak, 1978) or even 20% (Maruszaczak, 1985) of the total country area, and
the population density was 4–5 per km^2 (Zak, 1978). By AD 1000, gradual
human transformation of the environment caused the main river of Poland –
the Wisla – to transport about 12 t/km^2 of soluble compounds from its
drainage basin (Maruszaczak, 1985). However, areas of surviving forest,
bog, and meadow limited erosion. About the 16th century, the arable land

area constituted roughly 50% of the country acreage, almost the same as now. The population density was still small and amounted to about 20 per km². Grain crop production was intensified, subjecting loess soil areas to increasing gully erosion. At the beginning of the 19th century, row crops became widespread in Poland, particularly potatoes, and this significantly raised the rate of denudation (Szumanski, 1977). One of the consequences of increasing erosion in the 19th and 20th centuries could be the change in flood frequency. In the 19th century, the mean period between two floods on the Wisla River was 4.2 years, and in the first half of the 20th century 3.3 years. A similar time gap between floods (3.2 years), existed in the period 1901–39 on the other great Polish river – the Odra. Considering the fact that meteorological data do not indicate a rise in precipitation, the increased flood frequency must have been caused by an intensification of the rate of surface runoff, together with greater load of eroded soil silting up riverbeds (Ziemnicki, 1968). In the 20th century, besides agriculture, another source of intensified processes of matter translocation was added: the rapid development of industry. Industrial wastes released into rivers and saline mine water, are examples of such intensification. Maruszaczak (1985) estimated that at present the Wisla carries 47 tonnes (t) of dissolved material from each square kilometer of its basin. A considerable part – 23 t – of the transported dissolved chemical compounds is from sewage and air pollution. Soil erosion yields 24 t/km²/year, i.e. twice as much as the amount estimated by Maruszaczak (1985) for dissolved compounds transported by the Wisla from its watershed in the year 1000. According to data from the Chief Central Statistical Office, about 600 000 000 m³ of saline water is released into rivers from various mines in Poland. Erosion of a mine dump, particularly at strip mines, is another example of increasing erosion caused by industry. Also construction of large housing projects, highways, etc., increases erosion. As quantitative recognition of stimulation of erosion processes by industry is, however, very poor, it will be neglected in further analyses.

Natural conditions of erosion in Poland

Polar, arctic, and tropical masses of air which meet over Poland result in frequent changes of weather. These changes in air mass movements cause winter thaws which lower snow storage, and possibly prevent excessive spring thaws. The alternate occurrence of dry spells and excessive rainfall during the vegetative period is a characteristic feature of Polish climate (Schmuck, 1969). For example, rather frequent March dry spells can promote wind erosion due to poorly developed plant cover. Fortunately, low speed winds predominate over the country. Among the 26 European countries, Poland has the lowest precipitation, with a mean annual rainfall of about 600 mm. The structure of components in natural water balances of the country, particularly in its central parts, are shifted more towards evapotranspiration than in most other European countries. For instance, the Wisla River carries 24% of rainfall and the Odra 23%, as compared, for example, to the Rhine which carries 44%. Thus, river discharges in relation to normally low rainfall are further lowered due to relatively intensive evapotranspiration processes. The common effect of these climatic conditions is lower meteorological conditioning of erosion processes than in other parts of the globe, and their high variability with time. For example, thaw flows from snow in the Lublin uplands (which are covered with easily leached loess soils) were very low within the period 1970–77 and did not occur in other parts of Poland (Ziemnicki, 1978). During 18 years (1950–67), only two storms of local significance occurred in the Polish Lake District. The most fertile soils, such as loess and rendzina, which are most subject to erosion, constitute 4.3% of all Polish soils. The loose sandy soils, eroded to a lesser degree, occur on 3.47% of the acreage. The following regions have physical conditions favoring erosion (Ziemnicki, 1968):

1. Mountainous areas of the Karpaty and Sudety on the southern border, with serious erosion on steep slopes;
2. Areas of the Lublin and Malopolska uplands, with high erosion hazards causing severe agricultural losses;
3. Glaciated uplands of the Pomorze and Mazury Lakelands.

In the last-mentioned area, considerable forest areas, meadows, and pastures can be found on slopes. This causes a lower intensity of erosion there (Niewiadomski, 1968).

A very important characteristic of the Polish geographical environment with a significant effect on intensifying erosion processes, is that arable lands constitute 46% of the country area, and are a kind of surrounding and connecting tissue joining all types of Polish ecosystems. A rural management system, therefore, is most important for the conservation of the country's soil resources.

Occurrence and rates of soil erosion in Poland

Range of erosion

The first attempt at evaluating the potential range of water erosion in Poland was made by Reniger (1950). According to her estimation, around 1950, 20.1% of Poland was endangered by erosion (Fig. 9.1). A more precise analysis carried out 15 years later reduced this area to about 10% – approximately 30 000 km^2 – of the territory (Ziemnicki, 1968). The latter author recognized strong erosion hazard on about 1.5% of the country area, concentrated in the Lublin and Malopolska uplands. A recent assessment of potential occurrence of erosion forms showed their much larger range. Josefaciuk and Josefaciuk (1988) estimated that 39.3% (122 710 km^2) of the area is potentially threatened by water erosion. Gully erosion is found on 22% of the land, and wind erosion is observed on 10.8% of the land. Moreover, 8500 landslides were noted, half of which were rather small ones. In the lowlands area, 2360 slides were reported, 460 of which threatened buildings. Evaluating all forms of erosion, Josefaciuk and Josefaciuk (1988) estimated that almost 50% of the country area is endangered by erosion of varying intensity. Better mapping of the country and a broader set of criteria facilitating indication of a potential erosion hazard undoubtedly contributed to this increased estimate over that provided by Reniger (1950). The acreage

Fig. 9.1. Areas of potential erosion (after Ziemnicki, 1978).
1-medium, 2-strong in mountains, 3-strong in arable lands.

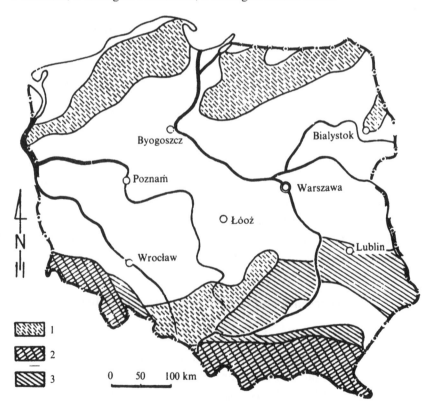

of arable fields in Poland constitutes 46.6% of the country and the area under meadows, pasturelands, and orchards is 13.8%. If, according to Josefaciuk and Josefaciuk (1988), nearly half of Poland's area is potentially threatened with erosion, one can presume that it occurs throughout arable land.

One can also suppose that during the last 30 years the area of land potentially threatened by erosion grew larger. The increase in territorial range of erosion is related to the intensification of agricultural production and particularly to its mechanization and the increase in arable land area. Industry also played a considerable role in increasing erosion by reducing vegetation cover.

Erosion rate

(a) Water erosion

Due to the lack of a station network measuring erosion rates in the whole country, it is only possible to present an estimation. In evaluating the extent of erosion, the variability of meteorological conditions in Poland must be considered. For instance, the measurements carried out in one place over 14 years concerning the denudation rate indicated enormous variability. In some years, erosion did not occur or the amount of eroded soil was slight, in other years 336 t/km^2 of soil washed by water was observed (Table 9.1). As it is well known, the topography of terrain has a very significant effect on the intensity of erosion. For example, Prochal (1973), while analyzing 11 mountain watersheds located close to each other, found considerable differences in the quantities of eroded material (Table 9.2), and indicated that the greatest influence on erosion rate (Y) was longitudinal inclination (X) of the stream ($Y = -90.3 + 0.97\,X; r = 0.89$, with erosion rate expressed in t/km^2/ year). Another important factor influencing the erosion rates in the streams studied by Prochal (1973), is the degree of afforestation. Plant cover and soil type have a major influence on limiting erosion. Complicated interactions of all factors affecting the erosion rate make estimates very difficult. Therefore, the evaluations of erosion rates for the whole country given below should be considered as approximate, and not as precise values. Preliminary evaluation of water transport of materials for various regions of the country were given by Ziemnicki (1978) using Branski's data from 1956 to 1970. The results presented by Ziemnicki can be expressed by the percentage of the country area corresponding to different erosion classes (Table 9.3). In order to estimate an average denudation rate for the whole country, water erosion rates for areas where the transport rates were unknown were approximated on the basis of physiographic similarity. Then, taking median values of distinguished erosion classes and a value of 120 t/km^2/year for the highest erosion class, the average water transport rate for the whole of Poland was calculated by multiplying median values of each class by the contribution of an area having similar denudation intensity in to the total area of the country. So the calculated average denudation rate is 26 t/km^2/year, and recalculation into the whole country area yields the estimate of water transport of soil of 8.1 million tonnes per year. The material transported in streams and rivers constitutes a part of the soil which was removed from fields by water erosion. Probably, at least half of the material subjected to water erosion is transported to rivers and one can assume that 16.2 million tonnes per year, or 52 t/km^2/year, to be a mean denudation rate for the whole country.

A large part of the area with intense erosion is located in southern Poland, either in the mountains or the Lublin and Malopolska uplands. Before water reaches the Baltic Sea, a high proportion of the eroded material will settle out due to the decrease in water velocity. Even fine material suspended in water is subject to sedimentation in large lowland rivers when their flow rates

Table 9.1. *On-side variation of water erosion rates on loess soils (after Ziemnicki, 1968). + means traces detected*

Watershed 1 (4.75 km^2)		Watershed 2 (6.22 km^2)	
Year	Erosion rate (t/km^2/y)	Year	Erosion rate (t/km^2/y)
1950	168		
1951	28		
1952	28		
1953	14		
1954	28		
1955	+		
1956	336	1956	154
1957	+	1957	+
1958	+	1958	+
1959	0	1959	0
1960	+	1960	+
1961	+	1961	+
1962	+	1962	+
1963	14	1963	+
		1964	56
		1965	+
mean	44	mean	21

decrease. For example, in the 394 km long segment of the Wisla River between Warszawa and Tczew, about 0.4 million tonnes per year of suspended material was sedimented in alluvial deposits (Debski, 1961). The Wisla carries 1.84 million tonnes per year of deposits from its river basin to the Baltic Sea. Using these data, Debski (1961) estimated that, from the whole country area, 4.8 million tonnes of deposits are removed. Comparing this value to a very imprecise evaluation of 16.2 million tonnes as the quantity of materials subjected to water erosion in the whole country, one can calculate that about 30% of the eroded soil is transported from the country.

Table 9.2. *Differences in mean erosion rates of mountain streams*

Stream	Watershed area (km^2)	Period of estimation (years)	Erosion rate (t/km^2/y)*
1	3.3	11.5	70.8
2	2.2	13.5	81.2
3	2.5	13.5	83.7
4	3.2	11.5	97.5
5	2.9	13.5	127.4
6	8.4	13.5	128.2
7	1.1	13.5	132.2
8	2.4	13.5	146.7
9	0.6	14.5	172.0
10	0.9	10.5	207.9
11	0.3	15.5	260.3

* The erosion rate presented by Prochal (1973) in m^3 was converted to tonnes using the factor of 1.56 expressing assumed bulk density of eroded material

Table 9.3. *Water erosion rates in Poland for 1956–70. Calculated from map presented by Ziemnicki (1978)*

Erosion rate (t/km^2/y)	Percentage of Poland area
<5	34.5
5.1– 10.0	15.5
10.1– 30.0	9.1
30.1– 50.0	1.8
50.1–100.0	12.7
>100	5.6
Lack of information on erosion rates	21.8

Most of the soil is transported to depressions, contributing to gradual change of the land profile, or forms alluvia. Ziemnicki (1968) estimated that on the loess Lublin uplands with a very high erosion rate, a slope with 10% inclination loses on average about 7000 t/km^2 of soil a year and, at the same time, valley bottoms are lifted. Also, the research carried out by Mazur and Pałys (1985) on changes in cultivated loess soils showed that, during 20 years, between 5600 and 23 800 t/km^2 was washed away annually. During that period, the slopes were lowered on average by 21 cm. The maximal lowering of the slope with the highest inclination was 34 cm. At the slope foot, the soil on the valley bottom was lifted on average by 12 cm. From the total amount of eroded soil, 44 to 76% of the material was deposited on the way down. The remaining part was transported further, mostly to water courses (Mazur and Pałys, 1985).

In Poland, soil particles deposited by the Wisla formed the fertile soils of the Wisla delta, or the fertile alluvial soils of some other river valleys (Ziemnicki, 1972). The intensity of this process is shown in the annual increment of the Wisla delta reaching a mean of 60 ha a year (Ziemnicki, 1968). Analyzing water transport of the materials, it should be kept in mind that some of the material originates from bank erosion and from the riverbed. Due to the removal of water mills from numerous smaller Polish rivers, the flow rates increased, which led to deepened riverbeds (Podwinska, 1970; Pałys, 1971; Szumanski, 1977; Łos, 1985).

The effects of erosion are not limited just to changes in the water network, but have great importance for the hydrobiological conditions of watersheds. Riverbed erosion can cause lowering of ground water in a valley and it increases the intensity of transporting eroded material in a river. Bank erosion creates losses in adjacent soils due to riverbed widening and meandering. In the studies carried out in the Lublin uplands, it was found that, at the beginning of the 20th century, there were about 150 installations damming up water. The distances between the dams varied from 2 to 5 km, which protected about 500 km of the river course against river erosion due to limiting the rate of water flow. Up to 1980, of the total number of 150 old installations, 120 were destroyed. Pałys (1985), evaluating the consequences of removing river dams, estimated that bed erosion increased on about 200 km of the studied rivers. Lowering of the riverbeds caused lowering of the ground-water level in the valley. This resulted in the drying of adjacent areas, which turned the meadows into arable fields because of low soil moisture (Pałys, 1985). Besides riverbed erosion, intensification of bank erosion was also found. Serious bank erosion occurs along 150 km of the studied rivers and about 60 km^2 of soils are lost each year. The soil of the banks contains about 4% humus, while the soil material deposited by water is almost totally devoid of it (Pałys, 1985).

To end the discussion of water erosion hazards, one should devote a few words to gully erosion. These phenomena have been studied in great detail by many Polish scientists (Ziemnicki, 1968, 1978; Ziemnicki and Naklicki, 1971; Maruszaczak, 1973; Buraczynski, 1977). One of the effects of advanced water erosion is gully formation on loess areas of the Lublin uplands where gully erosion is very serious. Here, the greatest concentration of gullies reaches the value of 10.5 km/km^2 in the neighborhood of Szczebrzeszyn (Buraczynski, 1977). Serious gully erosion with a gully density of 2 km/km^2 takes about 0.3% of the country. According to these authors, the total length of gullies in Poland is about 40 000 km and their total area is 1000 km^2. In the Lublin uplands, Maruszaczak (1973) estimated that, on average, 250 000 t of loess were removed per square kilometer due to gully erosion during the whole period of gully formation. Besides transforming soils into wastelands, gully erosion also causes a considerable fall in yields on adjacent slopes, and at gully mouths in places of accumulation of the transported material. Moreover, on the area cut with a gully network, the fields are small with irregular shapes, and this makes transport and cultivation difficult.

(b) Wind erosion

The evaluation of a wind erosion rate for the whole country has not been carried out yet. In spite of the fact that scientists have not been interested in wind erosion, we can suppose that this phenomenon occurs with increasing frequency. Intensification of wind erosion is related to cutting down shelter-belts, removal of boundary strips with creating large fields, and intensification of cultivation procedures (Josefaciuk and Josefaciuk, 1987).

In Polish conditions, wind erosion as compared to water erosion causes fewer soil transformations, though in extreme cases it can lead to large wastelands – dune areas. On loess and loose sandy soils, with strong winds blowing when the fields are not yet covered with vegetation, dust storms can be encountered. From measurements and observations it can be seen that wind erosion occurs with higher intensity in winter and early spring (Gerlach and Koszarski, 1967; Repelewska-Pakalowa and Pekala, 1988).

The wind erosion intensity is highly variable in Polish climatic conditions. When snow cover remains on the fields for a short time and during the vegetative period, if it is dry and very windy, then the wind erosion rate may exceed the value of 200 t/km^2/year. In years with higher rainfall and light winds it can be no more than 7.5 t/km^2/year in the same area (Repelewska-Pakalowa and Pekala, 1988). The mean annual erosion rate for endangered areas calculated from 15 measurements (Table 9.4) taken in various places in Poland during five years (Repelewska-Pakalowa and Pekala, 1988) is 93.1 t/km^2/year. According to Josefaciuk and Josefaciuk (1988), potential wind erosion hazards occur on an area of 88 430 km^2, which is 28.2% of the whole country area. In the opinion of these authors, serious wind erosion threatens 1%, medium erosion 9.8%, and slight erosion 17.4% of the land area. Assuming that the long-term mean value of the potential serious erosion corresponds to the mean value of measurements by Repelewska-Pakalowa and Pekala (1988), i.e. 93.1 t/km^2, and that the class of mild potential hazard has the mean value of 30 t/km^2, and the class of the slight one 5 t/km^2, then one can calculate the annual mean rate of wind erosion on the threatened areas to be 16.8 t/km^2.

Expressed in terms of the whole country area, the annual wind erosion rate is 4.8 t/km^2. As compared to mean annual water erosion, estimated to be 52 t/km^2, this is a small value. Such data support present opinion that, while at times and in certain local situations wind erosion may be strong and cause economical losses, wind erosion generally causes much smaller losses than water erosion in Poland. At present, because of the poor recognition of the wind erosion processes, it is impossible to evaluate how much of the soil material is transported outside the country's boundaries by wind. It should be pointed out that part of the material carried by water was blown there from land.

Erosion and organic matter in soil

Blowing off or washing of soil surface layers and blending deeper parts of subsoil into arable layer contributes to considerable differences in composition and content of humic compounds. Niewiadomski and Borenska

Table 9.4. *Annual wind erosion rate (t/km^2) in seriously threatened areas (after Repelewska-Pakalowa and Pekala, 1988)*

Area	Years				
	1981	1982	1983	1984	1985
Lublin upland	28.4	185.2	128.9	70.1	128.0
Mazowiecka lowland	–	112.7	87.6	31.9	47.0
Polesie	–	90.1	66.4	47.7	78.7
Roztocze	–	–	–	74.7	219.4

(1977) found a fall in humus concentration because of erosion from 0.88% on top to 0.75% in the middle of a slope, and then a rise to 1.07% at the foot for light soils. Increased fertilization doses (from 150 kg NPK per ha to 300 kg NPK per ha) on the most eroded slopes, facilitated uniform yielding of the whole slope.

Similarly, Mazur (1985) found a fall in humus concentration in loess soil on a slope with 10% inclination, from 15 to 30% as compared to the top. Higher fertilization doses equalized yields on eroded slopes to the level of the top area (Mazur, 1985). Similar changes in humus content in non-loess soils were described by Skrodzki (1978).

Water erosion has an influence on the composition of organic compounds in soil. Degraded by erosion, loess soil shows a narrower ratio of carbon to hydrogen in humic acids (Table 9.5), and a higher contribution of aliphatic chains combined with aromatic nucleus (Turski and Dobrzanski, 1968). Humus from eroded soils also shows a decrease in humic acid content and an increase in fulvic acid content, as well as a lower degree of humidification, and the humus is more loosely bound to soil minerals (Turski and Wincenciak, 1969). All these results indicate a higher susceptibility for the leaching of humus in eroded soils.

Long-term investigations carried out in the Institute of Agrobiology and Forestry of the Polish Academy of Sciences, showed high concentrations of soluble organic compounds in aquatic ecosystems. Concentrations of dissolved organic compounds reached values from 4.5 to 9 mg per litre (Zyczynska-Baloniak, 1980). Variability in the concentration of these compounds depended on rainfall and surface runoff intensity and on agrotechnical treatments. It was indicated that a considerable portion of organic matter dissolved in field water reservoirs constituted organic compounds of soil

Table 9.5. *Carbon and hydrogen content (%) in humic acids from upper layer (0–20 cm) of eroded soils. (Modified from Turski and Dobrzanski, 1968.)*

Type of soil	C	H	C:H
Non-eroded podzolized	51.07	6.18	8.26
Eroded podzolized	46.02	5.96	7.72
Non-eroded brown	54.87	5.80	9.46
Eroded brown	50.89	6.15	8.27
Non-eroded chernozem	55.43	4.87	11.38
Eroded chernozem	50.27	5.65	8.89

origin with fulvic acids dominating (Szpakowska *et al.*, 1986). In a drainage canal situated along cultivated fields, the concentration of soluble humus substances varied from 1.3 to 5.7 mg/l. Comparative research made in the Wisla River mouth by the same authors, showed that the concentration of dissolved organic compounds amounted to 0.9 mg/l. Assuming this concentration of dissolved organic matter to be representative, the quantity of dissolved organic compounds carried out to sea by all Polish rivers during a year (58.6 km^3, Debski, 1961), amounts to 9100 t.

Effects of erosion on the hydrobiological regime of soils and watersheds

Washing off the humus layer and uncovering rocky subsoil with lower permeability for water, or depositing fine particles blocking soil pores are among the mechanisms of the erosion responsible for decreasing the permeability and water capacity of soils (Prochal, 1984). The soils of slopes, often with a destroyed humus horizon and an arable layer made up of transient level or of subsoil, have decreased moisture content (Czerwinski, 1985). The moisture content of soil deposited at slope foot depends on soil type and on accumulated material. If fine particle material is deposited on compact soils, they become wet, and dry very slowly. If sand is deposited on compact soils, then, due to cultivation and mixing of the layers, one can observe improvement of soil absorbability. Similarly, when loam material is deposited on loose soils, one can expect an increase in water retention in the soil. The more varied the terrain relief, the greater the spatial differentiation of soil water content and so the greater intensity of erosion processes. Erosion can result in periodic excessive drying off or watering of the soils. Drying off occurs most on sloping soils where water deficit depends on angle of slope, aspect, and erosion of the soil. Conversely soil deposits are excessively wet at the foot of the slope (Mazur, 1988). Large differences in soil moisture content between slope foot and the remaining elements of the terrain relief were notable in the spring. These differences significantly delay the beginning of field cultivation in the valleys (Orlik, 1971). Disturbances in river flows result from decreased permeability and water-storing capacity of eroded soils, which increases surface runoff.

Deep cutting of slopes and valleys by gully erosion drains adjacent soils. In winter, snow from neighboring fields is blown into the gullies which lowers the field's supply of water at the beginning of spring (Ziemnicki, 1968). On loess soils, Naklicki (1971) found that a gully 3.5 m deep significantly changed the water regime in an adjacent soil belt of 20 m width. This was reflected in yield, which was between 5 and 20% lower than that obtained on the area beyond the gully effect zone.

Effects of erosion on vegetation and crop yield

Due to the climatic conditions of Poland, water erosion causes the greatest plant losses during spring thaw and during the heavy showers which occur in summer in central and southern Poland, and in spring in northern regions (Ziemnicki, 1968, 1978; Skrodzki, 1978; Prochal, 1984). The losses result from burying the plants in eroded material, uncovering root systems and washing off the soil, or from various indirect effects leading to changes in soil fertility. Field crops are most often threatened, particularly root plant crops that are strongly susceptible to erosion. Generally, direct losses have a local range. Losses due to the indirect effects of erosion are as widespread as areas with potential erosion hazard. Weak negative effects can be eliminated by means of appropriate agrotechnical treatments, e.g. higher level of fertilization. This applies to both loose and compact soils (Niewiadomski and Baranska, 1977; Niewiadomski and Grabarczyk, 1977; Mazur, 1985). As is well known, meteorological conditions play a decisive role in shaping the yield. Keeping this in mind, we can state that higher crop yields are obtained from top areas without eroded soils. In valley bottoms and slope bases with deposited soils rich in nutrients, one can observe high yields in dry years. In wet years, grain crop yields are low due to water logging, but root crops yield relatively well. On the slopes, the yields are generally the lowest, particularly under conditions of soil water deficit (Mazur and Orlik, 1985; Mazur, 1988). Detailed investigations on the development of root biomass by plants on soils with various degrees of erosion, indicate a decisive effect of soil moisture content which depends on meteorological conditions. The degree of soil erosion is not significant for root biomass development, but influences surface plant biomass (Orlik and Czerwinski, 1985).

Taking into account criteria of susceptibility to erosion, Ziemnicki (1968) stated that strong water erosion affects 10 000 km^2 of arable fields, and yield losses recalculated into wheat grain range from 0.5 to 1.0 million tonnes a year, depending on erosion intensity and on interplay of erosion effects with influences of meteorological conditions. Josefaciuk and Josefaciuk (1988) considered all kinds of erosion and applied a more versatile system for assessing possibilities of appearance of erosion hazards. They estimated that the losses can reach 3.5 million tonnes a year. In relation to the yields obtained in the whole country, these losses amount to 7%. High losses will occur when meteorological conditions stimulating erosion appear on the whole area. Such an unfavorable coincidence of conditions is only of small probability and it seems therefore that Ziemnicki's (1968) evaluation better reflects actual losses. Assuming Ziemnicki's evaluation, yield losses due to erosion range from 1 to 2% of yields.

To end the discussion on erosion effects on plant production, it should be pointed out that both water and wind erosion negatively affect soil fertility by removing nutrients. The quantity of mineral fertilizers taking part in erosion processes in Poland is not known. Only the amount of the fertilizers carried out to sea from the country in a year has been estimated. Ziemnicki (1968) estimated that in the water flowing out of the country 17 500 t of nitrogen, 4300 t of phosphorus, 25 000 t of potassium, and 43 000 t of calcium in elemental form are carried off. This constitutes from 0.5 to 2.1% of the amount of these components introduced with mineral fertilizers into arable field soils in 1984/85 (Table 9.6).

Prognosis of erosion in Poland

The prognosis given below is based on analysis carried out by Josefaciuk and Josefaciuk (1988). The rise in water and wind erosion intensities result from the intensification of agricultural, industrial, and mining production. The erosion processes are magnified by actions aiming at increasing the area of each field by removal of shelter-belts, stretches of meadows, and small field water reservoirs. Mechanization of cultivation, increased transport, and decline in organic fertilization of soil also have a negative effect. Mechanization without appropriate tractors and tools adopted for terraced soil cultivation, will cause an increase in water and wind erosion. It is estimated that intensification of surface runoff and wind erosion

Table 9.6. *Annual losses of mineral fertilizers from Poland into Baltic Sea (after Ziemnicki, 1968)*

Element	Loss (t)	Amount (t) contained in mineral fertilizers applied in 1984/85	Percentage of fertilizer elements removed
N	17 500	1 239 000	1.7
P	4 300	885 000	0.5
K	25 000	1 156 000	2.1
Ca	43 000	2 610 000	1.6

on large fields where anti-erosion protection has been neglected is about ten times higher than in small farms. Each kind of erosion will be intensified differently. The highest increase will concern surface-water erosion. Wind erosion intensification on arable soils will be related to drying off soils and increasing field size. Further, the development of mining and industry will favor the intensification of water and wind erosion. Deep mining causes very dangerous collapses of terrain which lead to higher intensity of erosion processes. Strip mining, which causes considerable excavations and dumps, is also a factor intensifying erosion. Considering expected transformations in spatial structure of arable fields and in mechanization, and also transformations due to mining and industrial activities, one should expect further intensification of erosion in Poland. Therefore, starting preventive actions protecting soil against erosion is an important element of nature conservation in Poland.

Program of protecting soil against erosion

The program of protecting soil against erosion was already developed in Poland by the end of the 1940s by Professor Stefan Ziemnicki. Several hundred papers presenting the causes of erosion and its various effects, and pointing out soil reclamation treatments protecting the soil against erosion, were published. A survey of this activity is contained in the books by Ziemnicki (1968, 1978) and Prochal (1984). Some methods for soil conservation were introduced into practice, particularly the treatments limiting river and gully erosion (Prochal, 1984). The outline of the rules of soil conservation given below is based on suggestions presented in the works of Ziemnicki (1968, 1978), Prochal (1984), and Josefaciuk and Josefaciuk (1988). Anti-erosion treatments should be adapted to natural conditions, and in Poland, three main regions with different specific conservation measures are identified. The northern region of lakelands is mildly threatened with erosion (Niewiadomski, 1968). The preventive measures consist of maintaining forests and shelter-belts covering 35% of the area; meadows and pastures should cover up to 20%, and arable land should not exceed 40% of the area. In crop rotation, cereals should not exceed 45%, and root crops 15% of the arable land area.

The second region is the Malopolska and Lublin uplands. This area is very seriously threatened by erosion. Since much of the area is upland, ideally a maximum of 60% of the region should be farmed using anti-erosion agrotechnics. At present, the fields constitute 80%. It is recommended that orchards should be planted on the slopes.

The third region with a specific strategy of soil conservation is the mountain area, which also has the highest annual rainfall. The main directions of soil conservation aim to cover 50% of the area with forest; arable soils should constitute no more than 25%, and it is recommended that the acreage of meadows and pastures be enlarged, and anti-erosion embankments of streams and water reservoirs be constructed. Slope-transverse cultivation is recommended on hillsides.

Agrotechnology has a great protective role to play in all the regions. It includes slope-transverse, or terraced plant and soil cultivation, differentiated fertilization between top, slope, and valley, and protective plant rotation. These treatments are very effectively conserving soil in erosion hazard areas with slope inclination up to 6%. In plant rotation, the most protective species are grasses and their mixes with legumes and perennial legumes. Cereals have lesser protective abilities. Root crops stimulate erosion intensity. It is not our goal to discuss detailed preventive measures, which are well described in numerous widespread publications. Only the importance of plant cover in limiting erosion will be stressed. As is well known, the greater the proportion of perennial plants in an area, the greater its resistance to erosion. Forests have the best erosion-preventing effects. The runoffs from forested areas are small and spring thaws are slow. Forests should be located on seriously eroded soils where strip tillage does not yield good effects, and on infertile ones. Also, field shelter-belts play a great conservation role, as they not only stop wind erosion but also water erosion. Field shelter-belts serve many purposes in agricultural landscapes, and therefore their propagation is of great importance in protecting rural areas. Permanent grasslands are also important. Their effect is connected not only with strong reinforcement of soil by a well-developed root system, but also with their soil forming role. Appropriate introduction of the system of soil conservation can effectively eliminate losses due to erosion. It should be stressed, at the same time, that many problems related to limiting erosion in its many forms have already been solved by scientists in Poland. The success of a conservation program depends mainly on society's understanding of the importance of this problem and on the practical introduction of theoretical work.

10

Soil erosion and conservation in the humid tropics

S.A. El-SWAIFY

Introduction

Soil erosion is perhaps the most serious mechanism of land degradation in the tropics generally and the humid tropics in particular (El-Swaify *et al.*, 1982). In the humid tropics, erosion by water, rather than by wind, assumes primary importance. The following discussions, therefore, will be restricted to rainfall-induced erosion.

Opinions differ as to which forms of water erosion are most serious, although gully erosion and mass wasting are the most catastrophic and visually impressive forms. However, any 'accelerated' erosion, i.e., erosion with a rate which exceeds the natural rate (prevailing geologically prior to human disturbance) or the rate of soil genesis from underlying parent strata is likely to produce detrimental impacts on the productivity and quality of the ecosystem. As such, even the insidious and visually least impressive sheet (inter-rill) and rill erosion can be seriously detrimental. A complete picture of erosion effects cannot be painted without a total consideration of onsite and offsite impacts. Since runoff water (including floods) and runoff-borne sediments recognize no political boundaries, erosion is always a watershed-based and often a multi-nation problem. Assessing the problem, and the planning and implementation of effective counter-measures, therefore, may well require the institution of not only national commitments and policies but also regional and international ones as well (FAO, 1982a).

The setting

Climatic classifications do not agree on a uniform definition for the humid tropics. Fig. 10.1 shows Troll's classification of tropical climates (El-Swaify and Dangler, 1982). Rainfall excess over evaporation (or evapotranspiration) is the usual criterion for defining wet or dry months. Those regions

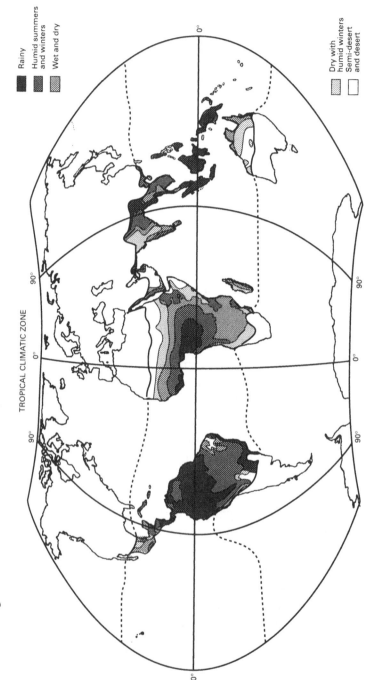

Fig. 10.1. Troll's classification of tropical climates as modified by El-Swaify and Dangler (1982).

TROPICAL CLIMATIC ZONE

Rainy

Humid summers
and winters

Wet and dry

Dry with
humid winters

Semi-desert
and desert

in the tropical zone with more than half the year (often 7.5 months) with a wet regime are normally considered as humid tropics. Strict imposition of this definition results in the exclusion of tropical regions with less than 1200–1500 mm of mean annual rainfall. A major consequence of total and seasonal water abundance is the establishment of dense forests. Undisturbed, natural forests are capable of imparting tremendous protection against erosional forces despite the high aggressiveness of rainfall.

Such a climate is categorized in the USDA Soil Taxonomy as a 'udic' moisture regime. Prevailing soils in the humid tropics are the Alfisols, Oxisols, Ultisols, and Inceptisols. Andisols prevail in volcanic ash dominated areas. In all these soils leaching is excessive and nutrient loss is accelerated when the nutrient-cycling natural forest is cleared. Prevailing minerals in these highly weathered soils generally display low activity, i.e. ability to retain basic nutrients in an exchangeable form or to associate strongly with organic matter and thus protect it from rapid decomposition in these warm and humid environments. Most have a vast capacity for immobilizing certain nutrients, particularly phosphates; soil acidity (primarily aluminum-derived) and infertility are primary constraints to the production of food crops. Physical properties appear to be generally favorable.

Harsh topographies often characterize the landscapes of the uplands and highlands in the humid tropics. Population expansion forces progressive encroachment on steeper lands where the erosion hazard is extremely high in the absence of natural vegetation. Traditional farming in these regions consists of slash-and-burn or shifting cultivation systems in which the all-important fallow period is becoming absent or very brief due to land scarcity. The unaffordability of inputs necessary to establish healthy cropping systems on exhausted soils leads to a vicious circle of erosion, fertility decline, and ultimate degradation.

Table 10.1 shows that estimated values for potential erosion from the major soils in the humid tropics are in the range of hundreds of tonnes per hectare per year. Protective land uses, therefore, are critical for preventing these hazards from becoming reality.

Erosion seriousness: extent and impacts

Qualitative and quantitative assessments of erosion magnitude
Erosion assessments have long been plagued by the diversity of adopted methods and scales; this has resulted in a wide range of reliability in estimates which should be recognized when data from different sources are reported. Qualitative descriptions of erosion status have long been standing

Table 10.1. Potential erosion from typical soils of the humid tropics under various climatic regimes. RI = rainfall index, SL = soil loss (Mg/ha/yr). The figures assume moderate steepness and length of slope (20% and 10 m, respectively), and soil erodibility values of 0.15, 0.2, 0.3, and 0.46 (metric units) for Ultisols, Oxisols, Andisols, and Alfisols, respectively. Rainfall data sources: Lo and El-Swaify et al. (1985)

Soils	Location									
	Cameroon		Hawaii		Indonesia		Philippines		Sri Lanka	
	RI	SL	RI	SL	RI	SL	RI	SL	RI	SL
Alfisols					2000	1840			619	570
Andisols			1466	880	1630	978	2708	1625		
Oxisols			524	210					694	278
Ultisols	1437	431			2381	714	1852	556	833	250

components of soil surveys and inventories (e.g. Soil Conservation Service (USDA), 1948, and Soil Conservation Service of NSW, 1982). Semi-quantitative methods include the periodic use of aerial photography and satellite imagery; both are rapid, useful for identifying sources with large features (e.g. gullies or landslides), but are too expensive and inaccurate for planning purposes (El-Swaify *et al.*, 1982). Workers have also attempted to translate the abundant available data on sediment loads of rivers and streams, and those deposited behind dams and reservoirs, etc., into soil-loss data from respective sources. This translation is difficult to accomplish because of uncertainties on sediment re-deposition, delivery ratios, re-suspension, the stability of stream banks during channel water flow, and the dependence of all these factors on the scale of measurement, i.e. size of catchment under study.

An increasingly available tool is the prediction of soil erosion rate from the climatic, soil, topographic, and land-use data by use of empirical or process models (e.g. Wischmeier and Smith, 1978, and the National Soil Erosion Research Laboratory, 1987). Empirical models, in particular the universal soil-loss equation, have received considerably more adaptation to the tropics than process models. However, the site-specificity of empirical models requires that considerably more work be done before they are considered reliable for erosion prediction in all of the humid tropics. Clearly, direct measurements of erosion at the source provide the best quantitative information on the magnitude of the problem.

Available geologic data on erosion of different continents indicate that Asia leads the way with 166 Mg/km^2/yr, followed by South America, North and Central America, Africa, Europe, and Australia with 93, 73, 47, 43, and 32 Mg/km^2/yr, respectively (various authors, cited by El-Swaify and Dangler, 1982). These data are derived directly from sediment loads in major rivers; no attempt is made to convert these data to field soil losses (i.e. by assuming sediment delivery ratios). While specific conclusions may be difficult to reach from these data, there is an apparent association between overall erosion rates and population densities in these areas. This is corroborated by the fact that the heavily populated regions of Asia possess the highest global sediment loads in their major rivers. Examples, presented as an average sediment removal from respective drainage basins (using appropriate sediment delivery ratios), are 550, 480, 430, 270, 217, and 139 Mg/ha/yr, respectively, from the Yellow River (China), Kosi River (India), Damodar River (India), Ganges River (India, Bangladesh, Nepal, Tibet), Red River (China, Vietnam), and Irrawady River (Burma). A similar picture is painted for other continents indicating the seriousness of erosion magnitudes in all of the tropics and in the humid tropics in particular. El-Swaify *et al.* (1982) assembled a review of the extent of water erosion in the tropics. Proceedings of numerous conferences and workshops include more up-to-date country reports many of which emphasize the continuing critical nature of the problem (e.g. Kussow *et al.*, 1982, and El-Swaify *et al.*, 1985).

No truly worldwide efforts have been undertaken to map soil losses and associated degradation in recent years (FAO, 1971). In the absence of such absolute data, it is constructive to provide illustrative available examples of erosion rates associated with various land uses in these regions.

Table 10.2 confirms the high potential erosion hazard in the tropics and shows the inferiority of managed systems as compared to natural forests in imparting adequate protection against erosion. Table 10.3 represents an attempt to quantify the relative effects of various human activities on erosion. The wide range of effects expected from forest conversion to annual crops reflects the extent to which various cropping systems and management practices may enhance conservation effectiveness. It is because of the potential benefits of tree-based systems and other perennial vegetation that agroforestry systems and agro-silvi-pastoral systems are perceived as being potentially more conservation-effective than strictly annual conventional cropping systems.

The specific examples of soil loss data from various locations (Table 10.2) reveal that harnessing the aggressive rainfall of the humid tropics requires good vegetation protection, particularly where topography is steep. These data show that leaving soils bare and unprotected even for a short period or within the existing cropping system (e.g. in clean till) is the most significant source of the problem. The data for bare fallow soils confirm the projection of high erosion potential in these environments as presented in Table 10.1. However, to appreciate the significance of these data and implications to the overall seriousness of the problem, it is important that the impacts of erosion also be reviewed.

Impacts of erosion on resource productivity and environment

Onsite impacts of erosion relate primarily to the soil's ability to support plant growth. The attributes which control this ability are the soil depth, root zone water-storage and transmission capacities, chemical/ nutritional qualities, and biological qualities. If the ideal rooting characteristics of a crop and its soil-based requirements for optimum production are known, erosion impacts on productivity can be predicted with some certainty from soil-loss data. Several predictive tools have been designed to accomplish this task but their applicability remains largely restricted to temperate

Table 10.2. Examples of erosion rates in tropical regions

Land use	Location (site–soil (slope))	Soil loss (Mg/ha/yr)	Reference
Bare soil (fallow)	Java–Oxisol (14%)	512	Suwardju* (modified)
	Java–Ultisol (4%)	115	
	Java–Alfisol (11%)	428	
Eucalyptus (blue gum)	India	0.1	Dhruva Narayana and Sastry*
Pomegranate	India	1.4	
Forest, montane undisturbed	Java, Indonesia	0.3	Weirsum*
Forest, no undergrowth or litter	Java, Indonesia	43.2	
Maize, clean till, along slope	Jamaica (30%)	127	Sheng (pers. comm., 1982)
Yams, clean till, along slope, rain 3300 mm	Jamaica (30%)	133	
Banana, clean till, along slope	Jamaica (30%)	183	
Maize, tilled, 612 mm rainfall	Trinidad (22%)	2.54	Gumbs et al.*
Cowpea, tilled, 612 mm rainfall	Trinidad (22%)	1.66	
Maize, untilled	Trinidad (22%)	1.98	
Cassava	Nigeria (11%)	125	Lal (1982)
Tea, seedling	Sri Lanka	40	Krishnarajah in Kusgow et al. (1982)
Tea, improved	Sri Lanka	0.30	
Sugar cane	Hawaii–Oxisol	7.16	
	Hawaii–Ultisol	4.26	
	Hawaii–Andisol	1.2	El-Swaify and Cooley (1980)
Pineapple	Hawaii–Oxisol	8.15	
	Hawaii–Inceptisol	10.2	

* In El-Swaify et al. (1985).

areas (Larson *et al.*, 1985; Williams, 1985). Recent research has shown that erosion impacts on the productivity of highly weathered tropical soils are more severe than for temperate soils. The primary factors leading to this conclusion are the higher erosion rates in the humid tropics, the severe changes in nutritional/chemical qualities as a result of erosion, and the inability of the typical resource-poor farmer to provide the inputs necessary for restoring those qualities to a respectable level.

Table 10.4 shows a typical example of erosion's effects on important characteristics of an Oxisol from southern Africa (El-Swaify *et al.*, 1982). Throughout the profiles, significant reduction is noted in pH, organic carbon (a reflection of total organic matter and nitrogen content), and major crop nutrients. While a significant part of the decline results from enhanced leaching in erosion-prone areas, surface erosion can account directly for much of the loss (Fig. 10.2). The N and P losses shown in Fig. 10.2 ignore the fact that erosion processes are frequently selective and that removed fine sediments are generally finer than the overall soil mass and are, therefore, enriched by higher nutrient loads than the bulk. It is possible, therefore, that 50–100 Mg/ha of soil erosion can result in the loss of N and P quantities which are comparable to those normally applied as maintenance fertilizers sufficing to produce a full crop of maize annually. Lal (1976) reported total nutrient

Table 10.3. *Some reported quantitative effects of human activities on surface erosion. (El-Swaify* et al., *1982)*

Initial status	Type of disturbance	Magnitude of impact by specific disturbance*
Forestland	Planting of row crops	100–1000
Grassland	Planting of row crops	20–100
Forestland	Building logging roads	220
Forestland	Woodcutting and skidding	1.6**
Forestland	Fire	7–1500
Forestland	Mining	1000
Row crops	Construction	10
Pastureland	Construction	200
Forestland	Construction	2000

* Relative magnitude of surface erosion from disturbed surface assuming an initial status of 1
** This low figure may be characteristic of the practice in the USA. It is essential to note that skidding is likely to cause more severe damage in tropical countries, particularly where no precautions against surface soil disturbance are required

Table 10.4. *Analytical data from some eroded and non-eroded Ferrasols (Oxisols) in Tanzania (El-Swaify et al., 1982)*

Depth (cm)	Clay (%)	Silt (%)	Sand (%)	pH	Organic C (%)	P (ppm)	Exchangeable cations (meq/100 g)					CEC*	Bases (%)	Zn (ppm)	Cu (ppm)
							Ca	Mg	K	Na	H				
Profile D1 (eroded plot)															
0–15	26.6	3.1	70.3	4.51	1.4	6	1.37	0.95	0.15	0.10	12.2	14.8	17.6	1.0	5.7
15–30	27.3	4.2	68.5	4.59	1.34	<2	2.08	1.16	0.11	0.06	11.3	14.7	23.1	0.9	4.1
30–60	34.2	3.0	62.8	4.5	—	<2	1.5	0.70	0.10	0.06	11.8	14.2	16.9	1.1	3.5
60–90	37.4	2.6	60.0	4.48	—	<2	1.21	0.48	0.09	0.05	11.9	13.5	11.9	1.0	3.3
90–150	38.6	2.3	59.1	4.46	—	<2	0.79	0.20	0.14	0.18	12.4	13.7	9.5	0.9	3.3
Profile D2 (non-eroded plot)															
0–15	21.1	3.5	75.4	5.29	2.59	33	5.21	2.05	0.32	0.16	8.3	16.0	51.4	3.5	31.0
15–30	21.7	4.2	74.1	4.78	1.22	5	2.28	1.79	0.18	0.18	8.4	12.8	34.6	0.9	4.4
30–60	21.4	3.5	75.1	4.49	0.98	3	1.76	1.32	0.19	0.22	8.1	11.5	30.2	1.9	4.2
60–90	24.9	4.1	71.0	4.29	0.53	3	1.2	0.58	0.14	0.12	6.6	8.6	23.6	5.5	8.2
90–150	30.6	2.3	67.1	4.51	0.49	<2	1.63	0.44	0.1	0.06	6.7	8.9	25.1	1.6	2.9

* CEC = cation exchange capacity

242   *S.A. El-Swaify*

and organic carbon losses of several hundred kilograms per hectare per year in sediments from Alfisols in Nigeria.

Crop responses to erosion of tropical soils are illustrated by data from long-term experiments on an Oxisol from central Oahu, Hawaii. Figure 10.3 and data from similar research in the tropics show that:

1. Erosion effects on productivity differ for different crops. Evidence so far is that crop sensitivity is in the order grain cereals > grain legumes > root crops.
2. Erosion effects are modified by the level of inputs (fertility, tillage, supplemental irrigation, etc.) provided to the growing crop; the lower the inputs the more severe is the impact. Clearly, resource-poor farmers in the humid tropics are more vulnerable to erosion-induced declines in productivity than their rich 'western' counterparts.
3. The replenishment of soil nutrients (as added fertilizers) is by itself insufficient to restore original soil productivity. Physical and biological rehabilitations are often also required. Table 10.5 shows

Fig. 10.2. Potential soil and nutrient loss from an unprotected tropical Oxisol (clayey, kaolinitic, isohyperthermic Tropeptic Eutrostox) in relation to steepness of slope. Calculations are based on experimentally measured soil erodibility, on a standard slope length of 23 m, and 1200 mm annual rainfall with a corresponding erosion index of 350 (Wahiawa, Hawaii). (El-Swaify and Dangler, 1977).

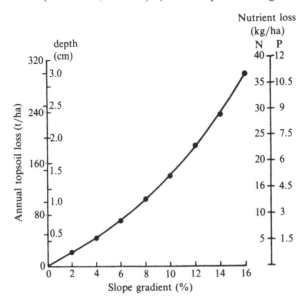

Fig. 10.3. Effects of erosion and added fertility on the yield of maize, Irish potato, sweet potato, and soybean grown on Wahiawa silty clay, an Oxisol in central Oahu, Hawaii. Erosion levels E0, E1, and E2 represent topsoil losses of 0 (control), 10, and 35 cm. Fertility levels F0, F1, and F2 represent nutrient replenishments of 0 (control), intermediate, and optimum. The yield of maize in the severe erosion, no-fertilization treatment (E2F0) was zero (original data).

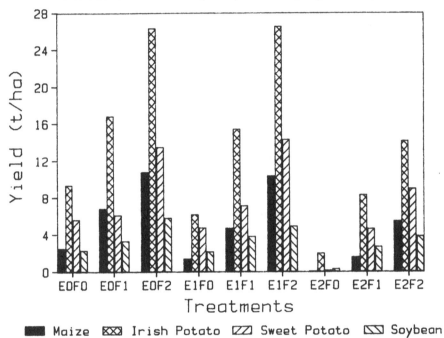

Table 10.5. *The influence of simulated erosion on indigenous populations of rhizobia nodulating cowpea, leucaena, or sesbania. (Unpublished data, M. Habte, 1987)*

Depth of topsoil removal (cm)	MPN* of rhizobia/g of soil		
	Leucaena	Cowpea	Sesbania
0	2.4×10^4a	3.5×10^4a	9.2×10^4a
7.5	1.6×10^4a	3.5×10^4a	2.8×10^4b
15	2.4×10^4a	2.4×10^4a	3.5×10^3c
25	3.4×10^2b	2.2×10^3c	1.5×10^2d
37.5	4.9×10^2b	1.1×10^3c	1.7×10^2d

* Most Probable Number (MPN) figures followed by different letters are significantly different at the 1% probability level

the significant decline in *Rhizobium* spp. populations due to erosion of the same Oxisol used for the above erosion–productivity experiments. Similar evidence exists for changes in the populations of *Mycorrhiza* sp. as a result of erosion (Yost *et al.*, 1985).

4. Highly weathered (tropical) soils are more vulnerable to erosion effects than temperate soils. This is partly because of their nutrient-depleted status and the tendency of beneficial nutrients and organic matter to exist in a concentrated manner at the very top of the epipedon and to display less uniform distribution with depth than in soil with active clays. As discussed above, these inherent tendencies are complemented by the very high erosion rates encountered in agricultural systems in the tropics.

5. Rooting depths in eroded tropical soils are considerably more restricted than in uneroded soils. These restrictions are due to both unfavorable physical and chemical characteristics. Subsoils of Oxisols, for instance, display limiting resistance to root proliferation at lower bulk densities than do temperate soils (Fig. 10.4).

6. Unfavorable root proliferation and distribution patterns (Table 10.6) and unfavorable water transmission properties combine to reduce the efficiency of water use by crops on eroded soils. Table 10.7 shows that even if comparable yields are obtained from eroded soils, the fertilizer and water inputs required to achieve

Fig. 10.4. Comparison of soil bulk density sufficiency curves (expressed as relative productivity indices, PI) for a tropical Oxisol and a temperate soil with comparable clay content (F. W. Chromec, pers. comm.).

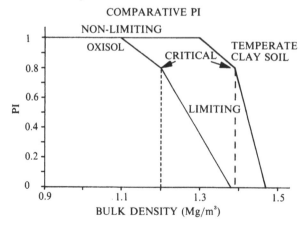

Table 10.6. *Maize root weight (grams fresh weight/2 kg soil) at two depths as influenced by soil erosion and added nutrients in an Oxisol. Figures in parentheses represent the corresponding percentages of roots in the respective soil layers. (Yost et al., 1985)*

Erosion level (cm)	Depth of Sample (cm)	Fertility level			Roots in each depth (%)
		Low	Intermediate	Optimum	
0	0–15	11.8 (74)	11.1 (67)	15.4 (81)	75
	15–30	4.1 (26)	5.5 (33)	3.5 (19)	25
10	0.15	11.9 (78)	13.2 (78)	18.7 (85)	80
	15–30	3.3 (22)	4.0 (23)	3.4 (15)	20
35	0–15	2.33 (97)	12.8 (96)	10.6 (96)	96
	15–30	0.07 (3)	0.51 (4)	0.48 (4)	4

LSD (0.05) for 0–15 cm is 6.32 and for 15–30 cm is 2.2

Table 10.7. *Changes of water and fertilizer use efficiency by maize as a result of erosion and restorative fertilization on an Oxisol (two crops)*

Erosion level (cm)	Fertility level	Water-use efficiency (kg stover/dm^3 of water)*	Fertilizer/amendment use efficiency (kg stover/ kg of elemental added amendment)*
0	Low	0.42	40
	Intermediate	0.48	8.6
	Optimum	0.60	2.6
	Average	0.5	17
10	Low	0.25	24
	Intermediate	0.36	5.6
	Optimum	0.51	2.3
	Average	0.37	11
35	Low	0.07	3.8
	Intermediate	0.17	4.0
	Optimum	0.37	1.9
	Average	0.20	3.2

* Stover is the vegetative component of yield (aside from the grain)

such yields far exceed inputs in uneroded soils. Clearly, this is another 'cost' of erosion. The implications for rainfed farming systems are important, not only in the long term as often accepted, but also in the short term since excessive erosional water loss by runoff represents lacking root-zone replenishment and, thus, can make the difference between a productive and a failing cropping season.

In recent years, offsite impacts have received increasing attention in the tropics. This, in large part, is due to increased encroachment of human populations on the upper reaches of river basins and watersheds with subsequent enhancement of sediment production from these areas as a result of introduced or increased human activity. Such disturbances induce excessive water losses which combine with sediment losses as partners in causing downstream damages. Effects caused by the *quantity* of runoff and erosional sediments include siltation of waterways, dams, and reservoirs, damaging the efficiency and equipment of hydroelectric power generating plants, burial and flooding of low-lying lands (and life), farmlands, property, and shoreline fisheries and reefs; destruction of roads, terraces, and other structures; and upsetting the balance involving sediment removal and deposition in stream beds and banks. Naturally, reduced reservoir life causes not only a loss in investment (cost of construction) but also a loss of water otherwise destined for storage and profitable use. The presence of sediment and associated nutrients in yielded watershed water also induces *quality* changes in waterways and at the destination. Effects caused by the quality of runoff water and borne sediments include the decline in optical water quality due to turbidity and enrichment of downstream receiving areas, particularly water bodies, by chemicals (including pesticides, fertilizers, metals and salts), microorganisms, and other substances. In developing tropical countries all these impacts have more than just aesthetic implications as the erosional disturbances of ecosystems generally have severe negative impacts on overall productivity, stability, and people's livelihoods.

Table 10.8 provides a sample of available data (various sources cited by El-Swaify *et al.* 1982) on the reservoir siltation problems caused by erosional sediments. It is truly intriguing that the actual useful life of a reservoir generally falls significantly shorter than that estimated by designers. This may be due to designers' optimism, or faulty assumptions of sedimentation rates either for lack of knowledge or to please politicians who may be keen on proceeding with 'beneficial' (and highly visible) public works. Invariably, and in all fairness to designers, there is often an increase in the rate of siltation with time reflecting increasing population encroachment and human activity in the erosion-prone upper reaches of dammed waterways. As stated earlier, a major additional dimension of the diminished storage capacity is the loss of valuable water resources for agriculture, power generation, or other uses.

The benefits of silt-laden river water to the productivity of soils in low-lying areas have been frequently documented and historically emphasized, particularly for ancient, low-input farming systems (e.g. Egypt is the Gift of the Nile). As an example, there have been recent concerns that the decline in silt load in Nile waters due to deposition behind the Aswan High Dam poses a threat to the productivity of irrigated agriculture in lower Egypt unless fertilizer use is proportionally increased. El-Swaify *et al.* (1982) argue that such benefits may not be forthcoming if the erosional sediments are derived from impoverished highly weathered soils (or subsoils). Furthermore, frequent short-term sedimentation damages, and burial of land and property as a result of catastrophic events, represent serious economic blows to farmers using these lands. In the long term, considerable cost and energy must be invested if such deposits are to be transformed into productive 'soils.'

The quality of stream water for human consumption, fish production, navigation, recreation, etc., is determined by its physical, chemical, and biological characteristics. A major distinction between tropical and temperate settings is the nature of the sediments derived from the soils in each. Aside from the inactivity of 'tropical' sediments, they seem to impart more dramatic effects to the optical characteristics of water than do those from 'temperate' soils. Fig. 10.5 shows that equal concentrations cause significantly higher turbidities with the first than the latter.

Table 10.8. *Siltation of selected storage dams and reservoirs. (El-Swaify et al., 1982)*

Reservoir	Annual siltation rate (10^3 m^3)	
	Assumed	Observed
Bhakra (Punjab)	28.4×10^3	41.6×10^3
Tungabhadra (Karnataka, S. India)	12.1×10^3	50.6×10^3
Matumbula (Tanzania)	7.9	13.2
Kisongo (Tanzania, 1960–69)	NA	4.0
Kisongo (Tanzania, after 1969)	NA	5.8
Ambuklao (Philippines)	1.3×10^3	2.5×10^3

All the above impacts must be considered, together with other factors, when the seriousness of an erosion problem is being judged. For management purposes, this information is usually translated into soil loss tolerances, i.e. soil erosion rates which are considered acceptable for preserving the stability, sustainability, and overall quality of the ecosystem.

Loss tolerances for tropical soils

Purists shudder to think that any soil loss should actually be deliberately permitted. In reality, however, accepting the inevitability of some erosion and determination of an acceptable level for it (soil loss tolerance), is a major aspect of land-use planning in erosion-prone areas. Such tolerances, also known as '*t*-values,' serve as management targets for designing conservation-effective agricultural systems and erosion control practices. Unfortunately, no simple formula can be provided for determining such values since management targets can vary alternately between onsite impacts, offsite impacts, and a host of other criteria such as natural erosion (geologically driven in the absence of human disturbance), soil renewal and profile development rates, intended land use, and socio-economic–political considerations pertaining to the costs, feasibility, and acceptability of the measures necessary to reduce erosion to a predetermined level. Indeed, a different value for acceptable soil loss may be derived for every criterion selected as a management target (Schmidt *et al.*, 1982). Since the costs of reducing erosion to meet conservative standards may be quite high, it is

Fig. 10.5. Standard turbidity curves for an Oxisol and Ultisol from Hawaii and an Aridisol from Colorado. Sediments derived from highly weathered 'tropical soils' are more detrimental to optical water quality than those derived from 'temperate soils'.

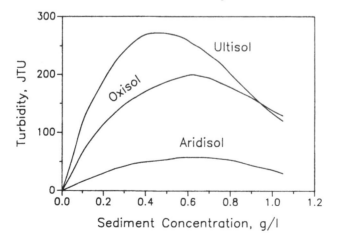

necessary to develop optimization procedures to achieve maximum overall benefits with the least cost by reconciling multiple criteria from the most important management targets.

A tolerance value which achieved common use in the USA and 'found its way' to the tropics, was derived from the estimated rate of soil renewal under intensive cultivation, i.e. 1 cm/30 years or 12 Mg/ha/yr (5 t/acre/yr). Recently, this value has been more often considered as a maximum soil loss allowable only for deep soils (< 1–2 m) with uniform favorable profile characteristics. The smallest tabulated *t*-value of 2 Mg/ha/yr is considered appropriate for shallow soils (10–25 cm) underlain by limiting layers, clay pans or bedrock (Mannering, 1981). Soils lying between these two extremes of depth and differentiation would be expected to tolerate intermediate losses ranging from 4 to 10 Mg/ha/yr, depending on their depth and profile characteristics.

The preceding discussions of onsite and offsite impacts indicate that the seriousness of these impacts in the topics exceeds that in temperate areas. These conclusions are based on soil-based factors, particularly infertility and nutrient distributions, the special optical characteristics of tropical soils and derived sediment, as well as resource-scarcity factors imposed by population pressures on cultivable lands, and the unaffordability of restorative measures by typical tropical small farmers. The severity of erosion impacts in the tropics, on the other hand, provides more incentive for and lends more feasibility to the implementation of preventive/corrective measures and the expectation of short-term benefits from such measures.

Planning, policy, and economic issues

Global concerns with erosion problems and increasing recognition of the importance of assured commitment from all relevant institutions and governments to adopt wise and long-term land management policies have culminated in the adoption of a World Soils Charter by the Food and Agriculture Organization of the United Nations. The charter sets principles for wise, productive, and protective land use to assure the welfare of future generations (FAO, 1982a).

Steps which facilitate the application of such principles and ultimate implementation of conservation-effective agricultural systems in the tropics are discussed below.

Importance of understanding erosion causes

A clear understanding of erosion causes is a requisite to any planning aimed to protect the viability of erosion-prone agricultural enterprises. The biological and physical factors which induce high erosion rates in the humid tropics (Table 10.2) and must be considered in selecting alternative land uses are:

1. High intensity and long-duration rainfall.
2. High rates of overland flow from adjacent uplands.
3. Poorly structured, non-cohesive soils with low infiltration and drainage rates.
4. Soils that lack coherence between top and subsurface layers (e.g. loose land fill, soils piled during construction activity, or soils with lithic contact).
5. Slopes with high or moderate steepness.
6. Long slopes with a non-rough surface.
7. Tillage and/or planting in rows directed with, rather than across, the prevailing slope direction.
8. Absent, sparse, or non-protective vegetative canopies, ground cover, or other surface protection cover as a result of inadequate land husbandry or excessive grazing.
9. Destructive and/or poorly timed land preparation or harvesting practices (including forest clearing and improper use of farm machinery).

Models, both empirical and process-based, are available for quantifying these factors and for use in predicting and controlling erosion (see El-Swaify *et al.*, 1982). Mass movements (e.g. landslides) are common in hilly terrain but are less predictable than surface erosion phenomena.

Socio-economic and policy issues

Unfortunately, the socio-economic–cultural–political factors which compel land users to use unsuitable lands and to practice ineffective preventative measures are neither quantifiable nor well understood. Frequent mention is made of the following 'human causes' of erosion which also act as barriers to erosion prevention and control in the humid tropics:

1. Lack of suitable land for expanding population to practice cultivation without high erosion risk.
2. Lack of recognition, awareness, or education of land users of the causes, sense of urgency, seriousness, and consequences of erosion.
3. Lack of will to accept 'improved' (but often high-risk) farming technology at the expense of proven (low-risk) traditional technology.

4. Lack of secure land tenure systems.
5. Increased fragmentation of lands with progressive generations and inheritance. Patterns of fragmentation often result in narrow slices oriented along the slope and are frequently blamed for increasing erosion potential.
6. Lack of technical services to assist land users with *appropriate* designs for erosion prevention. Shortage of trained personnel is a major bottleneck in many countries.
7. Lack of resources or necessary inputs for improving land husbandry and/or building conservation structures.
8. Lack of or inefficient enforcement of overall national and institutional commitment, policy, or incentives for erosion prevention and soil conservation implementation for the benefit of all society, and particularly future generations.
9. Non-involvement of grass-root levels of society (e.g. farmers) in the development of conservation projects to assure commitment and acceptability of erosion control measures.
10. Lack of sustained monitoring or periodic critical inventory of land resources to assure that accelerated erosion is not emerging as a threat.
11. Lack of systematic land resource inventory and policy for optimal land-use planning and designation of conservation lands.

Erosion cost considerations

Sensitization of policy-makers to the need for natural resource conservation may be assisted by attempting to place a price tag on the full costs of erosion. Two examples of such costs and their components are:

1. *Costs associated with onsite impacts.* Quantifiable onsite costs are those responsible for productivity changes on farmland. These may be measured, alternatively, as the costs of inputs necessary to maintain farm productivity at a viable level and to harness moving water and sediments for safe disposal. Components include fertilizer and tillage (to compensate for lost infiltration capacity and the quality in the exploitable root zone), water (to compensate for excessive runoff and inadequate root zone recharge), and installing land surface configurations, stabilization measures, and soil treatments to enhance infiltration, minimize runoff, and dispose of excess runoff safely so as not to endanger low-lying areas. A major (semi-quantifiable) cost in tropical uplands and highlands is the reduced value of eroded areas for water resources development by

virtue of declining quality for catchment/watershed purposes. This damage extends to both surface and underground water resources.

Non-quantifiable costs are those primarily due to long-term changes in soil quality (degradation) and depth, as the soil for all practical purposes is a non-renewable resource. Such changes generally cause ultimate abandonment of the land and cultivation of new lands where available. Barren lands such as these abound in the tropics. Other equally important changes arise from the severe disturbance or elimination of biological species in the original natural ecosystems which are transformed into agricultural lands.

2. *Costs associated with offsite impacts.* Quantifiable offsite costs are those associated with damage to low-lying lands, downstream life, property, structures, and the environment. These include the costs of flood and burial damage, reduced or eliminated value of silted water-storage structures (or the costs of regular dredging to maintain storage capacity), lost water which can no longer be stored in silted reservoirs, loss of water quality for human and animal consumption, loss of quality and productivity of fisheries, contamination and eutrophication due to sediment and runoff-borne chemicals, effects on the drainage efficiency and navigability of streams and waterways, and changes in other elements of ecosystem quality which are often considered aesthetic but do in fact translate into economically important impacts on society.

Non-quantifiable costs are primarily those arising from secondary changes in regional hydrology which are not easily predictable. An example of this is the salinization of many areas of northeast Thailand as a result of increasing water-table levels following deforestation.

No actual costing estimates have been made for either of the above categories. This, in part, explains our inability to mount and sustain convincing, non-emotional arguments with policy-makers on the urgency of enacting erosion-preventive land-use policies.

Elements of wise project planning and implementation

Two levels of concern arise when land-use issues and problems of soil resource instability are addressed (El-Swaify, 1988). The first level relates to specific technological issues aimed at generating necessary information then bridging the gap between results of research and actual application in the field. Technical guides are necessary, and will continue to be quite useful for practitioners and field planners who already possess expertise in analyzing and sorting out complex field problems. Most guides contain either broad generalizations or erosion/conservation principles to assure universal applicability, or strict site-specific features to assure relevance for local application. None are updated frequently enough to incorporate new information and few place sufficient emphasis on socio-economic considerations. These shortcomings are understandable because the amount of existing and new erosion/conservation knowledge is so large, and the databases which support this knowledge are so vast that conventionally written guides and manuals must be tailored for specific audiences, sites, and technical emphasis.

El-Swaify (1988) indicated that computer-based decision support systems are looming as increasingly powerful, useful, and available vehicles for structuring, storing, and retrieving large amounts of scientific data and knowledge for targeted applications. These systems, because they are systematic and interactive, also facilitate the identification of knowledge gaps and, thus, the targeting of research activities to essential needs. For instance, expert systems, which are special applications of artificial intelligence, can be easily updated substitutes for, or complements to, guide manuals and handbooks. They can also be useful as versatile aides to policy-makers, non-specialists, sponsoring agencies, and scientists for diagnosing erosion problems, selecting appropriate conservation recommendations, identifying important knowledge gaps, and overcoming socio-economic bottlenecks in the application of conservation technologies. It is expected that these new tools will expand very rapidly in the coming years.

The second level of concern deals with the broader considerations relating to establishing effective development projects in erosion-susceptible areas, particularly the steeplands of the humid tropics. El-Swaify (1988) identified the following elements as critical for assuring the success of such projects:

1. Sound conceptualization – farm productivity, rather than conservation *per se*, should be the primary target of the agricultural development enterprise; soil and water conservation is but one crucial requirement for sustaining and stabilizing the productivity of land resources on lands with high erosion hazard. Concerns of government and other agencies with 'offsite' impacts are consistent with farmer-based emphasis on sound management of soil resources and should be met without detracting from such emphasis.
2. Sound design – critical components of sound design include systematic matching of site characteristics with conservation-effective

and sustainable land uses, balancing both social and physical aspects of conservation by use of appropriate site surveys, involving recipient farmers in identifying problems and potential solutions, emphasizing simple, incremental, and gradual improvements in farming systems management, avoiding expensive structural conservation measures at the expense of the less visible but more important land husbandry (agronomic) measures, and emphasizing farmer education and awareness of the importance of continued maintenance and consequences of neglect of conservation measures.

3. Deliberate implementation – successful implementation of conservation measures requires true partnership (in both work and cost) between land users and the implementing agency. It is not uncommon for the agencies to subsidize, share, or facilitate in financing the cost of implementation with farmers, even in developed countries. This is particularly necessary when expensive downstream investments, stabilization structures, or common watershed (multi-farm) features (e.g. drainage ways or grassed waterways) are needed and a demonstration of long-term government commitment to project stability is necessary. Other important elements include the efficient delivery of important knowledge to all sectors of the farming community through an effective extension service, the maintenance of considerable flexibility in deliverable technological packages to allow continued updating or upgrading of management recommendations, and the assurance of periodic long-term maintenance of installed conservation measures.

4. Monitoring of conservation effectiveness – few projects whose mission includes erosion and runoff control in vulnerable tropical regions do in fact determine whether this objective is achieved and, if so, monitor the longevity of desired effects. Dependence on 'implied' or inferred applicability of experiences from elsewhere or results from small research plots may be erroneous. Erosional processes express themselves very differently at various sites and at operational field or catchment scales. The need for effective maintenance of existing conservation measures or for modifying their design in future projects can only be determined by quantitative monitoring of their effectiveness. Lack of monitoring allows design or implementation flaws to go unchecked, thus causing even more serious field erosion and defeating the purpose of

expensive investments. A balanced monitoring program should also include a continuous record of crop performance and soil characteristics which determine farming system productivity.

5. Coordination of project activities – the identification or perception of erosion problems and subsequent project conceptualization, design, implementation, and monitoring often involve several separate institutional entities in the country. An even more detrimental scenario, also not uncommon, is where one or more of these functions are not assigned to an accountable institution or executing agency. It is critical, therefore, that a 'Coordinating Board' be assigned to administer, oversee, and monitor all phases of the conservation/development scheme in the targeted region. This will ensure the integration of all project components and phases into a successful outcome. Coordinating Boards, while they may be charged with the responsibilities for specific projects, should not be intended to substitute for higher level agencies responsible for setting land-use policies at the national or regional level. Such policies are requisite to the formulation of systematic strategies for the orderly, effective, and safe use of all the nation's land resources, particularly those that are vulnerable to excessive erosion (e.g. steeplands). The Coordinating Boards, rather, would be responsible for designated target areas. Within these areas the multi-agency and multi-disciplinary Boards would assemble available land inventory data, prioritize land and soil suitability for development and conservation uses, coordinate the collection of information necessary to design and implement conservation-effective land-use systems, facilitate feedback between land users and project designers, designate pilot project areas to ensure effective diffusion of successful management technologies, and act as the clearing house for receiving and interpreting all monitoring information to safeguard the long-term conservation objectives and resource stability within the target area.

The seriousness of erosion extent and impacts in the humid tropics requires that full understanding of the problem be conveyed to policy-makers and land-use planners. This will assure that priority is given to the stabilization of existing and future agricultural lands.

11

The management of world soil resources for sustainable agricultural production

E.T. CRASWELL

Introduction

The thin layer of soil covering most of the earth's land surface supports a rich and wide diversity of terrestrial life. Formed slowly by physical, chemical, and biological weathering of parent materials, soil particles are moved and sorted by wind and water until, in most natural ecosystems, the counteracting processes of soil formation and loss reach equilibrium (Greenland, 1977a). The equilibrium established, and hence the soil type, is critically influenced by the nature of the climax vegetation which holds the soil in place and plays a central role in the processes of nutrient and organic matter cycling. The ultimate inherent fertility of the soil depends on the factors of soil formation – climate, parent material, and vegetation (Jenny, 1941).

Concern about the state of the world's soil and other resources has grown over the past two decades, culminating in the Brundtland Commission Report – Our Common Future (WCED, 1987a). Major problems – such as the greenhouse effect, caused by increasing levels of atmospheric carbon dioxide, effects of industrial atmospheric pollutants on forest growth, the extinction of irreplaceable animal and plant germplasm, and the impact of deforestation and land degradation – present mankind with hard choices between economic growth, poverty alleviation, and conserving the resource base for future generations. The need for economic development which is sustainable has never been clearer. For most developing countries this means meeting the food needs of a growing population by expanding food production without irreparably damaging the soil which is a key resource for agriculture.

Exposure of soils to erosion, the subject of this book, is accelerated when vegetation is removed by fire, grazing animals, or clearing and cultivation, which are the major means, for better or for worse, by which man has transformed the land. Early hunter–gatherers and pastoralists generally lived in harmony with the ecosystem, although their use of fire to clear undergrowth for hunting and other purposes severely perturbed the environment (Hallsworth, 1987b). Early agriculture, which began 10–12 000 years ago in Southeast and Southwest Asia, utilized swidden or shifting cultivation systems in which crops were grown after very rudimentary land clearing and in which long periods of bush fallow were used to restore soil fertility (Simmons, 1987). Modern agriculture, requiring the complete removal of native vegetation, transformed vast areas of forest and grassland in the so-called developed regions of Europe, Asia, Australia, and North America (Grigg, 1987).

The clearing and ploughing of soils for cropping exposes them to loss through wind and water erosion, and accelerates organic matter decomposition and the leaching of nutrients. Loss of the surface soil through erosion removes the most fertile soil layer, at the same time exposing subsoils which often are acid or have some other major physical or chemical properties which limit plant growth (Stocking, 1988). The removal of trees also alters the hydrology of the landscape leading to secondary salinity and waterlogging. Leaching, acid rain, and fertilizers can acidify soils under some circumstances (Overrein, 1978, Burch *et al.*, 1987). Degradation of soils by salinization, waterlogging, acidification, and related problems of compaction and crusting limit plant growth and vegetative cover, thus exposing the soil to wind and water erosion. These land degradation processes are quite well understood but their extent has not been assessed quantitatively (Burch *et al.*, 1987).

In this chapter, an overview of the state of the world's soil resources and the problem of erosion is presented. The focus is particularly on research to develop systems of soil and crop management which sustain agricultural production by conserving the soil resource. The associated policy needs are also discussed.

Population growth and agricultural intensification

It is useful to consider current problems with deforestation and soil erosion in the world in an historical perspective. Dregne (1982) differentiates three major historical periods of extensive soil erosion:

1. 1000–3000 years ago – the expansion of cultivation in China (including fragile loess soil areas), the Middle East and the Mediterranean.

2. 50–150 years ago – the migration of Europeans to develop colonies in other parts of the world where they took the best land and often forced the native inhabitants onto marginal lands which are difficult to manage.
3. The last 30 years – the expansion of agriculture in Latin America, Asia, and Africa due to growing population pressure. This expansion, to some degree caused by the use for urban purposes of good agricultural lands, has forced agriculture onto marginal lands unsuitable for conventional continuous cultivation.

Recent population growth and the expansion of agriculture is shown in Table 11.1. The rapid rate of population growth in the developing world was greatest in Africa (89%) and lowest in Asia (63%) but the absolute increase was highest in Asia where the population increased by around a thousand million people over the 23-year period. Increasing production to meet the consequent rapid increase in demand for food, fuel, and clothing has presented mankind with what is its greatest challenge (Swaminathan, 1983).

To some extent, man has so far been able to meet the challenge of increasing agricultural production to match population growth (Table 11.2). The increases in production of cereals are largely due to increased yields per hectare rather than overall expansion of cropped area; in the case of cash crops the expansion in area is the major contributor. The intensification of cereal production is the result of increased use of fertilizers, largely applied to responsive modern crop varieties, and the expansion in irrigated areas (Table 11.3). Much of the increase in cereal production has occurred in developing countries and is the result of the so-called green revolution. Nevertheless, cereal production in many industrialized countries has also intensified due to heavy government subsidies and protection (WCED, 1987a). Enough food can be produced to meet the rising global demand but it is not available where

Table 11.1. *Population growth and the expansion of arable land (adapted from Dudal, 1987)*

	Population (millions of people)			Arable land (Mha)		
	1961	1984	% increase	1961	1984	% increase
Developing countries	2158	3651	70	698	800	15
Developed countries	978	1202	25	654	676	3
World	3136	4853	55	1352	1476	9

it is needed. Uneven and inequitable distribution of food production capacity is a serious problem not only between developed and developing countries, but also between different regions within developing countries.

Valuable information about the prospects of feeding the burgeoning global population comes from an FAO-UNFPA/IIASA study which was under-taken to consider future population growth and to analyse the degree of agricultural intensification needed to satisfy future demand for food (Table 11.4). Projected rates of population growth are much higher in developing

Table 11.2. *Growth in area and production of staple cereals and cash crops 1961–84 (Dudal, 1987)*

	Area increase (%)	Production increase (%)
Cereals		
Wheat	14	131
Rice	26	117
Maize	24	118
Cash crops		
Soybean	120	236
Sugar cane	78	106
Tea	77	125
Coffee	5	15
Oilpalm	44	215

Table 11.3. *Growth in irrigation and fertilizer use 1961–84 (Dudal, 1987)*

	1961	1984	% increase
Fertilizer use			
$(N, P_2O_5, K_2O$ Mt$)$			
Developing countries	3.8	48.4	1100
Developed countries	27.3	82.3	200
World	31.1	130.7	320
Irrigation			
(10 Mha)			
Developing countries	101	158	56
Developed countries	37	62	67
World	138	220	60

countries, and it is estimated that 64 developing countries – 29 in Africa – will be unable to feed their population in the year 2000 if improved technologies such as fertilizers, pesticides, improved seeds, conservation measures, and improved cropping patterns are not introduced and adopted by farmers.

In addition to arable agriculture, the development of land for pasture, and overgrazing, logging, and fuelwood collection are all major causes of devegetation of soils in developing countries. Detwiler and Hall (1988) estimated that, in 1980, 3.1 and 2.9 million hectares of closed and open tropical forest, respectively, were converted to pasture and permanent agriculture. The area under shifting cultivation was estimated to be 34.3 million hectares. Overgrazing is a major cause of desertification in semi-arid rangelands, particularly in Africa and West and South Asia (Mabbutt, 1984). In developing countries, fuelwood production from the equivalent of 0.6 hectares of land per person is required annually to satisfy demand for cooking and heating (Pimentel *et al.*, 1986a). The unfortunate consequence of rapid population increase is an expansion in the demand for fuelwood with the result that large areas of land are denuded and exposed to soil erosion.

Global soil resources

Before moving on to a consideration of the extent of soil degradation in the world, it is useful to review the information on constraints to plant production associated with soils currently in use or potentially utilizable. The soil resources of the different continents vary widely in their suitability for agriculture (Fig. 11.1). Europe has the largest area of soils with no serious limitation to agricultural production and North and Central Asia the least. These trends in soil limitations are reflected in agricultural production and the general level of prosperity of the populations of the

Table 11.4. *Projected increases in population (FAO, 1984)*

	Population (millions)		
	1980	2000	2025
Developed countries	1131	1272	1327
Developing countries	3413	4969	6938
Latin America	364	566	865
East Asia	1175	1475	1712
Africa	470	853	1542
South Asia	1404	2075	2819

different regions. In humid tropical areas of South America, Africa, and Asia, mineral stresses to plant growth are the major limitation to agriculture, mostly due to the leaching effects of rainfall in removing nutrients and lowering soil pH (Sanchez, 1987). By contrast, Fig. 11.1 suggests that drought is the major limitation in large areas of South Asia, Africa, and Australasia. Like all assessments on a global scale, Allen's summary of soil constraints to agricultural production is a gross generalization. Nevertheless, such analyses of the current state of the world's soil resources and the trends

Fig. 11.1. Regional distribution of soils with or without limitations for agriculture. Source: Allen (1980).

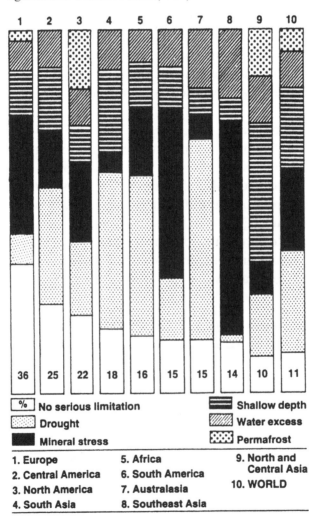

% No serious limitation	Shallow depth
Drought	Water excess
Mineral stress	Permafrost

1. Europe   5. Africa   9. North and
2. Central America 6. South America   Central Asia
3. North America   7. Australasia   10. WORLD
4. South Asia   8. Southeast Asia

in soil productivity make an extremely valuable contribution to efforts to quantify future prospects for growing enough food in different regions. Land use in different land classes in 1975 and the year 2000 have been estimated by Buringh and Dudal (1987) (Table 11.5). Their data suggest that 25% of all highly productive land will be lost during this period due to erosion and salinization or alkalization. There will be a concomitant decrease of 24% in the total area of forest and grassland but the reserves of highly productive land will decrease by 33%. Total forest area will decrease by 15%, but the area of forest on productive agricultural land (High, Medium, and Low Land Classes) will decrease by 55%. On a global scale, these developments are likely to cause a loss of 20% of the land's productivity in developing countries if appropriate soil conservation measures are not employed, and the new land brought into cultivation will merely replace lands degraded by erosion (Dudal, 1982). His analysis paints a bleak picture of the future capacity of some developing countries to produce sufficient food for their burgeoning populations unless urgent action is taken to arrest land degradation.

Table 11.5. *Land use (Mha) in different land classes in 1975 and 2000 (Buringh and Dudal, 1987)*

Land use	Land classes*				
	High	Medium	Low	Zero	Totals
1975					
Cropland	400	500	600	0	1500
Grassland	200	300	500	2000	3000
Forest land	100	300	400	3300	4100
Non-agricultural land	0	0	0	400	400
Other land	0	0	0	4400	4400
Totals	700	1100	1500	10100	13400
2000					
Cropland	345	745	710	0	1800
Grassland	170	320	510	2000	3000
Forest land	30	100	230	3140	3500
Non-agricultural land	0	0	0	600	600
Other land	0	0	0	4500	4500
Totals	545	1165	1450	10240	13400

* Land classes indicate the potential productive capacity of all land not covered by ice

264   *E.T. Craswell*

Soil degradation

As discussed above, soil erosion is but one of a number of processes of soil degradation which reduce the productive capacity of agricultural lands and lead to desertification. The term desertification, which has been adopted by a major UN programme to combat land degradation, has been defined as the diminution or destruction of the land leading ultimately to desert-like conditions. Mabbutt (1984) has presented the estimates of desertified areas made in this UN study (Table 11.6). Land at least moderately desertified was defined as areas where up to 25% of the productive potential was lost; and for severely desertified land there was in excess of 50% loss of productivity. The study concluded that desertification of rainfed croplands was the greatest threat because of the high potential for severe desertification and the large numbers of people dependent on these areas, viz. 85% of the rural population in dryland areas. South Asia, Africa south of the Sudano-Sahelian region, South America, and parts of South Asia are considered to be most at risk. The impact of desertification in these areas has serious implications because rainfed croplands in dry regions produce most of the world's wheat, barley, sorghum, and millet (Dregne, 1987). The major causes of desertification in the rainfed croplands are soil erosion by wind and water, soil compaction, and dryland salt seepage.

The global extent of soil degradation by various processes is the subject of a major study by FAO/UNEP/UNESCO (Riquier, 1982). The WMO (World Meteorological Organization) and the International Society of Soil Science have been involved and the main product of this global assessment is a World Map of Soil Degradation, Oldeman *et al.* (1990). The results of the study

Table 11.6. *Global extent of desertification* of dryland areas in relation to land use (Mabbutt, 1984)*

	Area (Mha) and % of total area	
Land use	At least moderately desertified	Severely desertified
Rangeland	3100 (80%)	1300 (35%)
Cropland		
Rainfed	335 (60%)	170 (30%)
Irrigated	40 (30%)	13 (10%)

* Desertification is defined as the destruction or diminution of the biological potential of land due to largely man-made processes.

show that water erosion, nutrient decline, wind erosion, waterlogging, and salinization are widespread on all continents. The most widespread problem is water erosion which moderately or severely affects an estimated 562 million hectares, 315 million hectares of which is in Asia. Wind erosion affects 204 million hectares. The rest of this section of the paper reviews current information on soil erosion processes and their impact.

Wind erosion

The process of wind erosion is not so well defined as that of water erosion (Burch *et al.* 1987). Vegetation protects the soil from wind erosion by increasing the threshold velocity for soil particle movement as well as reducing the wind velocity inside the canopy. Removal of vegetation in dryland areas is the major underlying cause of soil loss by wind erosion.

Global estimates of wind erosion are difficult to make and hence are not easy to find in the literature. Nevertheless, wind erosion is the cause of much of the desertification in rangelands and rainfed croplands referred to in Table 11.6. Tang Keli *et al.* (1988) estimated that wind erosion affects 1.3 million square kilometres of the People's Republic of China, a fact attested by the extremely dusty conditions in cities like Beijing in spring. Pimentel *et al.* (1976) quote Free (1911) who estimated that 850 million tonnes of soil per year were removed by wind in the western region of the United States. In developing countries the most serious threat to soil productivity from wind erosion is in arid rangelands such as those in the Sahara in Africa and in West Asia (Dregne, 1984). In some of these areas wind erosion is exacerbated by salinization which reduces the vegetative cover, exposing the soil surface to wind action.

Soil erosion by water

The erosion of soil by water is primarily affected by rainfall intensity, the degree and length of slope, the inherent properties (erodibility) of the soil, the crop or vegetation type, residue management, and tillage practices. In the United States of America, these factors were studied in detail at a number of locations to provide data for the Universal Soil Loss Equation (USLE) (Larson *et al.* 1983). For the cropped areas of the USA, the concept of a soil loss tolerance (*t*) value was developed to denote the highest rate of soil erosion that permits a sustainable economic level of crop productivity. This concept is quite controversial because quantitative data to support the *t* values are generally not available. Furthermore, Johnson (1988) has pointed out that the terms 'soil erosion' and 'soil loss' are not interchangeable since eroded soil particles are deposited downslope. Johnson suggested that, for this reason, the USLE seriously overestimates net erosion in the USA.

The USLE has been tested in other parts of the world and found to require considerable modification, particularly in the humid tropics where rainfall erosivity is generally much higher and soil conditions are quite different from the USA (El-Swaify and Dangler, 1982). An alternative approach to the empirical USLE is the development of soil erosion process models which have been discussed by Stocking (1988). An example is the model of Rose (1987) which includes deposition as well as detachment functions and is currently being evaluated in Southeast Asia (Craswell, 1987). The great value of these models is that they provide a means for predicting the potential soil loss through erosion at different sites, given the rainfall and soil characteristics, and also for understanding and improving soil and crop management technologies which reduce runoff and erosion.

The extent and impact of soil erosion can be estimated from onsite or from offsite measurements. Using a combination of methods, Brown and Wolf (1984) estimated that the world's soils are being depleted at the rate of 0.7% per year. Stream sediment yield data from major, largely agricultural, catchments have been used by Abernethy (1987) to estimate median values for gross soil erosion rates in Africa, Asia, and Europe and North America of 4170, 3890, and 830 $t/km^2/year$, respectively. The much higher values in Africa and Asia compared with Europe and North America reflect the erosive impact of tropical rain and the greater capacity of developed countries to mount soil conservation programmes. Onsite measurements of soil erosion reviewed by Pimentel *et al.* (1987) support the conclusion that average soil erosion rates in North America (USA) and Europe (East Germany) (18.1 and 13 t/ha/year, respectively) are much lower than in Asia and Africa where average erosion rates in some countries/regions are of the order of 40–100 t/ha/year.

Estimates of global or country-wide average rates of soil erosion may be misleading because erosion rates vary greatly from place to place depending on the slope, rainfall, soil characteristics, and agricultural practices. For example, in India, average erosion rates are estimated to be 25–30 t/ha/year whereas 40–100 t soil/ha/year are lost from the black soils of the Deccan (Pimentel *et al.* 1987). In some areas of the USA, combined wind and water erosion exceeds 240 t/ha/year (OECD, 1985). One of the best uses of measurements of erosion rates in different areas is to provide policy-makers with information on the size and distribution of erosion problems so that soil conservation efforts can be targeted on areas which are under the greatest threat.

Most global analyses of the soil erosion problem suggest that it is worst in the humid tropics (Dudal, 1982). This is a combined effect of high rainfall intensities, soil properties, and agricultural practices. Much less is known about soil erosion processes and the extent of the problem in the tropics than in temperate developed areas (Lal, 1986) (Table 11.7). Rates of erosion are particularly high in steepland areas (slope $> 10°$). In these areas, the USLE and other empirical models cannot be used reliably to predict erosion rates (El-Swaify *et al.* 1988). It could rightly be argued that steepland areas simply should not be cultivated. Nevertheless, many governments in developing countries are unable to prevent cultivation of steeplands by landless small-scale farmers, forced into these areas by the scarcity of alternative, less erosion-prone land (Ahmad, 1988).

High rates of soil erosion have marked effects on soil productivity, particularly in tropical areas where deep fertile soils are relatively rare and farmers may not be able to afford fertilizer inputs to substitute for the nutrients lost through erosion. Lal (1986) reviews data showing that 23–90% of annual crop yield may be lost when tropical soils are eroded. Erosion induces a wide range of soil constraints to crop growth including reduced root depth, increased soil bulk density, drought, nutrient deficiency, and aluminium toxicity (Stocking, 1984). Some tropical soils have shallow A horizons underlain by acid B horizons which, when exposed, are very intractable both chemically and physically. Because the surface soil contains much of the soil organic matter and nutrients, the soil lost by water erosion is that which is relatively enriched in nutrients (Rose *et al.* 1988). Some

Table 11.7. *Rates of soil erosion in selected areas of the tropics (after Lal, 1986)**

Country	Site characteristics	Field erosion rate (t/ha/year)
Tanzania	Bare soil	38–93
Nigeria	Bare soil	230
Bangladesh	50% slope	520
Indonesia	*Imperata* grassland	345
Trinidad	10–20° slope	490
Guatemala	Cultivated steepland	200–3600
Colombia	Cropland	22

* Contains more details and data sources

economists argue that sediments from upland areas enrich the soils in low-lying areas and infer that the net effect of soil erosion on soil productivity in the landscape is negligible (for an excellent review of this argument see Blaikie, 1985). Data on the economic effects on onsite productivity are inadequate for a clear-cut general conclusion on this issue to be reached. Nevertheless, rates of soil loss in the tropics such as those quoted in Table 11.7 must have disastrous effects on soil productivity, especially in large upland areas with shallow infertile soils. In the USA alone, onsite and offsite annual costs of soil erosion have been estimated to total $43.5 thousand million (Pimentel *et al.* 1987).

Although reduced productivity due to soil erosion is a well-documented and serious problem in developed and developing countries alike, offsite effects of soil erosion are considered by some economists such as Crosson (1986) to be the major problem in agricultural areas in the USA. Crosson believes that erosion effects on the recreational and residential value of water are a major environmental cost which is as high as $8 thousand million per annum in 1985 dollars. In an economic analysis of soil conservation in Allora, Queensland, the major benefits of reduced erosion included reduced road maintenance costs, improved water quality, and flood mitigation, although the latter two benefits could not be quantified (Anon. 1978). In a number of developing countries, siltation caused by soil erosion in watershed areas is shortening the productive life of major dams such as the Aswan high dam in Egypt, the Matumbulu and Kisongo dams in Tanzania, the Ambuklao dam in the Philippines, and the Mangla dam in Pakistan to 100, 30, 15, 32, and 75 years, respectively (El-Swaify and Dangler, 1982). Such offsite effects of soil erosion in these areas, and in other similar areas in India and Sri Lanka, seriously reduce the water availability for lowland irrigated cereal crops which are the mainstay of efforts to produce enough food for the burgeoning population in those countries. The decline in hydroelectricity generation capacity is also very costly to the national economy of countries affected.

Other degradation processes

A number of land degradation processes contribute indirectly to soil erosion by reducing vegetative cover. Salinity and sodicity are serious problems in close to 7% of the world's land area (Dudal and Purnell, 1986). Salinity and the associated problem of waterlogging, are widespread in irrigated lands (Shanan, 1987) where the effects on agricultural production can be devastating. Deforestation in some areas, such as northeast Thailand (Limpinuntana and Arunin, 1986), has changed the hydrologic balance causing groundwater to bring salt to the soil surface in low-lying areas. Some of these areas have to be taken out of crop production. Dryland salinity is also a serious problem in the northern great plains of the USA and the southern prairies of Canada (Daniels, 1987) and in 5500 km^2 in Australia (Bettenay, 1986). Salt-affected soils are prone to severe erosion because the vegetation has been destroyed and because the deterioration in soil structure caused by salinity reduces infiltration, thus increasing runoff after rain and hence soil erosion.

Acidification of soils can be caused by a number of factors. Acid rain from industrial emissions in developed countries has had a dramatic impact on forest growth and soil pH (Overrein, 1978). Vast tracts of land may have been acidified beyond reasonable hope of repair (WCED, 1987a). Acidification of clover pasture lands in southeastern Australia is reducing the productivity of pastures and associated crops grown in rotation (Burch *et al.* 1987). In tropical areas, many soils are naturally acid because of the high leaching rates but, for soils which are not naturally acid, land clearing and agricultural development can lead to further acidification, particularly if nitrogen fertilizers which can acidify the soil are used (Pieri, 1987). Acidification increases the concentration of toxic aluminium and manganese in the soil solution, inhibiting plant growth and reducing the vegetative cover which protects the soil from erosion.

Soil fertility decline due to continuous cropping after land clearing, and soil structural deterioration due to poor soil management are also widespread and serious problems, but are not so well understood or easily quantified as other land degradation processes (Sanchez, 1984; Burch *et al.* 1987). Decline in soil fertility reduces vegetative cover exposing the soil to erosion. Soil structural deterioration in soils prone to compaction through excessive tillage also limits vegetative cover by restricting germination and root growth (Nicou and Charreau, 1980). At the same time soil compaction reduces rainfall infiltration, increasing runoff and hence erosion.

Sustainable production technologies

The information available, together with the more detailed country studies presented in other chapters of this book, suggests that the productivity of large areas of rangelands and lands developed for arable agriculture has been, and is being, irreparably damaged by soil erosion. Urgent action is needed to reverse this decline through the development and application of appropriate soil conservation measures.

Soil conservation technologies can employ either vegetative or mechanical methods or a combination of both (Fig. 11.2). For a detailed consideration of these technologies readers are referred to standard texts (e.g. Hudson,

Fig. 11.2. Soil erosion control practices (El-Swaify et al., 1982).

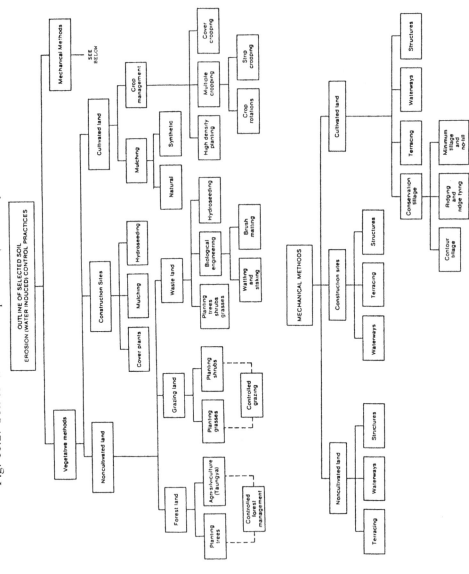

1971). In many countries soil conservation services have been dominated by engineers, with the consequence that mechanical measures have been used most extensively. Furthermore, insufficient attention has been paid to the social and economic factors influencing farmers' adoption and long-term acceptance (and maintenance) of soil conservation structures (Sanders, 1988). Sanders has recently emphasized the need for a fresh approach to soil conservation which takes account of the socio-economic environment of farmers and aims to provide them with direct, short-term benefits from the adoption of the conservation measures. For example, he cites a study which showed that 'Fanya Juu' terraces in semi-arid Kenya lead to substantially higher crop yields which is believed to be the main reason for their widespread adoption by farmers.

In the humid tropics the most widespread climax vegetation is rain forest. From an ecological point of view, trees should play a role in any development of such lands for agriculture (Wijewardene, 1984). One of the most successful and sustainable farming systems in these areas consists of plantation crops such as rubber and oilpalm with a legume ground cover which adds nitrogen to the system and stabilizes the soil surface (Zakaria *et al.* 1987). A wide variety of agroforestry (Nair, 1987) and sylvo-pastoral (Haque and Tothill, 1987) systems have been developed. For enhanced food crop production, the most promising systems utilize tree or shrub legumes which provide nitrogen and other nutrients recycled from the subsoil to the food crop while stabilizing the soil against erosion and giving the farmer a source of fuelwood and forage. For nearly 50 years the farmers of Flores, Sikka, and Timor in Indonesia have been utilizing *Leucaena leucocephala*, planted on the contour, to reduce erosion and provide fuelwood and forage for their cattle (Piggin and Pareira, 1985). In the southern Philippines, corn crops grown in the alleys between leucaena rows produce two- to three-fold higher yields than corn grown as a monocrop (O'Sullivan, 1985). In the same area, soil erosion was reduced by 60%. Research at the International Institute of Tropical Agriculture (IITA, 1985) has refined the alley cropping system, and programmes to evaluate and promote it in tropical Africa have been launched (Harrison, 1987). One of the serious difficulties to be overcome in some areas is the poor growth of leucaena on very acid soils, for which other shrub and tree legumes must be found (Craswell and Tangendjaja, 1985). Some nitrogen-fixing non-legumes such as *Casuarina*, or other trees such as *Azodiracter indica*, are very effective on acid soils (J. Turnbull, pers. comm. 1988).

The Tropical Forests Action Plan (WRI, 1985) has advocated the planting of trees, which provide fuelwood, fodder, and building poles, as part of a major programme to stabilize 160 million hectares of degraded upland watersheds. The planting of trees can be combined with mechanical soil conservation structures such as bench terraces in steep areas. Trees can also play an important role in shelter-belts to protect soil from the effects of wind erosion (Brown *et al*. 1988). Introduced *Eucalyptus* and *Acacia* species from Australia have proved to be particularly valuable sources of fuelwood and poles in semi-arid, saline, and low-fertility soil areas in many developing countries (Harrison, 1987; Thomson, 1987).

However, the greatest difficulties in incorporating such technologies in development schemes revolve around the social, political, and tenurial aspects of land use (Craswell *et al*. 1985; Bruce and Noronha, 1987; Hallsworth, 1987b). Squatters may be reluctant to plant trees because they do not provide sufficient immediate benefit or because improving the land with trees may prompt the resumption of land by the State or the landowners. Governments have a key role to play in reafforestation through land reform programmes and social forestry schemes which must take account of socio-economic factors at the grass-roots level.

Improper land clearing methods using bulldozers and root rakes have doomed many land development schemes in the humid tropics to failure (Lal, 1987c). Such methods lead to excessive soil erosion, lowered soil fertility, and abandonment of the land to unproductive grasses such as *Imperata cylindrica* which, for example, covers 18 million hectares of Indonesia (Sukmana, 1987). Mechanical clearing methods utilizing shear-blades, or manual methods, are much less destructive to the soil, and, combined with appropriate post-clearing soil management, can lead to sustainable agricultural production. For the acid soils of the humid tropics, a wide range of low-input cropping systems have been developed to serve as transition technologies between shifting and continuous cultivation (Sanchez and Benites, 1987). The key to reducing soil erosion is maintaining some vegetative cover through managed intercropping, especially with legumes, or the use of chemicals rather than cultivation to control weeds (von Uexkull, 1987).

Minimum tillage is now a widely accepted and profitable means of reducing soil erosion in temperate developed countries such as the USA (Pimentel *et al*. 1976). This technology also shows promise in the humid tropics (Greenland, 1975) and the semi-arid tropics (McCown *et al*. 1985). McCown *et al*. advocate the use of forage legumes undersown in sorghum or corn crops to stabilize the soil, provide feed for cattle, and provide a nitrogen input until the next crop is planted. Using intercrops or fodder banks is essential in semi-arid areas if overgrazing, which leads to desertification, is to be avoided (Tothill, 1987).

In temperate wheat-growing areas of southern Australia, soil erosion is effectively controlled by the rotation of legume leys with cereal crops (Puckridge and French, 1983). The medics or clovers used in the pasture leys fix nitrogen and boost soil organic matter levels which improve this soil structure and provide nutrients for the ensuing wheat crop. This system is being adapted to boost sheep and cereal production and to stabilize soils in desertification-prone areas of North Africa and West Asia (Cocks, 1988). Ironically, the medic species in the Australian pasture leys originated in West Asia and North Africa.

This brief review of selected technologies for sustainable agricultural production has focused largely on recent research developments for marginal upland areas of the tropics, where the soil erosion problem is most serious. Traditional methods of soil conservation, which have been in use for centuries, utilize mechanical or vegetative barriers to arrest soil movement, thus using the same principles as the more modern technologies (Hallsworth, 1987b). Unfortunately, many of these traditional methods are falling into disuse, possibly because they are too labour intensive and hence too expensive.

Policy and research needs

Debate among economists about the importance of the soil erosion problem is at the heart of many discussions of soil conservation policy (Blaikie, 1985). Some economists argue that gains in agricultural technology have in part led to the decline in the economic importance of agricultural land (Schultz, 1987). The essence of this argument is that induced technological innovation will substitute for loss of soil productivity. This is illustrated in developed countries such as the USA where the government, through the Conservation Reserve Program, pays farmers to cease agricultural production on highly erodible croplands (Taff and Runge, 1988). This and similar schemes in European countries would not be possible without – and indeed are the direct result of – modern crop production technology combined with agricultural price and other support policies which have led to massive overproduction of food and serious environmental pollution. Solutions to soil erosion problems, such as the Conservation Reserve Program in the USA and other similar programmes starting in some European countries, are most likely to be affordable only to developed countries which are highly industrialized and have a surfeit of good agricultural land to feed their own populations.

It could be argued that the green revolution gains in cereal yields in favourable, largely irrigated, production areas of many developing countries should provide the base for government schemes to take the marginal erodible areas out of production. However, the governments of developing countries generally do not have the financial resources to adopt such policies and would not be able to find land for the people displaced from the uplands. The main benefit of the cereal production gains in favourable areas will probably be to reduce the need to develop new lands for agricultural production (Pimentel *et al.* 1986a) and to keep down the price of staple cereals to urban and rural poor.

Economists who argue that soil erosion is a serious problem point to the increasing shortage of agricultural land as population grows, the impact of soil erosion on the productivity of existing lands, the sometimes devastating offsite effects of soil erosion, and the need for intergenerational equity, i.e. the responsibility to manage resources so that future generations are not disadvantaged (Blaikie, 1985; Crosson, 1986). To many, these are compelling arguments which do not require quantitative proof. Nevertheless, hard data are needed to convince some policy-makers, and one of the aims of this book is to gather together such information in a form accessible to them.

Education at all levels of society is one of the keys to success in saving soils, and is emphasized in the seven-point action plan of the World Commission on Environment and Development which stresses the right of every person to sustainable livelihood security (WCED, 1987b). The plan proposes that lands be divided into enhancement areas, where intensive but sustainable food production is possible; prevention areas, which are erosion prone and suited for less intensive sustainable uses such as forestry, grazing, etc; and restoration areas, which are already degraded and require re-vegetation and protection from further use. The action plan details the steps which national governments should take to implement the plan as well as indicating the international cooperation which will be needed. The plan builds on the World Conservation Strategy (Allen, 1980), the World Soils Policy (UNEP, 1982a), the UN Plan of Action to Combat Desertification (Mabbutt, 1984), and the Call for Action on Tropical Forests (WRI, 1985). An important common element in all of these plans is the emphasis on targeting the areas at most serious risk. In the case of soil erosion in the tropics, these are the marginal upland areas inhabited by poor subsistence farmers who have been the neglected clients of agricultural research (Blaikie, 1985; Craswell *et al.*

1985). The growing international awareness of these problems and the various schemes developed to remedy them is heartening, but implementing these plans will not be easy particularly if appropriate technologies are not available. The international agricultural research system which helped to generate the green revolution has recently undertaken a major study of the need to place more emphasis on the sustainability of agricultural production (TAC, 1988). The Consultative Group on International Agricultural Research (CGIAR) will continue to support research aimed at increasing food production in favourable environments of the tropics. However, additional research to develop technologies which prevent land degradation or restore degraded lands and which improve the productivity and sustainability of agriculture in the less favourable environments has been advocated. Research plays a vital role in developing the technologies which will be implemented through various national and international development plans. Thus a meeting in Bellagio recently considered research strategies to implement the tropical forestry action plan (Spears, 1988). The International Council for Research on Agroforestry (ICRAF) and the International Union of Forestry Research Organisations (IUFRO) will play a key role in this effort.

The past emphasis of the CGIAR-supported research centres on crop improvement led to the relative neglect of research on soil management (Craswell, 1986). This neglect has been addressed by the creation of the International Board for Soil Research and Management (IBSRAM) (Craswell and Isbell, 1984) which has been established in Bangkok. IBSRAM is dedicated to assisting and speeding applications of soil science in the interest of increasing sustainable food production in developing countries. IBSRAM has organized networks which currently are undertaking multidisciplinary adaptive research on soil management technologies such as those discussed in a previous section of this paper (Latham, 1988).

Conclusion

This chapter emphasizes technical aspects of the global soil erosion problem and the technological solutions emerging from research. Many such solutions have been developed but, especially in the tropics, the technologies need refinement and adaptation to local conditions along the lines pursued by IBSRAM. Actual implementation and adoption of these technologies by farmers is the most critical and most difficult step. Sociologists, economists, and extension workers potentially can play a key role at this stage. Political will and government support is also crucial to the success of soil conservation efforts. Policy-makers must be made aware not only of the cost of implementing soil conservation measures but also of the cost of not doing so. International awareness and concern about the problem must be translated into action by national governments.

12

Soil erosion and agricultural productivity

D. PIMENTEL, J. ALLEN, A. BEERS,
L. GUINAND, A. HAWKINS, R. LINDER,
P. MCLAUGHLIN, B. MEER, D. MUSONDA,
D. PERDUE, S. POISSON, R. SALAZAR,
S. SIEBERT AND K. STONER

D. Pimentel is a member of the faculty, and the other authors are graduate students at the New York State College of Agriculture and Life Sciences, Cornell University.

Introduction

Soil erosion is a major environmental problem that threatens world food production (UNEP, 1980; Dudal, 1981; Kovda, 1983). In the world today not only is the total population being fed greater than ever before in history, but more humans are malnourished (Swaminathan, 1983). At present, one thousand million people are malnourished and the problem is growing rapidly in severity (Latham, 1984). Many changes have occurred since the early 1900s when most nations were self-sufficient in food. Today, the great majority of the world's 183 nations are major food importers (FAO, 1983), underscoring a growing disparity in food resources (Swaminathan, 1983). The food supply problem has persisted, 'and in some cases worsened despite an increased pace of development' (Latham, 1984). Given these ominous trends, the control of soil erosion for a sustainable agricultural system is essential to any program to improve world food security and development.

Adequate food supplies depend on productive land. At present 97% of the food supply comes from land and only 3% from the oceans and other aquatic systems (CEQ, 1980). Hence, we must safeguard the productivity of the land to feed the ever-increasing world population. Just at a time when agricultural efforts are focused on increasing crop yields, land degradation is increasing throughout the world. Global dimensions of land destruction are alarming.

About 35% of the earth's land surface is affected (Mabbutt, 1984). The natural productivity of many soils has been reduced 8–100% because of erosion (Larson *et al.*, 1983; Lal, 1984a; Mabbutt, 1984; Langdale *et al.*, 1985). In some areas the productivity of eroded soils cannot be restored, even with heavy application of fertilizers and other inputs (Lal, 1984a; Sancholuz, 1984).

The relationship of soil erosion to the world food economy is addressed in this study. Specific factors analyzed are: patterns of soil degradation; the influence of soil erosion on food production; the offsite environmental consequences of erosion; erosion control measures for protecting agricultural land resources; the economic factors that cause and control erosion; and policies needed for soil and water conservation if soil erosion is to be controlled for a sustainable agricultural system.

Agricultural land degradation

The degradation of soil by erosion is of particular concern because soil reformation is extremely slow. According to T.W. Scott (pers. comm. 1985), under tropical and temperate agricultural conditions, from 200 to 1000 years are required for the renewal of 2.5 cm or 340 tonnes (t) of topsoil (Swanson and Harshbarger, 1964; Hudson, 1981; Larson, 1981; McCormack *et al.*, 1982; Lal, 1984a,b; Elwell, 1985). This is a reformation rate of 0.3 to 2 t/ha/yr. In the United States the accepted soil erosion tolerance level of 11 t/ha/yr (5 t/acre/yr) is much too high and should be reduced (Logan, 1977; OTA, 1982). Severely eroded soils can be restored in some cases with enormous investments, including adding either 2000 t/ha of quality soil or 500 t/ha (dry) of rotted cattle manure to the soil.

Serious soil erosion is occurring in most of the world's major agricultural regions (Table 12.1), and the problem is growing as more marginal land is brought into production. Data compiled by Posner (1981) indicate severe soil losses on slopes in Latin America, where instances of erosion exceeding 100 t/ha/yr have been documented. Clearly, soil loss rates on cultivated lands that range from 10 to 100 t/ha/yr are exceeding soil formation rates by at least ten times in most of the world's major agricultural nations (based on a soil formation rate of about 1 t/ha/yr).

Worldwide degradation of agricultural land by erosion and other factors is causing the irretrievable loss of an estimated six million hectares each year (UNEP, 1980; Dudal, 1981; Kovda, 1983). In addition, each year crop productivity on about 20 million hectares is approaching a negative net economic return due to land degradation (UNEP, 1980). Not included in the land degradation data is agricultural land being lost to urban areas and

Table 12.1. *Selected erosion rates in certain geographical regions (sources of information from Pimentel et al., 1986b)*

Country	Erosion rate (t/ha/yr)	Comments
United States	18.1*	average, all cropland
midwest deep loesshills (IA and MO)	35.6*	MLRA** No. 107, 2.2 million ha
southern high plains (KS, NM, OK, and TX)	51.5*	MLRA** No. 77, 6.2 million ha
China	3	average, all cultivated land
Yellow River Basin	100	middle reaches, cultivated rolling loess
India	28–31	cultivated land
Deccan black soil region	40–100	
Java, Indonesia	43.4	Brantas River Basin
Belgium	10–25	Central Belgium, agricultural loess soils
East Germany	13	1000-year average, cultivated loess soils in one region
Ethiopia	20	Simien Mountains, Gondor region
Madagascar	25–40	nationwide average
Nigeria	14.4	Imo region, includes uncultivated land
El Salvador	19–190	Acelhuate Basin, land under basic grains production
Guatemala	200–3600	corn production in mountain region
Thailand	21	Chao River Basin
Burma	139	Irrawaddy River Basin
Venezuela and Colombia	18	Orinoco River Basin

* Indicates combined wind and water erosion, all others are water only
** MLRA = Major Land Resource Areas

roadways. For example, from 1945 to 1975 an area of agricultural land similar to the total state of Nebraska was covered with concrete (Pimentel *et al.*, 1976).

Of the estimated 11.6 million hectares of forests cleared annually (FAO, 1982b), more than half is attributed to agricultural soil degradation and subsequent expansion of agriculture onto new land (Pimentel *et al.*, 1986a). Forest removal reduces fuelwood supplies and forces the poor in developing countries to rely more on crop residues and manure for fuel. This diversion of crop residues and manure compounds both the problems of erosion and declining productivity of agriculture by intensifying soil erosion and water runoff, as well as the loss of valuable nutrients that could be recycled in production.

Erosion affects crop production
Erosion adversely affects crop productivity by reducing water availability, nutrients (e.g. N, P, K, and Ca), and organic matter, and by restricting rooting depth as the topsoil thins (OTA, 1982; El-Swaify, 1985). Based on current worldwide soil loss, and projections for the period from 1975 to 2000, rainfed land degradation will depress food production by between 15 and 30% (Shah *et al.*, 1985).

Erosion reduces soil productivity primarily through the loss of water needed by plants (NSESPRPC, 1981; Follett and Stewart, 1985). Water is the major limiting factor for all crop production (FAO, 1981). In addition, both water and wind erosion reduce the available-water-holding capacity of soil by selectively removing organic matter and finer soil particles (Buntley and Bell, 1976). In an investigation of a broad range of soil types and textures in the US Midwest, water infiltration was most closely correlated with soil organic matter content (Wischmeier and Mannering, 1965). When soils have been exposed and degraded by erosion, water infiltration may be reduced as much as 93% (Lal, 1976). Because of cultivation, 20 to 30% of the rainfall runs off Zimbabwe cropland, resulting in water shortages even in years with good rainfall (Elwell, 1985).

Compacting the soil with tractors or livestock may reduce the infiltration rate of water into soil. For example, on North Carolina pasture before grazing, the infiltration rate was 76 mm/h, but after 74 animal-use days/ha, infiltration declined to only 16 mm/h (Wyman *et al.*, 1978).

Water shortages severely affect crops at most stages of development. Water deficits will reduce seed germination, seedling emergence, photosynthesis, respiration, leaf number and size, seed number, and seed filling (Jordan, 1983). The photosynthetically active leaf area is one of the most important factors affecting crop productivity, and it is also the component of growth that is the most sensitive to water stress (Jordan, 1983). Water stress also limits the nitrogen and other nutrient metabolism in crops. In leguminous crops, nitrogen fixation is much lower in stressed nodules (Pankhurst and Sprent, 1975), and water-stressed non-legumes show a decreased capacity to convert soil nitrogen to metabolically usable forms (Hanson and Hitz, 1983). An FAO (1979) study lists several major crops, such as corn, wheat, beans, and potatoes, as among the most sensitive to water stress, with rice and corn suffering the greatest yield losses.

After water, shortages of soil nutrients (N, P, K, and Ca) are the most important factors limiting crop productivity. One tonne of rich agricultural topsoil may contain a total of 4 kg of nitrogen, 1 kg of phosphorus, and 20 kg of potassium (Alexander, 1977); however, only a small portion of these would be immediately available to the crop – 18% of nitrogen and about 2% of the others (Larson *et al.*, 1983). With US cropland erosion rates of about 18 t of topsoil per hectare per year, estimates are that about half of the 45 million tonnes of fertilizers (USDA, 1985) that are applied annually are replacing the soil nutrients lost by erosion (see pages 288–290).

Erosion does not remove all the components of soil equally. Several studies have demonstrated that the material removed by erosion is commonly from 1.3 to 5 times richer in organic material than the soil left behind (Barrows and Kilmer, 1963; Allison, 1973).

Organic matter is important to soil quality because of its positive effects on water retention, soil structure, and cation exchange capacity. It is also the source of a large portion of the nutrients needed by plants (Allison, 1973; Volk and Loeppert, 1982). About 95% of the nitrogen in the surface soil and 15–80% of the phosphorus is found in soil organic matter (Allison, 1973). Reducing the amount of soil organic matter from 3.8 to 1.8% lowered the yield of corn about 25% (Lucas *et al.*, 1977). Conversely, raising organic matter by annual application of manure or by use of leguminous green manures increased yields of wheat, sugar beet, and potatoes beyond any that were achieved with equivalent inorganic fertilizers – no matter how much additional nitrogen was added (Johnston and Mattingly, 1976; Cooke, 1977).

Organic matter is a necessary component of soils because it helps formation of soil aggregates of desirable size, porosity, and stability, which are critical to good soil structure and tilth (Greenland, 1981; Tisdall and Oades, 1982; Chaney and Swift, 1984). Soil structure is closely related to nutrient availability because good soil structure does not limit root growth and aids the plant in obtaining nutrients (Allison, 1973). Therefore, when organic matter is lost by erosion both soil quality and crop productivity suffer.

Organic matter and soil biota are interrelated in maintaining soil quality and recycling nutrients in the crop/soil system. For example, earthworms and arthropods break up plant and animal residues, mix them into the soil, and make them available to microbes (Kuhnelt, 1976; Edwards and Lofty, 1977). Then microbes break down and recycle the active organic matter fractions, making nutrients available to plants and producing stable humus (Allison, 1973; Alexander, 1977; Jenkinson and Ladd, 1981). Substantial losses of organic matter limit populations of soil biota ranging from microbes to earthworms in the remaining soil and result in decreased soil quality, both in nutrient content and structure (Alexander, 1977; Edwards and Lofty, 1977).

Because reduced crop productivity from erosion is caused by a complex set of ecological factors including water, nutrients, organic matter, soil biota, and soil depth, serious confusion exists over the effects of erosion on crop productivity (Follett and Stewart, 1985). For instance, most economists and modellers assess soil erosion effects on productivity solely by reduced soil depth. The result of using only soil depth indicates that the annual effect on reduced productivity ranges from 0.1 to 0.5% (Crosson, 1985; Walker and Young, 1986). However, when the total effects of erosion are measured, reduced productivity ranges from 15 to 30% (Battiston *et al.*, 1985; Schertz *et al.*, 1985). Thus, although reduced soil depth is cumulative, the other factors, including loss of water, nutrients, and organic matter have a greater impact on productivity.

If reduced crop production on eroded land with decreased soil depth can be measured, then the fertilizer and other fossil energy inputs required to offset the reduced yields can be calculated. For example, suppose 7 cm of topsoil was lost and it is assumed that a 6% corn yield reduction occurs per centimeter of soil loss, then the production inputs to maintain yields can be determined. Normally there are diminishing returns for fertilizer and other production inputs; however, for these calculations we will assume a linear relationship. Based on average US corn yield of 6500 kg/ha, a 7-cm soil loss would reduce yields by 2730 kg. Fossil energy inputs amounting to an additional 2.5 million kcal/ha would be required to restore yields to pre-erosion levels (920 kcal of fossil energy inputs are expended per 1 kg of US corn produced [Pimentel *et al.*, 1981].)

Sometimes it is impossible to offset topsoil loss using fertilizers because the soils are too severely degraded (Engelstad *et al.*, 1961; Heilman and Thomas, 1961; Eck 1968, 1969; Mbagwu, 1982). This is especially true if soil water conditions make nutrients less available and in this way decrease the efficiency of nutrient utilization by plants. Clearly, the real costs of erosion-induced yield declines should include the costs of extra fertilizer and irrigation, as well as economic loss suffered when land is held fallow for soil restoration.

Offsite environmental effects

In addition to reducing the productivity of the land, erosion and water runoff can cause offsite environmental effects. Water runoff in the United States is 'delivering approximately 3 billion tonnes/year of sediment to waterways in the 48 contiguous states' (NAS, 1974). About 60% of sediments come from agricultural lands (Highfill and Kimberlin, 1977; Robinson, 1971), and these frequently have detrimental effects on industry, agriculture, and wildlife. The offsite effects of erosion sediments in the United States cost an estimated $6 thousand million annually (Clark, 1985). Just to dredge several million cubic meters of sediments from US rivers, harbors, and reservoirs costs an estimated $570 million each year (Clark, 1985). Another cost of sedimentation is the major reduction in the useful life of reservoirs. An estimated 10 to 25% of new reservoir storage capacity in the United States is 'currently built solely to store sediment' (Clark, 1985).

Offsite effects of sediments are equally serious in other regions of the world. For example, the Archicaya Dam of Colombia was one-quarter filled with sediments after only 21 months and three-quarters filled after only 10 years of use (UNEP, 1982b). A survey of 132 dams constructed just 30–50 years ago in Masvingo, Zimbabwe, indicated over half were more than 50% filled with silt (Elwell, 1985). Siltation of dam sites in India is reducing the capacity of reservoirs for irrigation, electricity, and flood control (CSE, 1982).

When soil sediments that include pesticides are carried into rivers, lakes, and reservoirs from agricultural lands, fish production is adversely affected (EPA, 1979; USDI, 1982). Sediments interfere with fish spawning, increase predation on fish, and destroy fish food (NAS, 1982). These sediments can also destroy fisheries in estuarine and coastal areas (Begg, 1978; Alexander, 1979; Day and Grindley, 1981). In US streams, damages caused from soil erosion on fish, other wildlife, water storage facilities, and navigation have been estimated to be about $4.1 thousand million each year (Clark, 1985).

Contrary to popular belief, deposits of sediment left by floods do not necessarily enhance the productivity of lowland agricultural lands. In the United States, where alluvial soils already are highly productive and usually are more fertile than sediment transported from less productive areas, deposited sediments cause about a 10% decline in productivity on flood-plains across the country (Clark, 1985). Rapid water runoff floods poorly managed agricultural land in lowlands. In India, for example, increased erosion and water runoff more than doubled the amount of valuable cropland that was flooded during the 1970s, and annual flood damages in these areas ranged from $140 to $170 million (CSE, 1982). In the United States, crop losses from flooding and thunderstorm activity reached $3.8 thousand million in 1982 and average at least $1 thousand million annually (USDC, 1983).

When water is not held in soils but runs off quickly, the total soil moisture is diminished for crop and forages. Add this problem to areas where droughts are a major concern and the effect is magnified for all nations of the world. As many as 80 nations, accounting for nearly 40% of the world population, experience frequent droughts (Kovda *et al.*, 1978). The rapid growth in the world population, accompanied by the need for more crops, will only intensify the water problem, particularly if soil erosion is not contained. With projected water needs expected to almost double in the next 20 years, water supplies emerge as a crucial aspect of crop production.

Erosion control technologies

The principal method of controlling soil erosion and its accompanying rapid water runoff is to maintain adequate vegetative cover. Plant cover intercepts and dissipates the energy in raindrops before they strike the soil, enabling the water to reach the soil without damage. Furthermore, plant stems, roots, and organic matter help control runoff and encourage water percolation into the soil. For example, water runoff rates have been measured to be as much as ten to 102-fold greater on cleared land than on vegetation-covered land (Bennett, 1939; Charreau, 1972; USDA-ARS and EPA-ORD, 1976). With greater water retention on the land, more water is available to sustain plant growth (Wischmeier and Smith, 1978).

Terracing, mulching, cover crops, contour planting, rotations, no-till, ridge planting, and combinations of these conservation technologies are effective in preventing or slowing soil erosion (Table 12.2). For example, in the Philippines, farmers have constructed terraces on slopes greater than 50%, and successfully cultivated irrigated paddy rice for several centuries without erosion problems (Conklin, 1980). Terraces have been shown to be an effective control measure for rainfed crops (Hatch, 1981; Hurni and Nuntapong, 1983). Many soil conservation technologies can be combined to reduce erosion rates about 1000-fold (IITA, 1973). To determine the best combination of appropriate technologies for crops, soil, slope, locale, as well as the source and amount of water available must all be calculated.

Table 12.2. *Erosion control technology examples (sources of information from Pimentel et al., 1986b)*

Technology	Treatment	Soil loss (t/ha/yr)	Slope (%)	Country
Rotation	corn–wheat–hay–hay–hay–hay	3	12	USA
	continuous corn	44		USA
Contour planting	potatoes on contour	0.2	—	USA
	potatoes, up-and-down-hill	32		USA
Rotation plus contour planting	cotton on contour and grass strips	8	—	USA
	continuous cotton planted up-and-down-hill	200		USA
Terraces	peppers on terraces	1.4	35	Malaysia
	peppers on slope	63		Malaysia
Manure	corn with 36 t/ha of wet manure	11	9	USA
	corn without manure	49		USA
Mulch	corn planted on land with 6 t/ha of rice straw	0.1	5	Nigeria
	continuous corn	148		Nigeria
Grass cover	grass	0.08	10	Tanzania
	plowed	13.6		Tanzania
No-till	corn	0.14	15	Nigeria
	conventional corn	24		Nigeria
Ridge planting–crop residues left in trenches on land surface*	corn	0.2	2	USA
	conventional corn	10		USA

* J. Krummel, 1985 (pers. comm.). Oak Ridge National Laboratory, Oak Ridge, TN

Causes of land degradation

Although many erosion control technologies are available that are well understood and effective, the world soil erosion persists at levels greatly in excess of soil formation rates in most major agricultural regions. Explaining this discrepancy and the causes of continued land degradation requires consideration of several factors related to agricultural technology and human populations and their behavior.

The most fundamental cause of land degradation in the world is the clearing and planting of land to increase the area necessary to feed the growing number of people. Because most of the good arable land in the world is already under production, the majority of new land that is put under cultivation is marginal (Buringh, 1984). This often means that newly planted land is either located on fairly steep hillsides and/or contains relatively poor soils. For instance, in many developing countries, such as the Philippines and Costa Rica, inequitable land tenure and poverty force many small family farms onto steep, easily eroded hillsides (Hartshorn *et al.*, 1982; Sajise, 1982; UNEP, 1982b). Similar patterns are found in more developed nations, such as the United States, where 'nationally, only 40% of cultivated cropland owned by the best conservation managers is classified as having an erosion hazard, while 59% of cultivated cropland owned by the lowest income group is labeled erosion prone' (Lee, 1980).

Erosion increases dramatically on steep, unprotected cropland. For example, in Nigeria when cassava was planted on land with only a 1% slope, the soil erosion rate was 3 t/ha/yr (Aina *et al.*, 1977). However, on 5 and 15% slopes, erosion rates increased to 87 t/ha/yr and 221 t/ha/yr, respectively. The fact that such land is typically farmed by small land-holders only compounds this problem. Since these farmers need immediate cash or food returns and have problems obtaining credit for conservation investments, marginal lands are farmed more intensively than land of larger operations is. Also, patterns of fragmented land ownership make it impossible to design and encourage farmers to implement conservation measures (e.g. El Salvador, Java, plains of Central India) (Hudson, 1982).

In the post-World War II period, US agricultural structure has undergone a dramatic shift toward fewer and larger farms, greater mechanization, and greater farm and regional specialization (USDA, 1981). The trend toward fewer and larger farms in the United States has been accompanied by increased use of heavier, more powerful machinery (OTA, 1982). Former terraces, shelter-belts, and hedgerows have been removed because they restrict the operation of large machinery. Since large machinery is unable to follow contours on sloping cropland as easily as smaller equipment, contour planting has been modified in order to adapt to larger farm machinery. In each case, all these changes have intensified soil erosion (Buttel, 1982; OTA, 1982).

Farm and regional specialization in the United States and elsewhere also has contributed to increased soil loss. Most of the major food crops are annual row crops that require the soil to be tilled and planted each season. The exposure of freshly tilled soil to rain and wind facilitates erosion (Wischmeier and Smith, 1978). Planting these crops with conventional till in larger and more continuous monocultures increases soil degradation by reducing crop rotations and crop diversity (see rotations in Table 12.2). The erosion potential of annual row crops is especially severe in the tropics because rainfall intensity is high (Lal, 1976, 1984b; Hudson, 1981; Quansah, 1981; El-Swaify *et al.*, 1982; Elwell, 1985).

On the level of the microeconomics of the farm, there are several constraints on soil conservation. Although the literature contains conflicting claims (see below) as to the private profitability of investment in erosion control technologies and practices, conditions of tenancy, debt, and inadequate income may preclude such investments even when it would otherwise be profitable. The problem of inadequate income is particularly relevant in the developing nations where government policies often favor 'low producer prices and subsidized consumer prices' (Brown *et al.*, 1985). Low farm prices discourage adoption of conservation technologies and force farmers to use low-cost and often poor management practices (IUCN, 1985). Although the political, social, and human costs of reversing such policies may seem prohibitive, in the long term such adjustments will be necessary to encourage both increased food production and economic development (Brown *et al.*, 1985). In the United States, for example, conservation tillage is now being practiced on nearly one-third of US cropland (Follett and Stewart, 1985).

Finally, on the level of individual perception, erosion is often ignored and exacerbated by its 'insidious nature.' Although farmers are aware of the added effort and cost to control soil erosion, the damage caused by erosion often goes unnoticed. In fact, the amount of soil that is eroded with each rain or windstorm is almost imperceptible to the observer. For instance, 15 tonnes of soil lost from a hectare of land during a single storm removes only about 1 mm of soil from the surface. Unless one actually observes the soil being washed or blown away, it is difficult to see that erosion has occurred by looking at the land after the storm has passed, especially after cultivation.

Economics of land conservation

Implementing soil and water conservation technologies would benefit both farmers and society as a whole. For individual farmers, reducing erosion from about 18 t/ha/yr to about 1 t/ha/yr would help preserve the productivity of the land, reduce the exaggerated need for fertilizers and other energy inputs, and decrease water stress on crop production.

Most assessments of the effect of erosion on productivity are based on removing a certain depth of topsoil and measuring the impact on crop yields (Crosson and Stout, 1983; Pierce *et al.*, 1984). This technique does provide an estimate of the effect of reduced soil depth on crop yields, but it tends to under-estimate the effects of erosion because it does not selectively remove soil components such as organic matter, as natural erosion does. In these kinds of studies, corn yields are reduced about 6% per centimeter of soil lost. Because some conservation technologies would increase annual production costs by about 10%, some studies have concluded that the costs of implementing certain conservation technologies are greater than the annual benefits they would produce in corn yields (Shrader *et al.*, 1963; Pimentel *et al.*, 1976; Berglund and Michalson, 1981). Several other studies, however, have indicated that certain soil conservation technologies provide benefits greater than their costs (Lee *et al.*, 1974; Pollard *et al.*, 1979; Pope *et al.*, 1983; Wijewardene and Waidyanatha, 1984). The reason for these results is that different techniques were used to assess erosion effects and to measure the costs and benefits of conservation.

The major benefits of erosion control, as mentioned, are conserving water and retaining soil nutrients and organic matter, as well as maintaining soil depth. In several experimental studies, for example, soil conservation techniques retain about 10 cm more water per growing season than conventional plow-plant technologies (Table 12.3). Five centimeters (50 mm) of water applied to corn, spring wheat, and sorghum crops, which normally experience transient drought periods during the growing season, have been calculated to increase yields 15%, 25%, and 23%, respectively, based on average yields (de Wit, 1958; Shalhevet *et al.*, 1979; Hanks, 1983). These calculations assume that the 50 mm of water lost because of conventional tillage systems would be fully utilized by crops grown under conservation tillage systems. The calculated cost of replacing 50 mm of water by pumping ground water for irrigation averages about $15/ha/yr in the United States (USDA, 1981).

Conserving soil nutrients is nearly as important as conserving water for crop production. Based on sediment loss from water and wind erosion

Table 12.3. *Water runoff in corn plantings as affected by conservation technology*

Technology	Treatment	Water runoff (cm)	Increased loss (cm)	Slope (%)	Reference
Till plant	till plant	14.5	2.5	3.4	Moldenhauer et al., 1971
	conventional	12.0			
Chisel plant	chisel plant	1.14	3.38	8–12	Romkens et al., 1973
	conventional	4.52			
Level terraced	terraced	0.94	7.2	2–18	Schuman et al., 1973
	contour planted	8.14			
Mulch	corn stover	0.06	1.24	7	Ketcheson and Onderdonk, 1973
	no stover	1.3			
Mulch	rye cover crop	3.9	13.5	2–4	Klausner et al., 1974
	residue burned	17.4			
Mulch	manure mulch	9.0	4.1		Musgrave and Neal, 1937
	no manure	13.1			
Disk-chisel (no-till)	disk-chisel	0.1	2.8	5	Oschwald and Siemens, 1976
	conventional fall plow	2.9			
No-till	no-till in sod	3.7	7.0	14	Spomer et al., 1976
	conventional	10.7			
No-till	no-till	8.1	16.5*	7	Lal, 1984c
	conventional	24.6			
Rotation	corn–oats–hay–hay	0.58	2.50	7–10	Ketcheson, 1977
	conventional continuous	3.08			
Coulter-chisel	coulter-chisel (no-till)	9.45	5.08	1–2	Siemens and Oschwald, 1978
	conventional fall plow	14.53			
Disk-chisel	disk-chisel (no-till)	2.8	5.6	5	Moldenhauer, 1979
	conventional fall plow	8.4			
Average			10.28		

* Ibadan, Nigeria

(Larson *et al.*, 1983) and nitrogen in water runoff (Lal, 1976), the available nutrients lost annually worldwide were calculated to be 8.2 million tonnes of nitrogen, 0.06 million tonnes of phosphorus, and 2.0 million tonnes of potassium.

At 1988 fertilizer prices (nitrogen = $53/kg; phosphorus = $51/kg; potassium = $27/kg), this loss of nutrients is the equivalent loss of nearly $5 thousand million. Earlier estimates of fertilizer nutrient losses range from $7 to $18 thousand million annually (USDA, 1965; Beasley, 1972; Troeh *et al.*, 1980; Hartwig, 1984). Since US farmers currently spend about $10 thousand million annually for fertilizer (USDA, 1985), retention of these nutrients through conservation technologies would substantially reduce fertilizer costs.

When conservation technologies are employed, increased crop yields result because water, nutrients, and soil organic matter are retained. For example, in Texas, yields from cotton grown on the contour were 25% greater than from cotton grown with the slope (Burnett and Fisher, 1954). Yield increases from planting on the contour have also been reported for corn (12.5%) in Missouri (Smith, 1946), and for corn (12%), soybeans (13%), and wheat (17%) in Illinois experiments (Sauer and Case, 1954). On land with a 7% slope, yields from cotton grown in rotation were increased 30%, while erosion was reduced nearly one-half (Hendrickson *et al.*, 1963). In Nigeria, yields from corn grown by no-till under favorable soil and climatic conditions were 61% greater than from corn grown with conventional tillage (Wijewardene and Waidyanatha, 1984). In an experiment comparing tillage practices used on 22 consecutive corn crops grown on highly erodible Nigerian soils, the average grain yields from no-till plots were 20% higher than from conventional plots. This was attributed to the accumulated effects of erosion-induced degradation of the unprotected soil. Using a mulch in corn grown in Mexico, erosion was reduced 99% and corn yields were increased nearly 30% (Maass *et al.*, 1988).

Assuming a minimum of 10% reduced annual crop yield and fertilizer loss of $5 thousand million, the onsite costs of erosion total $18 thousand million annually in the United States. (Conventional production results in erosion and water runoff and yield reductions from 12 to 61% compared with conservation technologies.) These are only short-term losses. The long-term losses would be much greater. For example, when an average erosion rate of 18 t/ha/yr is assumed to occur for 10 years, about 1.3 cm of soil would be lost. Based only on soil depth, this would reduce corn yields about 8% on soils less than 30 cm in depth, and is equivalent to a loss of 520 kg/ha/yr. Assuming this reduced yield can be offset by fertilizer and other production inputs, then about $20/ha/yr is required to offset reduced soil depth (478 000 kcal additional energy required for the 520 kg reduced yield).

While this reduction in the productivity of the soil took place in only 10 years, from 110 to 400 years are required to replace the 1.3 cm of topsoil. Therefore, even if conservation measures are adopted after the 10th year, added fertilizer and other inputs would have to be applied for 110 to 400 years in order to maintain the original productivity of the land. If fuel prices increase ten-fold during this period, then costs would rise to $200/ha/yr. (Note, fuel prices increased about seven-fold during the 1970s, and were projected to rise two to three-fold just during the following 20 years (DOE, 1981).) Thus, wasting a renewable soil resource that requires hundreds of years to replace by substituting a non-renewable resource is economically and environmentally unsound because of the long-term costs.

These long-term, onsite costs should be added to the annual short-term, onsite costs of $18 thousand million. The offsite environmental costs of erosion are estimated to contribute at least an additional $6 thousand million annually for sediment and flood effects (Clark, 1985). Thus, the total short-term cost of both onsite and offsite environmental effects of soil erosion and water runoff in the United States is estimated to be about $24 thousand million annually.

Clearly, significant returns are possible to both farmers and society when soil and water conservation technologies are employed on land susceptible to erosion. However, several important impediments to the implementation of sound soil and water management practices exist, as mentioned.

Socio-economic factors that force poor family farmers onto marginal land in many countries and that encourage US farmers to increase large-scale mechanization and specialization interfere with implementing soil management policies. In addition, the insidious, gradual effects of erosion on soil productivity and reservoirs discourage the implementation of conservation practices.

Conclusions

Severe soil erosion and rapid water runoff problems are a worldwide crisis that is seriously affecting the world food economy. High rates of soil loss are causing declines in soil productivity worldwide, and most nations do not have sound land-use policies to protect their valuable soil and water resources. In fact, low-cost food policies by most governments encourage farmers to use low-cost, poor management practices in agriculture (Brown *et al.*, 1985). Soil degradation has serious implications for the economy and security of nations. In the past, several nations and civilizations failed in large measure because of soil erosion and environmental degradation (Jacks and Whyte, 1939; Lowdermilk, 1953; Carter and Dale, 1974; Troeh *et al.*, 1980).

Because of the insidious nature of soil erosion and the complexity of environmental effects, few measurements have been made of the total costs of erosion and water runoff to farmers and society as a whole. Our study found that soil erosion and associated water runoff cost the United States about $43.5 thousand million annually in both direct and indirect effects. In the long term, the environmental and social costs may be several times this level. Clearly, it would pay society to invest in soil and water conservation since the payoff is so great when this serious environmental problem is controlled.

The increasing demand for food from growing numbers of people is placing increasing pressure on limited cropland and water resources. Poverty and inequity in some nations is forcing subsistence farmers to expand their cultivation onto erosion-prone marginal lands. The limited availability of total fossil energy resources and their future high cost make it unlikely that increased fertilizers and other inputs can offset severe land and water degradation problems, especially for impoverished nations. Fuel prices are projected to rise again, which will further reduce the opportunity to use fossil energy as a substitute for productivity lost due to land degradation.

Clearly, the results of this assessment not only underscore the serious nature of the environmental and social costs of soil erosion worldwide, but emphasize the need for immediate implementation of soil erosion control technologies. Sound soil and water resource management policies would be a substantial benefit to all nations now and in the future.

13

Vetiver grass for soil and water conservation: prospects and problems

M.V.K. SIVAMOHAN, C.A. SCOTT, AND
M.F. WALTER

Introduction

Although a great deal of interest in watershed conservation has been generated recently, schemes which address resource use within the watershed focus on either arable or non-arable lands. Rarely is the watershed viewed as an integrated resource unit. Most watershed-based planning has treated 'natural resources,' such as forests, soil, and water, as if these were somehow in an entirely different domain from agriculture. Even in high mountain environments, agriculture as one of multiple resource use sectors often plays a crucial role in sustaining the local economy. In semi-arid regions, where water is the input constraining agricultural production, sound management of water resources is a critical need. Water management comprises both surface and ground-water sources for irrigation, and soil moisture from precipitation. When dealing with integrated soil and water management, the watershed does appear to be the logical unit for resource management.

Where sufficiently high population densities in the semi-arid zone necessitate intensive land use, rainfed agriculture is the predominant means of livelihood. While animals are an important part of many systems of rainfed cropping, cultivated agriculture remains the primary land-use strategy. Farmers exercise a range of management decisions over various processes of agricultural production. They change cropping patterns in response to a variety of factors: climate, soil conditions, the market, and their own subsistence needs. Simultaneously, they seek to judiciously distribute resources to meet their own objectives. We refer here to labor and capital resources, which are combined to achieve the farmers' perceptions of maximum benefit. This goes equally for the use of such inputs as it does for tillage, weeding, and a range of conservation operations.

Soil and water conservation using a variety of techniques has always been an important management process in tropical rainfed farming. While we will not attempt to document the rich array, both traditional and modern, of soil and water conservation practices from around the globe, we would like to note the increasing attention being given in agricultural research and development to vegetative techniques for this purpose. One technique which has received considerable attention is the planting of live, grass contour hedges on sloping land.

In this chapter, we are concerned with the application of one vegetative soil and water conservation technique in particular, namely vetiver (*Vetiveria zizanioides*) hedges, for integrated watershed management. The use of vetiver grass ('khus') is said to solve a range of soil and water related problems (Holden, 1989; World Bank, 1990; Yudelman *et al.*, 1990; Subramanya and Sastry, undated). Intended to replace contour field bunds, vetiver is planted in furrows at a spacing of 15 to 20 cm along contours with a vertical interval of 1 to 2 m. Once established, vetiver contour hedges increase in-field sedimentation, while simultaneously reducing the erosivity of runoff.

Vetiver grass

The World Bank is actively promoting vetiver as a soil and water conservation tool in a number of countries including Madagascar and India. Based on three decades of experience with the grass in Fiji plus observations of its use in several other locations, agricultural scientists of the World Bank emphasize that vetiver increases sedimentation, reduces the velocity of runoff, and 'binds the soil ... to a depth of up to 3 meters' (World Bank, 1990). In on-station research trials in India, vetiver is reported to reduce both surface runoff and soil loss, while increasing crop productivity (World Bank, 1991). Its 'roots prevent rilling, gullying, tunneling' (World Bank, 1990), while their aromatic nature makes the 'grass unpalatable to rodents and other pests'. Further, the roots keep weeds out of the fields, and the 'leaves keep snakes away.' Reportedly, most cultivars of vetiver will not spread, weed-like, by themselves; rather they must be propagated by planting slips. The leaves of mature vetiver plants are reportedly not palatable to animals, though young shoots can be used as fodder. Finally, vetiver 'is a climax plant, and therefore even when all surrounding plants have been destroyed by drought, flood, pests, disease, fire, or other adversity, the vetiver will remain to protect the ground from the onslaught of the next rains.' Summarizing the virtues of vetiver, R.G. Grimshaw (World Bank, 1990) writes, 'In every case this unique grass has displayed the same extraordinary characteristics that make it an ideal low-cost, non-site-specific system for controlling soil loss and improving soil moisture.'

Yudelman *et al.* (1990) assess vetiver's performance and potential as a soil and water conservation measure, comparing it to a range of alternative grass species. Vetiver scores high marks for its resistance to adverse conditions, grows in 'all climates,' and is easy to establish. In comparison, only *Atriplex* spp. and *Cymbopogon nardus* (citronella) are close to but still below vetiver's performance. Perhaps the most emphatic recommendation given to vetiver by Yudelman *et al.*, is its economic benefits. In a comparative study of vetiver and earthen bunds for soil conservation, they attribute the net present value of vetiver to be greater than twice that of field bunds, while vetiver's internal rate of return is said to be 95%, compared to field bunds' IRR of 28%. Although their methodology differs somewhat from the World Bank, on whose project these figures are based, their results indicate that vetiver far outperforms earthen bunds. They conclude that, 'while every effort should be made to learn more about the conservation potential of vetiver and other grasses, it needs to be appreciated that they are not panaceas. Rather, they are a very promising component of what will inevitably be a very complex solution to the challenge of sustainable agricultural development.'

While the World Bank, and to some extent Yudelman *et al.*, maintain that vetiver has near universal application, Subramanya and Sastry (undated) endorse the use of vetiver only in particular locations. They document the use of vetiver in Mysore district, Karnataka, India, where it has been used for the past century. Employing an interesting etymological analysis of local names for the grass, they speculate that it was perhaps introduced from nearby Kerala. In the villages studied, vetiver appears to have been used primarily as a field boundary, although its use within fields expressly for conservation purposes does not appear to be uncommon. Vetiver is amenable to management. It should be planted with the first monsoon shower. Regular cuttings provide fodder, while the existence of ecotypic variations of vetiver indicate that selection has been occurring for a considerable time.

Other researchers have cautioned that, while vetiver certainly plays an important role in conservation, it should be treated as one of a number of approaches to soil conservation. Kanodia and Singh (1987) have pointed out that vetiver has several shortcomings, specifically that it is not always deep-rooted, it can spread via seeds, it is palatable in some cases, its aromatic roots are in demand for the production of soaps, perfumes, flavorings, and mats, and its 'ecological niche ... is ... restricted to lowlying, marshy areas and illdrained lands where subsoil water table is quite high ...' Kanodia and

Singh, both of the Indian Grassland and Fodder Research Institute, instead recommend 'multi-purpose' species, and detail the applicability of various other grasses in a range of topographical, edaphic, and ecological conditions.

Harvesting vetiver entails uprooting the entire plant, with dire consequences for erosion. It is reported that Indonesia has prohibited vetiver for precisely this reason. Experience in Haiti suggests that while vetiver is often grown as a cash crop for its roots, it is rarely used for soil conservation. According to one soil conservation expert who has worked in both Haiti and Indonesia, the temptation on the part of poor farmers to harvest vetiver root is extremely high, even if it was originally planted to control erosion (Lester Stillson, United States Department of Agriculture (USDA), Soil Conservation Service, pers. comm.). The short-term returns must be immediately apparent for small farmers to adopt soil and water conservation. For example, farmers in the Dominican Republic favor hedges of elephant grass for its value as livestock feed over citronella, although the latter forms a better hedge (Thomas, 1988).

Lal (1990) emphasizes that the establishment of vetiver hedges without gaps is of utmost importance. He rates vetiver as 'relatively tolerant of moderate levels of drought or inundation, and damage by fire'. Lal's primary emphasis is to show that vetiver technology must be combined with other soil erosion control techniques for maximum effect.

Our own assessment is that vetiver is an appropriate technique for soil and water conservation in a limited set of conditions. While the emphasis on vegetative techniques is definitely sound, structural measures and improved management practices (e.g. the type of tillage) are often appropriate as well. In many instances, the best soil and water conservation systems use a combination of control and protection measures. The most practical approach to soil conservation will be based on the local availability of knowledge systems, and materials and techniques which are compatible with local agricultural and natural resource use practices. As such, the *inherent characteristics of vetiver are somewhat less important than its integration into local farming practices*. Therefore, whether its root system is invariably three meters deep or not is less important than whether farmers adopt vetiver, and *innovate with it to meet their own requirements*.

That vetiver is found throughout the world is hardly surprising at all. That it is used for soil and water conservation is also no surprise. However, the cases reported in the literature where vetiver has been used for erosion control are on ex-colonial cash crop plantations (sugar, rubber, tea, coffee, fruit) virtually without exception. Plantation agriculture can be characterized as management intensive, which suggests that *vetiver does require a great deal of maintenance to act as an effective hedge against erosion*. The expenditure of scarce resources required for the management of soil and water conservation systems are often not made by farmers with small land-holdings because they are forced to focus on short-term rather than long-term benefits. The problem of creating incentives for farmers to manage soil conservation systems is chronically apparent to professional conservationists.

Our investigation of the applicability of vetiver is based on its implementation in semi-arid watersheds in Andhra Pradesh, India. We will attempt to show that, like other conservation practices, the performance of vetiver is very much dependent on management. Nursery techniques, planting, and maintenance all are important factors in determining its performance as a soil and water conservation measure.

Physiology and ecological constraints

Vetiver is one of at least ten species of 'coarse perennial grasses' of the tribe Andropogoneae (World Bank, 1990). There are at least two known species of vetiver, *V. zizanioides* and *V. nigratana*. *V. zizanioides* originated in the Indian subcontinent, where it is found primarily in low-lying marshy lands. It has been described as a 'densely tufted, perennial, 1.2 to 1.5 m tall grass arising from a branching root-stock, with large tufts of spongy, aromatic, fibrous roots, leaves 30–60 cm long and ligule nearly obscure. Inflorescence an erect, conical panicle, often reddish brown or purplish, with two types of spikelets, one ... fertile, while the other is ... sterile' (Kanodia and Singh, 1987).

On the question of vetiver's sterility, there is some disagreement. The World Bank scientists consider it 'sterile outside its natural habitat of swampland' (World Bank, 1990). This is qualified by the observation that '*V. zizanioides* does not produce seeds that germinate under natural field conditions. *V. nigratana* (the Nigerian species) does seed, but the seedlings are easily controlled'. Exactly what constitutes 'natural field conditions' is not clear, and, for example, Wayne Hanna of the USDA Soil Conservation Service feels that vetiver could potentially pose a weed problem in the southern United States (pers. comm.).

Gil Lovell, of the USDA's Agricultural Research Service (ARS), National Plant Germplasm System, reports that 'All the Indian accessions we [the ARS] are working with are very fertile' (pers. comm.). He maintains that this contradicts World Bank statements, which are 'rather broad.' Noting that the Indian accessions with which he is working had been selected in India for their oil production characteristics, Lovell observes that vetiver is 'very vigorous' and holds that the north Indian and south Indian plant material could be different. To be sure of vetiver's fertility, he states, California has prohibited vetiver until it is determined that it will not spread as a weed. Lovell reports that the ARS is 'still looking for vegetative material from India that does not propagate by seed' and is also trying to procure Caribbean plant material. To his knowledge, disease is not a major problem for vetiver propagation. In summing up, Lovell states that vetiver 'must be used in the right situation, with the caution to use only the non-fertile variety.'

Vetiver apparently favors marshy land, though both the World Bank and Yudelman *et al.*, claim it is xerophytic in addition to being naturally hydrophytic. It is said to grow in areas of annual rainfall ranging from 200 to 6000 mm (World Bank, 1990). The World Bank scientists maintain that 'It has an exceptionally wide pH range, seems to be able to grow in any type of soil regardless of fertility, and has been found to be unaffected by temperatures as low as $- 9 °C$'. There is no consistent evidence, however, that shows vetiver to be frost resistant. (Earlington Jacobson of the USDA Soil Conservation Service reports that cold tolerance is a limiting factor to the application of vetiver in the United States (pers. comm.). Temperatures as low as $10 °F$ ($- 12 °C$) in December 1989 killed most of the vetiver which the SCS was working with in the southern and midwestern United States.) Additionally, the suitability of vetiver in semi-arid conditions, particularly in sloping land with a poor moisture regime, is questionable, as will be demonstrated by the subsequent analysis of field data.

For soil conservation purposes, it is important to distinguish between whether a grass grows and whether it forms hedges. As the experience from Andhra Pradesh will show, a significant number of gaps continue to persist after several seasons, although the height and number of tillers increases with increasing age. Replanting is required to fill the gaps. Additionally, 'during the first two seasons, and sometimes the third, the plants need protection ...' (World Bank, 1990). In semi-arid conditions such as in the Maheshwaram Watershed Development Programme in Andhra Pradesh, vetiver appears to take longer than three years to establish a hedge able to serve a soil and water conservation function.

Little is written on pest infestation of vetiver. Brown spot, black rust, and stem borers have been observed (World Bank, 1990), although little is known about their effects on vetiver's growth. From our own field experience, termites ('white ants') are known to be a problem in India.

Maheshwaram Watershed Development Programme
The critical areas of rainfed cultivation in India were recognized as those which receive erratic precipitation of about 750 mm annually, placing

them in the semi-arid region. Since the beginning of the planning era in 1951, the government sporadically initiated several development programs to conserve soil and water and to improve agricultural production in these areas.

One watershed development pilot program that is being implemented in the Maheshwaram area of Andhra Pradesh with World Bank aid, unlike earlier programs, seeks to develop both arable and non-arable lands in an integrated manner. The Maheshwaram Watershed Development Programme has the following components:

1. Soil and *in situ* moisture conservation activities,
2. improved crop production programs,
3. pasture and fodder development,
4. forestry and social forestry programs,
5. horticulture development.

The pilot project is located in Ranga Reddy district of Andhra Pradesh. The area is in the catchment of Phulandari Vagu, a small tributary of the Musi River, which in turn joins the Krishna River. The watershed is situated in the Deccan Plateau, between 17°5' and 17°15'N, and 78°24' and 78°37' E. It has an area of 25 330 ha covering 23 villages. Average annual precipitation is 750 mm, with most occurring in the southwest monsoon during June to October. The soil types in Maheshwaram watershed are medium to deep alfisols. Slopes range from 1 to 3% for cultivated land, and from 3 to 10% for pastures, permanent fallows, and forest lands. Elevation ranges from 200 to 620 meters above mean sea level.

Approximately 61% of the total area is arable; the principal crops grown are castor and sorghum (*jowar*). Being located 15 km from Hyderabad, the state capital, the population in the area is dependent on the city for services.

From 1984 to 1986, initial soil and water conservation activities in the project focused on graded bunds with waterways (Rao *et al.*, 1990). Subsequent changes were made by the World Bank supervisory teams. In 1986–87, bund cross-sections were reduced from 0.500 to 0.334 m^2. Also starting in 1986, it was decided to use vetiver hedges in place of bunds. In July–August 1986, 25 000 vetiver slips were procured from neighboring Karnataka state, a region with climatic and soil characteristics similar to Maheshwaram. Most slips were multiplied in nurseries, and initial planting started in 1987–88, when vetiver was planted along with the reduced section bunds. The planting technique was changed to contour furrows on land previously untreated with bunds. Full-scale implementation of vetiver hedges began in the 1988 planting season and has continued into 1990–91.

Organizational arrangements

The nodal agency responsible for implementation of the watershed development project is the Directorate of Agriculture, Government of Hyderabad. However, a state-level Watershed Development Council has been organized under the chairmanship of the Agricultural Production Commissioner. Secretaries of concerned departments are members of the council. The council determines policy and guidelines and reviews watershed development programs in the state. A Watershed Development Team is constituted with an Additional Director of Agriculture as the team leader. The organizational structure of the team provides for a multidisciplinary team with state-level officers drawn from the departments of forestry, crop production, extension, horticulture, animal husbandry, and so on, to work under one umbrella. However, the team at present is staffed with only Department of Agriculture officials and a statistician.

At the field level two Assistant Directors of Agriculture with their staff are posted to implement the conservation works and watershed development activities. The Assistant Agricultural Officer (AAO) is the grass-roots-level official to implement the works in the field. There are eight to ten AAOs under each Assistant Director. A Divisional Forest Officer (DFO) from the Forest Department has been deputed to work with the watershed develop- ment team at the field level. For the remaining activities, the team seeks cooperation of the concerned departments and agencies in the implemen- tation of the program.

Assessing vetiver's performance

AAOs were given survey forms to complete. Surveys from 60 plots of land planted with vetiver, comprising 82 individual farmers' holdings, and totaling 434 ha were conducted, resulting in quantitative data and qualitative observations regarding a number of aspects of vetiver technology. While the sample size is not large enough, nor the survey exhaustive enough, to claim any definitive results, the diversity of the sample in terms of the number of villages covered and the number of AAOs involved in conducting this survey, give us confidence that we can state some of the major trends in the performance, management, and acceptance of vetiver as a soil and water conservation measure in semi-arid conditions in Andhra Pradesh.

Qualitative observations are interpreted and presented first. The results indicate that vetiver technology has been accepted by farmers with varying degrees of enthusiasm (see Table 13.1). Whether the rate of acceptance is based on vetiver's own merits or because it is part of a large extension program is subject to some speculation.

It was deemed in 77% of the cases to have no effect on the crop; the remaining 23% of surveys gave no reply to this question. (We assume that farmers assessed its short-term effect on crop yields when answering this question. The effects of vetiver directly on soil moisture and erosion, and indirectly on yields, are clearly long-term.) Significantly, the introduction of vetiver technology appears to have had an impact on tillage practices, with 100% of the responses indicating that vetiver lines act as the key line for contour cultivation (no response was found on 3 of 60, or 5%, of the surveys). Participatory rapid rural appraisals in the field study area indicate that on smaller plots the vetiver lines made plowing very difficult. Vetiver was reported to have been plowed up in some instances.

On the other hand, it is interesting to note that browsing of vetiver by domestic animals is highly prevalent. In the study area, vetiver has been planted on agricultural land. On private holdings, grazing as a rule is limited to cows and buffaloes, with minimal watch kept over them. Grazing by goats and sheep is restricted to range and forest land. Thus, browsing of vetiver in the Watershed Development Programme area refers to cows and buffaloes. The degree of browsing in some cases was so severe that a number of AAOs had crossed out the word 'browsed' and written in 'grazed' or 'completely grazed' (see Table 13.2). It must also be emphasized that the cases where

Table 13.1. *Farmers' acceptance of vetiver technology*

Level of acceptance	% of observations
Below average	3
Satisfactory	60
Good	22
Very good	10
Excellent	0
No reply	5

Table 13.2. *Browsing of vetiver by animals*

Degree of browsing	% of observations
Browsed	48
'Grazed'	3
'Completely grazed'	5
Not browsed	38
No reply	6

either browsing or grazing took place were not limited to the younger vetiver plants, and 35% of the cases of browsing or grazing were of vetiver planted in 1988, 21% of vetiver planted in 1989, and 44% of vetiver planted in 1990.

Pests appear not to be a problem. A total of 85% of the surveys indicate that vetiver had no pest attack. Of the remaining 15%, the only identified pests were termites ('white ants').

There was a virtual complete lack of maintenance. In three instances, vetiver slips were reported to have been watered. Likewise, in three instances, hedges were reported to have been pruned. More importantly, although gap filling, i.e. maintaining the hedge by replacing vetiver plants that did not survive, should be a standard maintenance activity, this was not performed in 65% of the cases reported. Of the remaining cases where it was performed, 48% of gap filling occurred for slips planted in 1988, 38% for slips planted in 1989, and 14% for slips planted in 1990. Despite the initial program emphasis on gap filling, gaps continue to persist (see Fig. 13.1).

Of the 35% of cases where gap filling was done, the average number of gaps in 1990 was still between 7 and 8 per 30 meters of vetiver hedge. It is apparent that gap filling has scarcely kept pace with vetiver mortality. This has very important implications for vetiver's ability to act as a soil and water conservation barrier.

Perhaps more significant, however, is the fact that gap filling was carried out by the Agriculture Department, not by farmers. If the area planted with

Fig. 13.1. Percentage of hedges where gaps were filled and the average number of gaps persisting in 1990 vs. planting date.

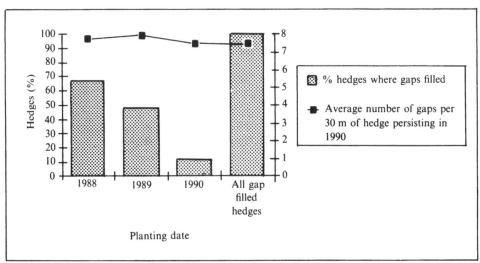

vetiver increases as projected, the department will be unable to cope with gap filling in a very short time.

Standard statistical regression techniques were used to interpret quantitative results. In the regression analyses, the performance of vetiver was assessed by two primary indicators (the maximum number of tillers per vetiver plant, and the number of gaps per 30 m of vetiver hedge) and one secondary indicator (the maximum height, in centimeters, of vetiver plants). In total, 30 regressions were run, investigating the effect of the following 'management' variables on the three performance indicators:

1. date of planting
   (a) 1988–90 as a set,
   (b) 1988,
   (c) 1989,
   (d) 1990;
2. size of holding on which vetiver was planted;
3. age of slips taken from the nursery and planted in the field;
4. time elapsed from picking up the vetiver slips at the nursery to finally planting them in the field;
5. slope of the plot on which vetiver was planted;
6. number of rows planted per hectare;
7. number of slips planted per rill (i.e. plants per 15 cm rill length).

The latter two variables, i.e. number of rows planted per hectare and number of slips planted per rill, were found to have no significant correlation (at the 95% confidence level) with any of the performance indicators, while the data for individual year of planting was limited by sample size. The remaining management variables were found to have significant correlation with at least one performance indicator (see Table 13.3), which suggests that management of vetiver, starting at the nursery and continuing through to follow-up maintenance, is crucial.

The date of planting within the season was not as important as initially believed. The differences seen from year to year in the time series data suggest that rainfall plays an important part in the performance of vetiver. As expected, there was high correlation between year of planting with number of tillers per vetiver plant (see Fig. 13.2).

An important result of the regression analysis has to do with the size of the holding planted with vetiver. An interesting feature of the implementation of vetiver in the Watershed Development Programme area is the range of holding sizes on which it was planted varying from 0.2 to 28.5 ha. As the holding size increased, performance decreased. This trend is evidenced by

Table 13.3. *Summary of regression analyses of management variables on performance indicators*

Management variables	Performance indicators		
	Max. no. tillers per vetiver plant	No. gaps/30 m of vetiver hedge	Max. vetiver height
Date of planting (1988–90)	$r = 0.68$ $p › 99.9\%$	not statistically significant	not statistically significant
Size of holding planted with vetiver	$r = 0.34$ $p > 99\%$	not statistically significant	$r = 0.31$ $p > 95\%$
Age of slips planted	not statistically significant	$r = 0.43$ $p > 99.9\%$	not statistically significant
Time elapsed from nursery to planting	not statistically significant	not statistically significant	$r = 0.42$ $p > 99.9\%$
Slope of plot planted with vetiver	not statistically significant	not statistically significant	$r = 0.33$ $p > 95\%$

Note: 'not statistically significant' refers to probability of correlation less than 95%

Fig. 13.2. Maximum number of tillers per vetiver plant vs. planting date. $n = 58$, $r = 0.68$, $p > 99.9\%$.

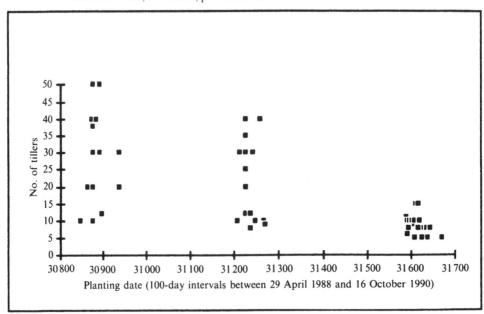

Planting date (100-day intervals between 29 April 1988 and 16 October 1990)

statistically significant correlation of this management variable with both the number of tillers per plant and vetiver height (see Figs. 13.3 and 13.4).

It was found that number of gaps/30 m of vetiver hedge decreased as the age of vetiver slips increased (see Fig. 13.5). Extending the length of nursery operations from six to ten months (and correspondingly the date of initiation of nursery operations from December back to September of the year preceding planting) has cost and labor implications that are crucial from a management perspective.

The time elapsed from picking up the vetiver slips at the nursery to finally planting them in the field was found to have considerable effect on the maximum height of vetiver (see Fig. 13.6). As expected, the longer the time from nursery to planting, the less the maximum height.

Slopes in the sample range from 0.23 to 3.30%, although most are in the range of 1.50 to 3.00%. (It should be noted that, with the limited range of slope values in the sample, assessment of the effect of slope on vetiver performance is tenuous.) There was statistically significant correlation of slope only with height. As slope increases, the maximum height of vetiver decreases (see Fig. 13.7).

In planting vetiver, slips are planted along the contour generally at a spacing of 15 to 20 cm. Results of the regression of the number of vetiver slips

Fig. 13.3. Maximum number of tillers per vetiver plant vs. holding planted with vetiver (ha). $n = 59$, $r = 0.34$, $p > 99\%$.

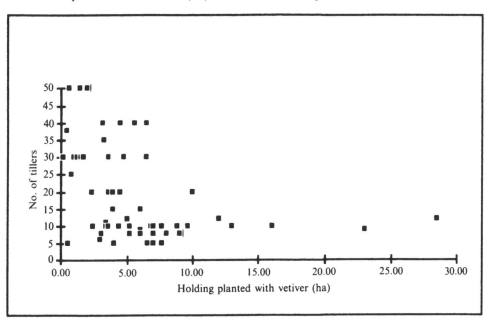

Fig. 13.4. Maximum vetiver height (cm) vs. holding planted with vetiver (ha). $n = 58$, $r = 0.31$, $p > 95\%$.

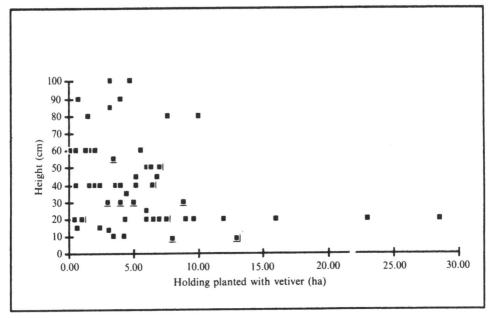

Fig. 13.5. Number of gaps/30 m of vetiver hedge vs. age of slips planted (months). $n = 58$, $r = 0.43$, $p > 99.9\%$.

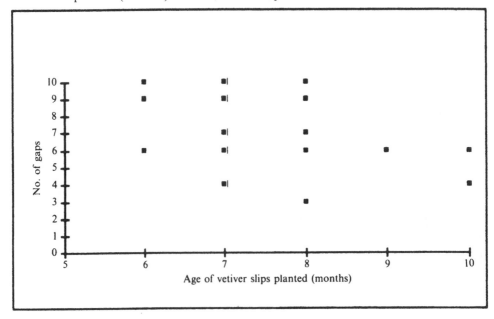

Fig. 13.6. Maximum vetiver height (cm) vs. time elapsed from nursery to field planting (hr). $n = 56$, $r = 0.42$, $p > 99.9\%$.

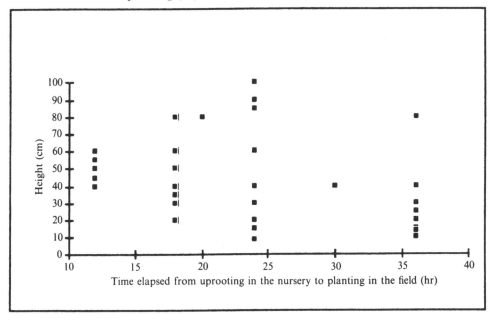

Fig. 13.7. Maximum vetiver height (cm) vs. slope of plot planted with vetiver (%). $n = 59$, $r = 0.33$, $p > 95\%$.

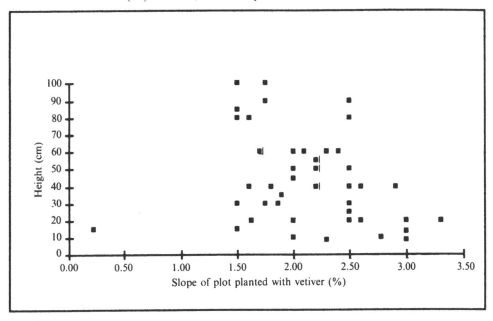

planted per rill are not presented because the sample was limited to either one or two slips planted per rill. However, it was found that with two slips planted per rill, both the number of tillers and the height tended to increase. No discernible effect on the number of gaps was observed.

Returning to the qualitative observations, our analysis takes up the effect of performance (as measured by the two primary indicators – the maximum number of tillers per vetiver plant, and the number of gaps per 30 m of vetiver hedge) on the suitability of vetiver as a soil and water conservation technique. The perceptions of the Agricultural Department staff vis-a-vis vetiver survival, hedge formation, and its acceptance by farmers have been investigated in relation to performance. In each case, the trends found in the data were compared with reference to the relationships assumed to prevail among performance and suitability (see Fig. 13.8).

While the data for the number of tillers showed no perceptible trend in two cases, those for the number of gaps indicated that our assumptions regarding suitability appear to be correct in all three cases. Perhaps the most significant relationship is that prevailing between the number of gaps and farmers'

Fig. 13.8. Summary of the relationships assumed and actually found prevailing among performance indicators and vetiver suitability.

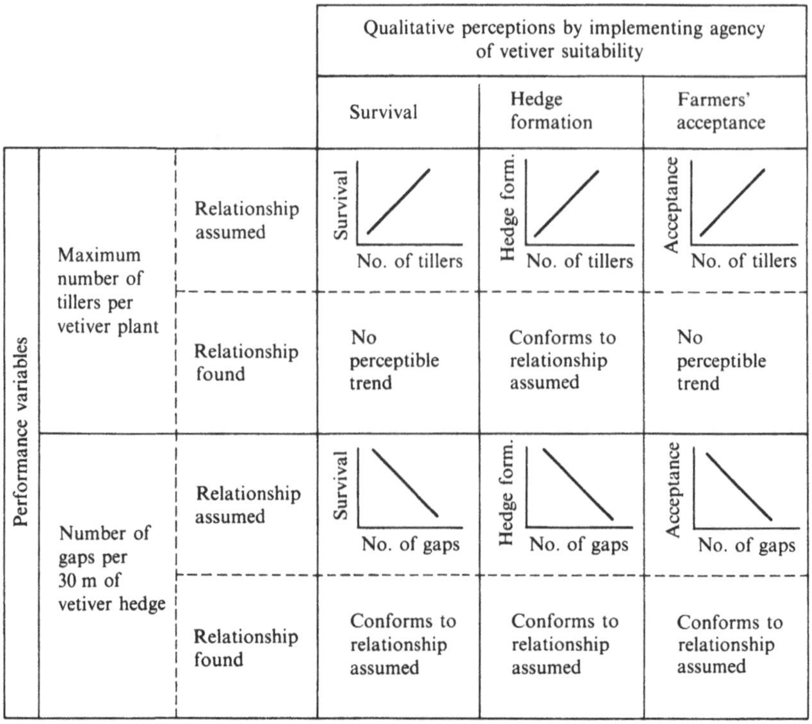

acceptance, which leads to our conclusion that the number of gaps is the most important performance indicator. Clearly, farmers' acceptance of vetiver technology is in part based on the number of gaps. Given what we have said about the persistence of gaps, it is important to look at management activities to reduce the number of gaps in the vetiver hedges.

Conclusions

The potential of vetiver grass as a soil and water conservation technique appears to be diminished in the arid and semi-arid regions. In such ecological conditions, establishing vetiver hedges is problematic and the persistence of gaps continues to raise questions about the suitability of vetiver technology. We do not reject the application of vetiver in semi-arid areas; however, we are confident that its effectiveness for soil and water conservation must be based on a high level of management.

Our survey population is too small and limited to draw any final conclusions for vetiver's performance in Maheshwaram, much less for other conditions. Further location-specific research to exhaustively establish vetiver's suitability is required. From our findings, it is apparent that a number of management lessons can be learned. Contrary to claims that vetiver will establish itself as a hedge in two to three years with minimal management, it is clear that in the semi-arid areas, nursery operations and holding size and slope are important determinants of performance. Gap filling, as carried out by government agencies, appears to have had limited success. Alternatively, involving farmers in participatory planning and implementation processes and in management is likely to have a beneficial influence on vetiver performance.

REFERENCES

AAC. 1970. *Modern Farming and the Soil*. Report of the Agricultural Advisory Council. HMSO, London.

Abawi, G.Y. and Stokes, R.A. 1982. *Wights Catchment Sediment Study 1977–81*. PWD Water Resources Technical Report No. 100., Public Works Department, Perth, WA.

Abernethy, C.L. 1987. Soil erosion and sediment yield. A review of data on river-borne sediment, and their use to indicate catchment-scale soil erosion rates. FAO Consultancy Report, November 1987. (Mimeo.)

Abrahams, A.D. 1972. Drainage densities and sediment yields in eastern Australia. *Aust. Geographical Studies* **10**:19–41.

Abrol, I.P. and Bhumbla, D.R. 1973. *Reclaiming Alkali Soils*. Central Soil Salinity Research Institute, Karnal, India.

Adamson, C.M. 1974. Effects of soil conservation treatment on runoff and sediment loss from a catchment in southwestern New South Wales, Australia. Proceedings Paris Symposium 'Effects of man on the interface of the hydrological cycle with the physical environment'. *IAHS-AISH Publication No.* **113**:3–14.

Aguilar, M. 1964. Promoting farmers' groups. *Ciencia e Investigación* (Buenos Aires) **20**:255–264. (In Spanish.)

Ahmad, N. 1988. *Steepland Agriculture in the Humid Tropics*. International Conference on Steepland Agriculture in the Humid Tropics. Kuala Lumpur, 17–21 August 1987.

Aina, P.O., Lal, R. and Taylor, G.S. 1977. Soil and crop management in relation to soil erosion in the rainforest of western Nigeria. *In: Soil Erosion: Prediction and Control*. SCS Spec. Publ. No. 21. Soil Conservation Society of America, Ankeny, Iowa.

Alchin, B.M. 1983. Runoff and soil loss on a duplex soil in semi-arid New South Wales. *J. Soil Cons. NSW* **39**:176–187.

Alexander, M. 1977. *Introduction to Soil Microbiology*. 2nd edition, Wiley, New York.

Alexander, W.J.R. 1979. *Sedimentation of Estuaries: Causes, Effects and Remedies*. 4th (S. Afr.) Natl. Oceanog. Symp. Cape Town, July. Council for Scientific and Industrial Research, Pretoria.

Allen, R. 1980. *How to Save the World. Strategy for World Conservation*. IUCN–UNEP–WWF, England.

Allison, F.E. 1973. *Soil Organic Matter and its Role in Crop Production*. Elsevier, New York.

Amartya Sen. 1987. *Hunger and Entitlements. Research for Action*. WIDER of UNU, Helsinki, Finland.

Anon. 1927. Forest destruction and its effects. *Nature* **119**:37–99.

Anon. 1978. *Economic Evaluation of a Soil Conservation Project in Allora Shire, Queensland. Commonwealth and State Government Collaborative Soil Conservation Study 1975–77.* Department of Environment, Housing and Community Development, Report 10. Australian Government Printing Service, Canberra.

Anon. 1986a. Dingxi Anjiagou Liuyu 1963 Nian, 6 Yue, 4 Ri Baoyu Jingliu Baogao. *Shuitu Baochi Shiyan Yanjiu Chengguo Ziliao Xuanbian 1956–1985* (Dingxi Area Soil Conservation Station). Pp. 67–75.

Anon. 1986b. Dingxi Zhuan Qu Qunzhong Xingxiu Titian Diaocha Yanjiu. *Shuitu Baochi Shiyan Yanjiu Chengguo Ziliao Xuanbian 1956–1985* (Dingxi Area Soil Conservation Station). Pp. 101–113.

Anon. 1986c. Handi Titian Kanghan Baoshang Yu Gaitu Peifei Shiyan. *Shuitu Baochi Shiyan Yanjiu Chengguo Ziliao Xuanbian 1956–1985* (Dingxi Area Soil Conservation Station). Pp. 138–147.

Arden-Clarke, C. and Hodges, R.D. 1987. The environmental effects of conventional and organic/biological farming systems. Soil erosion with special reference to Britain. *Biological Agriculture and Horticulture* 4:309–357.

Armstrong, C.L., Mitchell, J.K. and Walker, P.N. 1980. Soil loss estimation research in Africa: a Review. *In*: Deboodt, M. and Gabriels, D. (Eds), *Assessment of Soil Erosion.* J. Wiley & Sons, Chichester, UK. Pp. 285–294.

Arregui, J.C.T. and Bautista, I.D. 1965. *Pronostico de rendimientos de trigo en base a la humedad acumulada en el subsuelo en el momento de la siembra.* Numero Estadistico de la Bolsa de Cereales de Buenos Aires.

Atkinson, G. 1984. *Soil Erosion Following Wildfire in a Sandstone Catchment.* National Soils Conference, ASSSI, Brisbane, Australia.

Atkinson, G., Aina, P.O., Lal, R. and Taylor, G.S. 1977. Soil and crop management in relation to soil erosion in the rainforest of western Nigeria. *In: Soil Erosion: Prediction and Control.* SCS Spec. Publ. No. 21, Soil Conservation Society of America, Ankeny, Iowa. Pp. 75–82.

Aveyard, J.M. 1983. Soil erosion: productivity research in New South Wales to 1982. *Tech. Bull* 24, Wagga Wagga Res. and Tech. Serv. Centre, Soil Conservation Service, Sydney, Australia.

Aveyard, J.M. 1984. Report of working group on soil erosion and production prepared for the Standing Committee on Soil Conservation, Canberra, Australia.

Bai Jinian. 1987. The speech on the 2nd meeting of the soil and water conservation committee of the middle reach of the Yellow river. *Soil and Water Conservation in China. (Zhongguo Shuitubaochi)* 2:2–5.

Bali, J.S. 1975. Address delivered at Indian Photo-interpretation Institute, Dehradun, India. (Unpublished.)

Barber, R. 1984. *An Assessment of the Dominant Soil Degradation Processes in the Ethiopian Highlands.* Ethiopian Highlands Reclamation Study, Addis Abeba.

Barr, D.A. 1957. The effect of sheet erosion on wheat yield. *J. Soil Cons. NSW.* 13:27–32.

Barrows, H.L. and Kilmer, V.J. 1963. Plant nutrient losses from soils by water erosion. *Adv. Agron.* 15:303–315.

Battiston, L.A., McBride, R.A., Miller, M.H. and Brklacich, M.J. 1985. Soil erosion–productivity research in Southern Ontario. *In: Erosion and Soil Productivity.* Proc. Nat. Symp. American Society of Agricultural Engineers, St Joseph, MI. Pp. 28–38.

Beasley, R.P. 1972. *Erosion and Sediment Pollution Control.* Iowa State University Press, Ames.

Beckmann, G.G. and Coventry, R.J. 1987. Soil erosion losses: squandered withdrawals from a diminishing account. *Search* 18:21–26.

Begg, G.W. 1978. *The Estuaries of Natal.* Natal Town and Regional Planning Commission, Pietermaritzburg.

Bennett, H.H. 1939 and 1947. *Soil Conservation.* McGraw-Hill, New York.

Bennett, H.H. 1957. Soil conservation in Argentina. *Hombre y Suelo* (Buenos Aires) No. 6.

Berglund, S.H. and Michalson, E.L. 1981. Soil erosion control in Idaho's Cow Creek watershed: an economic analysis. *J. Soil Water Cons.* 36(3):158–161.

Bettenay, E. 1986. Salt affected soils in Australia. *In*: Barrett-Lennard, E.G., Malcolm, C.V., Stern, W.R., and Wilkins, S.M. (Eds), *Forage and Fuel Production from Salt Affected Wasteland.* Proceedings of a seminar held at Cunderdin, Western Australia, 19–27 May 1984. Elsevier Science Publishers B.V., The Netherlands.

Birch, P.B. 1981. Effects of wind erosion on crop yields on Geraldton sandplain soils. *Tech. Bull. No. 54*, Dept. of Agric., Perth, Australia.

Blaikie, P. 1985. *The Political Economy of Soil Erosion in Developing Countries.* Longman Scientific & Technical, England.

Blong, R.J., Riley, S.J. and Crozier, P.J. 1982. Sediment yield from runoff plots following bushfire near Narrabeen Lagoon, NSW. *Search* 13:36–38.

Blyth, M. and McCallum, A. 1987. Onsite costs of land degradation in agriculture and forestry. *In*: Chisholm, A.H. and Dumsday, R.G. (Eds), *Land Degradation: Problems and Policies.* Cambridge University Press and CRES, Canberra, Australia. Pp. 79–88.

Boardman, J. 1983. Soil erosion at Albourne, W Sussex, England. *Appl. Geog.* 3:317–329.

Boardman, J. 1984. Erosion on the South Downs. *Soil and Water* 12:19–21.

Boardman, J. 1986. Soil erosion in SE England. *Lond. Environ. Bull.* 3(4):5–7.

Boardman, J. 1987a. The context of soil erosion. *In*: Burnham, C.P. and Pitman, J.I. (Eds), *Soil Erosion*, see *Soil* 3:2–13.

Boardman, J. 1987b. A land farmed into the ground. *The Guardian* 18 Dec. 1987.

Boardman, J. 1988. Public policy and soil erosion in Britain. *In*: Hooke, J. (Ed.), *Geomorphology and Public Policy.* J. Wiley, Chichester, UK. Pp. 33–50.

Boardman, J. and Robinson, D.A. 1985. Soil erosion, climatic vagary and agricultural change on the Downs around Lewes and Brighton, autumn 1982. *Appl. Geog.* 5:243–258.

Bosworth, D.A. and Foster, A.B. 1982. *Approved Practices in Soil Conservation.* Pergamon Press, London.

Bradley, D.B. and Sieling, D.H. 1953. Effect of organic anions and sugars on phosphate precipitation by iron and aluminium as influenced by the pH. *Soil Sci.* 76:175–179.

Brown, L.R., and Wolf, E.C. 1984. *Soil Erosion: Quiet Crisis in the World Economy.* Worldwatch Paper 60.

Brown, L.R., Chandler, W.U., Flavin, C., Pollock, C., Postel, S., Starke, L. and Wolf, E.C. 1985. *State of the World 1985.* W.W. Norton, New York.

Brown, L.R., Chandler, W.U., Durning, A., Flavin, C., Heise, L., Jacobson, J., Postel, S., Pollock Shea, C., Starke, L. and Wolf, E.C. 1988. *State of the World. A Worldwatch Institute Report on Progress Toward a Sustainable Society.* W.W. Norton & Company, New York and London.

Bruce, J.W. and Noronha, R. 1987. Land tenure issues in the forestry and agroforestry project contexts. *In*: Raintree, J.B. (Ed.), *Land, Trees and Tenure.* Proceedings of an International Workshop on Tenure Issues in Agroforestry, Nairobi, 27–31 May 1985. Pp. 121–160.

Brühne, S. 1988. *Agrarian Development, Famine and Foreign Aid – The Ethiopian Experience.* Paper presented to the Xth International Conference of Ethiopian Studies, Paris, August 1988. 18 pp.

BTN (Bureau of the 'Three North' Protection Forest System Construction, Ministry of Forestry, China). 1986. *Work on the Project of 'Three North' Protection Forest System is Under Way.* Beijing. 15 pp.

Buenos Aires Stock Exchange. 1987. *Estadistic number*. Buenos Aires. (In Spanish.)

Buntley, G.J. and Bell, F.F. 1976. Yield estimates for the major crops grown on the soils of west Tennessee. *Tenn. Agr. Exp. Stn Bull*.

Buraczynski, J. 1977. Intensity of gully erosion and soil erosion in Roztocz Gorajskie. *Zeszyty Problemowe Postepcw Nauk Rolniczych* **193**:91–99.

Burch, G., Graetz, D. and Noble, I. 1987. Biological and physical phenomena in land degradation. *In*: Chisholm, A.H. and Dumsday, R.G. (Eds), *Land Degradation: Problems and Policies*. Cambridge University Press and CRES, Canberra, Australia. Pp. 27–48.

Buringh, P. 1984. The capacity of the world land area to produce agricultural products. Paper presented at Workshop Intecol and Institute of Mediterranean Agronomy, Zaragoza, Spain, 30 January 1984.

Buringh, P. and Dudal, R. 1987. Agricultural land use in space and time. *In*: Wolman, M.G. and Fournier, F.G.A. (Eds), *Land Transformation in Agriculture*, SCOPE 32. John Wiley & Sons, Chichester, England. Pp. 9–43.

Burnett, E. and Fisher, C.E. 1954. The effect of conservation practices on runoff, available soil moisture and cotton yield. *Proc. Soil Sci. Soc. Am.* **18**:216–218.

Burt, T.P. and Oldman, J.C. 1986. The implications of sediment inputs to upland reservoirs. *In: Current Geomorphological Research in the Northern Dark Peak, Field Guide*. British Geomorphological Research Group. Spring field meeting, Buxton 1986. Pp. 5–33.

Butcher, D.P., Labadz, J.C., Pemberton, T.J.L. and Potter, A.W.R. 1990. Erosion rates and reservoir sedimentation in the Southern Pennines. *Proceedings of the North of England Soils Discussion Group*, **24**:63–79.

Buttel, F.H. 1982. Environmental quality in agriculture: some observations on political–economic constraints on sustainable resource management. *Cornell Rural Sociol. Bull. Ser. Bull. No.* **128**. Cornell University, Ithaca, NY.

Butzer, K.W. 1981. Rise and fall of Auxum, Ethiopia: A geo-archaeological interpretation. *American Antiquity*, **46**(3):471–495.

Capelin, M.A. and Truong, P.N. 1985. Soil erosion within pineapple fields in SE Queensland. *Proceedings Fourth Australian Soil Conservation Conference*. Pp. 220–221.

Carpenter, R.A. (Ed.) 1983. *Natural Systems for Development: What Planners Need to Know*. Macmillan, New York.

Carter, V.G. and Dale, T. 1974. *Topsoil and Civilization*. Revised edition, University of Oklahoma Press, Norman, OK.

Catt, J.A. 1986. Soils and Quaternary geology. *Soil and Resources Surveys*, Monog. 11. OUP, Oxford.

CBIP. 1981. Sedimentation studies on reservoirs. *Tech. Report No. 20*, Vol. 2. Central Board of Irrigation and Power, New Delhi, India.

CEQ (Council on Environmental Quality). 1980. *The Global 2000 Report to the President. The Technical Report*. Vol. 2. US Government Printing Office, Washington, DC.

Chaney, K. and Swift, R.S. 1984. The influence of organic matter on aggregate stability in some British soils. *J. Soil Sci.* **35**:223–230.

Chankou Forestry Station. 1986. Huangtu Qiuling Goubao Qu Zaolin Jishu Yanjiu. *Zhongguo Ganhan Ban Ganhan Nongye Keji Ziliao Xuanji Di Si Ji*. Pp. 219–261.

Charreau, D. 1972. Problemes poses par l'utilisation agricole des sols tropicaux par des cultures annuelles. *Agron. Tropical (France)* **27**:905–929.

Charreau, D. and Nicou, R. 1971. L'amelioration du profil culturel dans les sols sableux et sablo argilleux de le zone tropicale seche ouest Africaine et ses incidences agronomique. *Agron. Tropical (France)* **26**:209–255, 565–631, 903–978, 1184–1274.

Chaturverdi, M.C. 1978. *Proceedings of symposium on Indian National Water Perspective, Technological Policy Issues and Systems Planning*. Physical Research Laboratory, Ahmedabad, India. Pp. 459–489.

Chaudhry, K. 1987. Wastelands and the rural poor: essentials of a policy framework. *FAO–Forest News* **14**(2):1–6.

Chen Yanhai. 1987. The trend of the population growth in China and the way to deal with it. *Geography and Territorial Research (Dilixue He Guetu Yanjiu)* **3**(2):5–10.

Chen Yongzhong. 1988. Some discussions about the land management in the Loess Plateau. *Journal of Natural Resources* **3**(1):1–8.

Chen Zhangling. 1985. Huangtu Gaoyuan Shuiping Titian Jianshe De Zhuyao Shiyan. *Tudi Liyong Gongcheng Cankao Ziliao* **12**:13–19.

Chen Zhilin. 1986. Preliminary analysis on variations of oncoming sediment load of Yellow river. *Soil and Water Conservation in China (Zhongguo Shuitubaochi)* **11**:12–16.

Chisholm, A.H. 1987. Abatement of land degradation: regulation vs. economic incentives. *In*: Chisholm, A.H. and Dumsday, R.G. (Eds), *Land Degradation: Problems and Policies*. Cambridge University Press and CRES, Canberra, Australia. Pp. 223–247.

Chittleborough, D.J. 1983. The nutrient load in surface waters as influenced by land use patterns. *In*: Holmes, J.W. (Ed.), *The Effects of Changes in Land Use upon Water Resources*. Adelaide, Australia. Pp. 42–53.

Chu Wenguan. 1981. Tianshui Diqu Nongye Jishu Cuoshi Dui Shan Qu Baochi Shuitu He Zengchan De Zuoyong, *Shuitu Baochi Shiyan Yanjiu Chengguo Huibian 1952–1980* **2**:62–84.

Chui Mengshen. 1985. The relations between rainfall, topography and vegetation in Chengde region. *Soil and Water Conservation in China (Zhongguo Shuitubaochi)* **8**:15–17.

Ciesiolka, C. 1987. *Catchment Management in the Nogoa Catchment*. Australian Water Research Project 80/128 – Completion Report. Department of Resources and Energy, Canberra, Australia. 204 pp.

Clark, E.H. II. 1985. The off-site costs of soil erosion. *J. Soil Water Cons.* **40**:19–22.

Clarke, A.L. 1986. Cultivation. *In*: Russell, J.S. and Isbell, R.F. (Eds), *Australian Soils: The Human Impact*. University of Queensland Press, St Lucia, Australia. Pp. 273–303.

Cocks, P.S. 1988. The role of pasture and forage legumes in livestock based farming systems. *In*: Beck, D.P. and Materon, L.A. (Eds), *Nitrogen Fixation by Legumes in Mediterranean Agriculture*, ICARDA, Martinus Nijhoff Publishers, The Netherlands. Pp. 3–10.

Colborne, G.J.N. and Staines, S.J. 1985. Soil erosion in South Somerset. *J. Agric. Sci.* **104**:107–112.

Conacher, J. and Conacher, A. 1986. Herbicides in agriculture: minimum tillage, science and society. *Geowest* No. 22., University of Western Australia, Nedlands, Australia.

Conklin, H.C. 1980. *Ethnographic Atlas of Ifugao: A Study of Environment, Culture and Society in Northern Luzon*. Yale University Press, New Haven, CT.

Constable. 1985. *Ethiopian Highlands Reclamation Study (EHRS): Summary*. EHRS, Addis Abeba. (Draft.)

Cooke, G.W. 1977. The roles of organic manures and organic matter in managing soils for higher crop yields – a review of the experimental evidence. *In: Proc. Internatl. Seminar on Soil Environment and Fertility Management in Intensive Agriculture*. The Society of the Science of Soil and Manure, Tokyo, Japan. Pp. 53–64.

Costin, A.B. 1980. Runoff and soil loss and nutrient losses from an improved pasture at Ginninderra, Southern Tablelands, New South Wales. *Australian J. Agric. Res.* **31**:533–546.

Craswell, E.T. 1986. Cherish the Earth. *Interdisciplinary Science Reviews* **11**(3):234–235.

Craswell, E.T. 1987. Soil management in Asia: research supported by the Australian Centre for International Agricultural Research. *In*: Latham, M. (Ed.), *Soil Management Under Humid Conditions in Asia, ASIALAND: Proceedings of the First IBSRAM Regional Seminar.* Khon Kaen, Phitsanulok, Thailand, 13–20 October 1986. Pp. 53–67.

Craswell, E.T. and Isbell, R.F. 1984. Proceedings of the International Workshop on Soils, Townsville, Australia, 12–16 September 1983. *ACIAR Proceedings Series No. 2.*

Craswell, E.T. and Tangendjaja, B. 1985. Shrub legume research in Indonesia and Australia: proceedings of an International Workshop held at Balai Penelitian Ternak, Ciawi-Bogor, Indonesia, 2 February 1984. *ACIAR Proceedings Series No. 3.*

Craswell, E.T., Remenyi, J.V. and Nallana, L.G. (Eds) 1985. Soil Erosion Management: proceedings of a workshop held at PCARRD, Los Banos, Philippines, 3–5 December 1984. *ACIAR Proceedings Series No. 6.*

Crosson, P. 1985. National costs of erosion on productivity. *In: Erosion and Soil Productivity.* Proc. Nat. Symp. American Society of Agricultural Engineers, St Joseph, MI. Pp. 254–265.

Crosson, P. 1986. Sustainable food production. *Food Policy*, May 1986:143–156.

Crosson, P.N. and Stout, T.T. 1983. *Productivity Effects of Cropland Erosion in the United States.* Resources for the Future, Washington, DC.

Crouch, R.J. 1987. The relationship of gully sidewall shape to sediment production. *Australian J. Soil Res.* **25**:531–539.

CSE. 1982. *The State of India's Environment 1982.* A Citizens' Report. Centre for Science and Environment, New Delhi.

CSWCRTI. 1981. Report on operational research project, watershed management. Central Soil and Water Conservation Research and Training Institute, Dehradun, India.

Cuba, P. 1949. *Rolo-facas.* Notas agricolas de la Secretaria de Agricultura de Sao Paulo (Brasil). (In: Portuguese.)

Cucullu, O.P. 1961. CREA, una verdadera cooperativa de ideas. *Hombre y Suelo.* Special number in homage to H.H. Bennett (Buenos Aires). (In Spanish.)

Cullen, P. and O'Loughlin, E.M. 1982. *Prediction in Water Quality.* Australian Academy of Science, Canberra, Australia.

Cummins, W.A. and Potter, H.R. 1967. Rate of sedimentation in Cropston Reservoir, Charnwood Forest, Leicestershire. *Mercian Geol.* **2**:31–39.

Cunningham, G.M. 1987. TCM – the recipe for survival in the Murray–Darling Basin. *Proceedings Soil Conservation Service NSW Annual Conference 1987.* Pp. 18–23.

Czerwinski, S. 1985. Effect of rolling loess terrain on the crop of some cultivated plants. *Zeszyty Problemowe Postepcw Nauk Rolniczych* **311**:37–51.

Daniels, R.B. 1987. Saline seeps in the Northern Great Plains of the USA and the Southern Prairies of Canada. *In*: Wolman, M.G. and Fournier, F.G.A. (Eds), *Land Transformation in Agriculture*, SCOPE 32. John Wiley & Sons, Chichester, England. Pp. 381–406.

Das, D.C. and Singh, S. 1981. Small storage works for erosion control and catchment improvement: mini case studies. *In*: Morgan R.P.C. (Ed.), *Soil Conservation: Problems and Prospects.* Wiley, Chichester, England. Pp. 425–450.

Das, D.C., Bali, Y.P. and Kaul, R.N. 1981. Soil conservation in multipurpose river valley catchments. Problems, programme approaches, and effectiveness. *Indian J. Soil Conservation* **9**(1):5–26.

Davies, D.B. 1983. Wind erosion in the UK. *In*: Prendergast, A.G. (Ed.), *Soil Erosion.* CEC, Brussels. Pp. 23–25.

Day, J.H. and Grindley, J.R. 1981. The estuarine ecosystem and environmental constraints. *In*: Day, J.H. (Ed.), *Estuarine Ecology*. A.A. Balkema, Rotterdam. Pp. 345–372.

Debski, K. 1961. *Charakterystyka hydrobiologiczna Polski*. Panstwowe Wzdawnictwa Naukowe. 159 pp.

Department of Environment, Housing, and Community Development. 1978. *A Basis for Soil Conservation Policy in Australia*. Report 1. Aust. Govt Pub. Serv., Canberra, Australia.

Detwiler, R.P. and Hall, C.A.S. 1988. Tropical forests and the global carbon cycle. *Science* **239**:42–47.

Dingxi Area Conservation Station. 1983. Daizhuang Pingbo Qilong Gengzuofa. *Quanguo Shuitu Baochi Gengzuo Xueshu Taolun Hui Ziliao Ji Bian*. Pp. 177–179.

DOE. 1981. *Energy Projections to the Year 2000*. Office of Policy, Planning and Analysis, Department of Energy, Washington, DC.

Donald, C.M. 1967. Innovation in agriculture. *In*: Williams, D.B. (Ed.), *Agriculture in the Australian Economy*, Sydney University Press, Sydney. Pp. 57–86.

Douglas, I. 1970. Sediment yield from forested and agricultural lands. *In*: Taylor, J.A. (Ed.), *The Role of Water in Agriculture*. Pergamon Press, Oxford, UK.

Dregne, H.E. 1982. Desertification: man's abuse of the land. *J. Soil Water Cons.* **33**:11–14.

Dregne, H.E. 1984. Combating desertification: evaluation of progress. *Environmental Conservation* **11**(2).

Dregne, H.E. 1987. Desertification. *In*: McLaren, D.J. and Skinner, B.J. (Eds), *Resources of World Development*, John Wiley & Sons Ltd, Chichester, England. Pp. 697–710.

Dudal, R. 1981. An evaluation of conservation needs. *In*: Morgan, R.P.C. (Ed.), *Soil Conservation, Problems and Prospects*. John Wiley & Sons, Chichester, UK. Pp. 3–12.

Dudal, R. 1982. Land degradation in a world perspective. *Journal of Soil and Water Conservation* **82**:245–249.

Dudal, R. and Purnell, M.F. 1986. Land resources: salt affected soils. *In*: Barrett-Lennard, E.G., Malcolm, C.V., Stern, W.R. and Wilkins, S.M. (Eds), *Forage and Fuel Production from Salt Affected Wasteland*. Proceedings of a seminar held at Cunderdin, Western Australia, 19–27 May, 1984. Elsevier Science Publishers B.V., The Netherlands. Pp. 1–9.

Dugue, M.J. 1986. The African functioning of production systems and use of environment in a village in Yatenga: Boukere (Burkina Faso). *Collection Documents Systemes Agraires Dept. Systemes Agraires du CIRAD*, Montpellier No. 1.

Duley, F.L. and Kelly, L.L. 1939. Effect of soil type, slope and surface conditions on intake of water. *Nebraska Agric. Exp. Sta. Res. Bull.* **112**.

Dumsday, R.G. 1973. The economics of some soil conservation practices in the wheat belt of northern New South Wales and southern Queensland – a modelling approach. *Farm Management Bulletin. No. 19*, University of New England, Armida.

Eck, H.V. 1968. Effect of topsoil removal on nitrogen-supplying ability of Pullman silty clay loam. *Soil Sci. Am. Proc.* **32**:686–691.

Eck, H.V. 1969. Restoring productivity on Pullman silty clay loam subsoil under limited soil moisture. *Soil Sci. Am. Proc.* **33**:578–581.

Edwards, C.A. and Lofty, J.R. 1977. *Biology of Earthworms*. Chapman and Hall, London.

Edwards, K. 1980. Runoff and soil loss in the wheat belt of New South Wales. *In*: *Proceedings, Agr. Eng. Conf. Geelong*. The Institution of Engineers, Barton, Australia.

Edwards, K. 1987. Runoff and soil loss studies in New South Wales. *Technical Handbook No. 10*. Soil Conservation Service of NSW, Sydney, Australia.

Edwards, K. 1988. How much soil loss is acceptable? *Search* **19**:136–140.

Edwards, K. and Wild, A.E.R. 1987. Drought and landholder attitudes and their effects on soil degradation. *In*: Pla Sentis, I. (Ed.), *Soil Conservation and Productivity* (Proc. IV Intl.

Conf. on Soil Conservation, Maracay, Venezuela, 1985), Sociedad Venezolana de la Cienca del Suelo, Maracay, Venezuela. Pp. 516–523.

EHRS. 1985. Ethiopian Highlands Reclamation Study. Various Reports. c/o Land Use Planning and Regulatory Department Addis Abeba.

Elliott, G.L. and Dight, D.C.G. 1986. *An Evaluation of the Surface Stability of Rehabilitated Overburden in the Upper Hunter Valley, NSW.* Report to NSW Coal Association, Sydney, Australia.

Elliott, G.L., Campbell, B.L. and Loughran, R.J. 1984. Correlation of erosion and erodibility assessments using caesium-137. *J. Soil Cons. NSW* **40**:24–29.

El-Swaify, S.A. (Ed.) 1985. *Soil Erosion and Conservation.* Soil Conservation Society of America, Ankeny, IA.

El-Swaify, S.A. 1988. Advances in water erosion research and conservation practices. *In:* Rimwanich, S. (Ed.).

El-Swaify, S.A. and Cooley, K.R. 1980. Sediment losses from agricultural watersheds in Hawaii. *Agricultural Reviews and Manuals* No. 17. USDA-SEA Western Series, USDA, Oakland, California.

El-Swaify, S.A. and Dangler, D.G. 1982. *Rainfall Erosion in the Tropics: A State-of-the-Art.* Soil Erosion and Conservation in the Tropics, American Society of Agronomy and Soil Science Society of America, Special Publication No. 43.

El-Swaify, S.A. and Dangler, E.W. 1977. Erodibilities of selected tropical soils in relation to structural and hydrologic properties. *In: Soil Erosion: Prediction and Control*, Special Publication No. 21, Soil Conservation Society of America, Ankeny, Iowa.

El-Swaify, S.A., Dangler, E.W. and Armstrong, C.L. 1982. Soil erosion by water in the Tropics. *Research Extension Series* 024, College of Tropical Agriculture and Human Resources, University of Hawaii.

El-Swaify, S.A., Moldenhauer, W.C. and Lo, A. 1985. *Soil Erosion and Conservation.* (Proc. of Malama Aina 1983.) Soil Conservation Society of America, Ankeny, Iowa.

El-Swaify, S.A., Garnier, C.L. and Lo, A. 1988. *Recent Advances in Soil and Water Conservation in Steep Lands in the Humid Tropics.* International Conference on Steepland Agriculture in the Humid Tropics, 17–21 August 1987, Kuala Lumpur, Malaysia.

Elwell, H.A. 1985. An assessment of soil erosion in Zimbabwe. *Zimbabwe Sci. News* **19**(3/4):27–31.

ENDS. 1984. Soil erosion: An Unsustainable Face of Modern Farming. *ENDS Report* **115**:9–10. Environmental Data Services Ltd, London.

Engelstad, O.P., Shrader, W.D. and Dumenil, L.C. 1961. The effect of surface soil thickness on corn yields: I. As determined by a series of experiments in farmer-operated fields. *Soil Sci. Soc. Am. Proc.* **25**:494–497.

EPA. 1979. *Effects of Suspended Solids and Sediment on Reproduction and Early Life of Warmwater Fishes: A Review.* Environmental Protection Agency, EPA 600/3–79–042, Environ. Res. Lab., Corvallis, OR.

Evans, R. 1971. The need for soil conservation. *Area*, **3**:20–23.

Evans, R. 1977. Overgrazing and soil erosion on hill pastures with particular reference to the Peak District. *J. Brit. Grassland Soc.* **32**:67–76.

Evans, R. 1980. Characteristics of water eroded fields in lowland England. *In*: De Boodt, M. and Gabriels, D. (Eds), *Assessment of Erosion.* J. Wiley, Chichester, UK. Pp. 77–87.

Evans, R. 1981a. *Potential Soil and Crop Losses by Erosion.* Paper given at conference on Soil and Crop Loss – Developments in Erosion Control. Royal Agric. Soc. Eng., Stoneleigh, UK.

Evans, R. 1981b. Assessments of soil erosion and peat wastage for parts of East Anglia, England. A field visit. *In*: Morgan, R.P.C. (Ed.), *Soil Conservation: Problems and Prospects*. J. Wiley, Chichester, UK.

Evans, R. 1983. Accelerated water erosion of soils in England and Wales. *In*: Prendergast, A.G. (Ed.), *Soil Erosion*. CEC, Brussels. Pp. 27–28.

Evans, R. 1985. *Soil Erosion – The Disappearing Trick*. Paper given at a conference on Better Soil Management for Cereals and Oilseed Rape, NAC, Stoneleigh, UK.

Evans, R. 1988a. Water Erosion in England and Wales. Report for the Soil Survey and Land Research Centre. Silsoe, Beds., UK. (Unpublished.)

Evans, R. 1988b. Some effects of erosion in Britain. *Geog. Rev.* 1(5):34–36.

Evans, R. 1990. Erosion studies in the Dark Peak. *Proceedings of the North of England Soils Discussion Group* 24:39–61.

Evans, R. and Catt, J.A. 1987. Causes of crop patterns in E England. *J. Soil Sci.* 38:309–324.

Evans, R. and Cook, S. 1987. Soil erosion in Britain. *In*: Burnham, C.P. and Pitman, J.I. (Eds), *Soil Erosion*, see *Soil* 3:28–59.

Evans, R. and Nortcliff, S. 1978. Soil erosion in N Norfolk. *J. Agric. Sci. Camb.* 90:185–192.

Evans, R. and Skinner, R.J. 1987. A survey of water erosion. *Soil and Water* 13(1/2):28–31.

Fang Zhengshan. 1960. *Terraced Fields*. Water Conservancy and Electrical Power Press, Beijing.

Fang Zhengshan. 1985. Several proposals on policies of harnessing Yellow river and speeding up the erosion controls in the Loess Plateau. *Bulletin of Soil and Water Conservation (Shuitubaochi Tongbao)* 6:14–17.

Fang Zhengshan. 1987. Tentative discussion on macroeconomic losses caused by soil erosion. *Bulletin of Soil and Water Conservation (Shuitubaochi Tongbao)* 7(1):66–70.

FAO (Food and Agriculture Organization). 1969. *Technical Conference on Soil Conservation in Latin America*. IDIA (INTA) Sup. No. 23, Pp. 1–172 (1970).

FAO. 1971. *Land Degradation*. Soils Bull. No. 13. FAO, Rome, Italy.

FAO. 1979. Yield response to water. *FAO Irrigation and Drainage Paper* 33.

FAO. 1981. *Agriculture: Toward 2000*. FAO, Rome, Italy.

FAO. 1982a. *World Soil Charter*. FAO, Rome, Italy.

FAO. 1982b. *1981 Production Yearbook*. FAO, Rome, Italy.

FAO. 1983. The state of food and agriculture 1982. *FAO Agricultural Series*. No. 15, Rome, Italy.

FAO. 1984. Land food and people. *Economic and Social Development Series*. No. 30, Rome, Italy.

FAO. 1986. *African Agriculture the Next 25 Years*. Main Report:25–39.

FAO/UNEP. 1978. *Methodology for Assessment of Soil Degradation*. Rome, Italy.

FAO, UNEP and UNESCO. 1980. *Provisional map of Present Degradation Rate and Present State of Soil*. FAO, Rome, Italy.

Fei Xiaotong. 1986. The concept of the poverty regions and their development. *Agricultural Modernization Research (Nongye Xiandaihua Yanjio)* 6:1–14.

Field, J.B. 1984. Erosion in a catchment in New England, NSW. *In*: Loughran, R.J. (Ed.), *Drainage Basin Erosion and Sedimentation*, Vol 2, University of Newcastle, Newcastle, Australia. Pp. 43–58.

Follett, R.F. and Stewart, B.A. 1985. *Soil Erosion and Crop Productivity*. American Society of Agronomy, Crop Science Society of America, and Soil Science Society of America. Madison, WI. 533 pp.

Fournier, F. 1967. Research on soil erosion and soil conservation in Africa. *African Soils* 12:53–96.

Freebairn, D.M. 1982. Soil erosion in perspective. *In: Div. Land Util. Tech. News*, **6**(1), Qld Dept Primary Ind., Brisbane, Australia. Pp. 12–15.

Freebairn, D.M. and Wockner, G.H. 1986a. A study of soil erosion on vertisols of the eastern Darling Downs, Queensland. I. Effects of surface conditions on soil movement within contour bay catchments. *Australian J. Soil Res.* **24**:135–158.

Freebairn, D.M. and Wockner, G.H. 1986b. A study of soil erosion on vertisols of the eastern Darling Downs, Queensland. II. The effect of soil, rainfall, and flow conditions on suspended sediment losses. *Australian J. Soil Res.* **24**:159–172.

Frost, C.A. and Speirs, R.B. 1984. Water erosion in SE Scotland – a case study. *Research and Development in Agriculture* **1**:145–152.

FTN (Fertilizer Test Network of Chinese Academy of Agricultural Sciences). 1983. The effects of fertilizers on crop yields, and the suitable fertilization methods in China. *Soil and Fertilizers (Tuyang Ha Feiliao)* **6**:13–18.

FTN. 1986. The variations about the effects of fertilizers on crop yields and its improvement. *Soil and Fertilizers (Tuyang Ha Feiliao)* **1**:1–8.

Fu Dexing. 1986. *Establishing Shelterbelt Systems According to the Viewpoint of Ecological Economy*. Forestry Institute of Heilongjiang Province, Harbin.

Fullen, M.A. 1985a. Erosion of arable soils in Britain. *Intern. J. Environ. Studies* **26**:55–64.

Fullen, M.A. 1985b. Compaction, hydrological processes and soil erosion on loamy sands in E Shropshire, England. *Soil and Tillage Res.* **6**:17–29.

Fullen, M.A. 1985c. Wind erosion of arable soils in E Shropshire during spring 1983. *Catena* **12**:111–120.

Gallais, J. 1979. La situation de l'elevage bovin et le probleme des eleveurs en Afrique occidentale et centrale. *Les Cahiers d'outre-Mer* **32**:113–138.

Gansu Hydrology Bureau. 1983. *Gansu Xiao Liuyu Pucha Chengguo Ziliao Huibian*. Soil Conservation Office.

Gao Qijiang. 1987. The current situation of soil and water loss in Shanxi province. *Bulletin of Soil and Water Conservation (Shuitubaochi Tongbao)* **7**(2):17–21.

Gao Zhu and Zhou Lie. 1988. The aggravation of soil erosion in southwestern region. *Science and Technology Daily (Keji Ribao)*, Jan. 22, 1988.

GBPUAT. 1982. *Integrated Natural and Human Resources Planning and Management in the Hills of UP*. Govind. Ballab. Pant. University of Agric. and Tech., Pantnagar, India. Vols I and II.

Gerlach, T. and Koszarski L. 1967. Wspçlczesna rola morfogenetycznawiatru na przedpolu Beskidu Niskiego. *Studia Geomorphologica Caraptho-Belcanica*, **2**:67–76.

Ghildyal, B.P. 1981. Soils of the Garwhal and Kumaon Himalayas. *In: The Himalaya: Aspects of Change*. OUP, Delhi, India. Pp. 120–151.

Gil, E. and Slupik, J. 1972. The influence of the plant cover and landuse on the surface runoff and wash down during heavy rain. *Studia Geomorphologica Carpatho-Balcanica* **6**.

Goedewaagen, M.A.J. and Schuurman, J.J. 1950. Root production by agricultural crops on arable land and on grasslands as a source of organic matter in the soil. *Trans. 4th Intl. Soil Sci. Cong.* **2**:28–31.

Gopalakrishna Kumar. 1987. *Ethiopian Famines 1973–1985: A Case Study*. WIDER Working Paper 26, WIDER of UNU, Helsinki, Finland.

Graesser, N.W. 1979. How land improvements can damage Scottish fisheries. *Salmon and Trout Magazine* **215**:39–43.

Greenland, D.J. 1975. Bringing the Green Revolution to the shifting cultivator. *Science* **190**(4217):841–844.

Greenland, D.J. 1977a. Soil damage by intensive arable cultivation: temporary or permanent? *Phil. Trans. Roy. Soc.* London **B281**:193–208.

Greenland, D.J. 1977b. The magnitude and importance of the problem. *In*: Greenland, D.G. and Lal, R. (Eds), *Soil Conservation and Management in the Humid Tropics*. John Wiley & Sons, Chichester, England.

Greenland, D.J. 1981. Soil management and soil degradation. *J. Soil Sci.* **32**:301–322.

Greenland, D.J., Rimmer, D. and Payne, D. 1975. Determination of the structural stability class of English and Welsh soils, using a water coherence test. *J. Soil Sci.* **26**:294–303.

Grigg, D. 1987. The Industrial Revolution and land transformation. *In*: Wolman, M.G. and Fournier, F.G.A. (Eds), *Land Transformation in Agriculture*, SCOPE 32. John Wiley & Sons, Chichester, England. Pp. 79–109.

Guan Junwei and Wang Lixian. 1987. Water and soil conservation. *In: Agricultural Encyclopaedia of China*, Water conservancy volume. Agricultural Press, Beijing. Pp. 729–731.

Guo Xiulin. 1986. Ningtiao Zaolin Jishu, *Zhongguo Ganhan Ban Ganhan Nongye Keji Ziliao Xuanji Di Si Ji*. Pp. 145–152.

Gupta, G.P. 1975. Sediment production: status report on data collection and utilization. *Soil Conserv. Digest* **3**(2):10–21.

Gupta, G.P. 1980. Soil and water conservation in the catchment of river valley projects, Status Report. *Indian J. Soil Cons.* **8**(1):1–7.

Gupta, P.N. 1981. *Integrated Watershed Rehabilitation and Development Project*. UP Forest Dept, Lucknow, India.

Gupta, S.K., Tajwani, K.G., Mathur, H.N. and Srivastava, M.M. 1970. Land resource regions and areas of India. *J. Indian Soc. Soil Sci.* **18**:187–198.

Hallsworth, E.G. 1987a. Soil conservation down under. *J. Soil Water Cons.* **42**:394–400.

Hallsworth, E.G. 1987b. *Anatomy, Physiology and Psychology of Erosion. The International Federation of Institutes of Advanced Study Monograph No. 1*. John Wiley & Sons, Chichester, England.

Hamel, P. 1986. Assessment of the risks of erosion in young oil palm plantations where food crop intercropping is practiced based on the Wischmeier and Smith equation. *Oleagineux* **41**:419–428.

Hamilton, A.P.F. 1935. Siwalik erosion. *Indian Forester* **36**:375–387.

Hamilton, G.J. 1970. The effect of sheet erosion on wheat yield and quality. *J. Soil Cons. NSW* **26**:118–123.

Han Yichi. 1987. The continued deterioration of ecological environment in China. *Science and Technology Daily (Keji Ribao)*, 13 Nov. 1987, p.2.

Hanks, R.J. 1983. Yield and water-use relationships: an overview. *In*: Taylor, H.M., Jordan, W.R. and Sinclair, T.R. (Eds), *Limitations to Efficient Water Use in Crop Production*. American Society of Agronomy, Crop Science Society of America, and Soil Science Society of America, Madison, WI. Pp. 393–411.

Hanson, A.D. and Hitz, W.D. 1983. Whole-plant response to water deficits: water deficits and the nitrogen economy. *In*: Taylor, H.M., Jordan, W.R. and Sinclair, T.R. (Eds), *Limitations to Efficient Water Use in Crop Production*. American Society of Agronomy, Crop Science Society of America, and Soil Science Society of America, Madison, WI. Pp. 331–343.

Haque, I. and Tothill, J.C. 1987. Forages and pastures in mixed farming systems of sub-Saharan Africa. *In*: Latham, M., Ahn, P. and Elliott, C.R. (Eds), *Land Development and Management of Acid Soils in Africa*, Proceedings of the Second Regional Workshop on Land Development and Management of Acid Soils in Africa, Lusaka and Kasama, Zambia, 9–16 April 1987. IBSRAM Proceedings No. 7. Pp. 107–131.

Harrison, P. 1987. *The Greening of Africa*. An International Institute for Environment and Development – Earthscan Study. Paladin Grafton Books, London.

Harte, A.J. 1984. Effect of tillage on the stability of three red soils of the northern wheat belt. *J. Soil Cons. NSW* **40**:94–101.

Harte, A.J., Enright, N.F. and Watt, L.A. 1984. Soil loss measuring simplified. *J. Soil Cons. NSW* **24**:62–63.

Harte, A.J., Thomas, E.C. and Packer, I.J. undated. Manipulation of soil structure for soil conservation – a review. *Soil Conservation Service NSW*, Sydney, Australia.

Hartley, R.E., Maschmedt, D.J. and Chittleborough, D.J. 1984. Land management – key to water quality control. *Water* **11**:18–21.

Hartshorn, G., Hartshorn, L., Atmella, A., Gomez, L.D., Mata, A., Mata, L., Morales, R., Ocamp, R., Pool, D., Quesada, C., Solera, C., Stiles, G., Tosi, J., Umana, A., Villalobos, C. and Wells, R. 1982. *Costa Rica. Country Environmental Profile, A Field Study*. Tropical Science Center, Costa Rica.

Hartwig, N.L. 1984. Crownvetch and no-tillage crop production for soil erosion control. *Penn State Univ. Coop. Ext. Serv. Publ.* USA.

Hatch, T. 1981. Preliminary results of soil erosion and conservation trials under pepper (*Piper nigrum*) in Sarawak, Malaysia. *In*: Morgan, R.P.C. (Ed.), *Soil Conservation, Problems and Prospects*. John Wiley & Sons, Chichester, UK. Pp. 255–262.

Hawtrey, K. 1987. Outlook for the rural economy. *In*: *Proceedings Soil Conservation Service NSW*. Annual Conference, SCS, Sydney, Australia. Pp. 224–237.

HCH (Hydrodynamic Committee of Haihe). 1984. Adding to the achievements and making further development. *Soil and Water Conservation in China (Zhongguo Shuitubaochi)* **9**:13–16.

Heilman, M.D. and Thomas, J.R. 1961. Land levelling can adversely affect soil fertility. *J. Soil. Water Conserv.* **16**:71–72.

Hendrickson, B.H., Barnett, A.P., Carreker, J.R. and Adams, W.E. 1963. Runoff and erosion control studies on Cecil soil in the southern Piedmont. *US Dept. Agric., Tech. Bull. No.* 1281.

Henrickson, B., Sultan Tilimo, Ross, S., Wijintie-Bruggeman, H.Y. and Fitssum Fitwe. 1983. Provisional soil depth map of Ethiopia. LUPRD, Addis Abeba.

Hensel, W. and Tabaczynski, S. 1978. *Rewolucja neolityczna i jej znaczenie dla rozvoju kultury europejskiej*. Ossolineum, Wroclaw. 157 pp.

Highfill, R. and Kimberlin, L. 1977. Current soil erosion and sediment control technology for rural and urban lands. *In*: *Proc. Natl. Symp. on Soil Erosion and Sedimentation by Water*. Palmer House, Chicago, IL. American Society of Agricultural Engineers. Pp. 14–22.

Holden, D.W. 1989. Miracle grass takes root. *Development Forum*, Nov.–Dec. 1989.

Hoogmoed, W.B. 1986. Crusting and sealing problems on West African soils. *In*: Callebaut, F., Gabriels, D. and De Boodt, M. (Eds), *Assessment of Soil Surface Sealing and Crusting*. University of Ghent, Belgium. Pp. 48–55.

Hooper, M.D. 1987. Conservation interest in plants of field margins. *In*: Way, J.M. and Greig-Smith, P.W. (Eds), *Field Margins*. BCPC Monog., British Crop Protection Council, Croydon, UK. **35**:49–52.

Hopkinson, J.W. 1927. Studies on the vegetation of Nottinghamshire. The ecology of the Bunter Sandstone. *J. Ecol.* **15**:130–171.

Hore, H.L. and Sims, H.L. 1954. Loss of topsoil – effect on yield and quality of wheat. *J. Agric. Victoria* **52**:241–250.

Hou Guangjun and Gao Huimin. 1982. *An Introduction to Agricultural Soils in China*. Agricultural Press, Beijing. 541 pp.

House of Lords Select Committee on the European Communities. 1984. *Agriculture and the Environment*. HMSO, London.

House of Lords Select Committee on Science and Technology. 1984. *Agricultural and Environmental Research Vol. 2, 4th Report, Session 1983–84*. HMSO, London.

Huang Yong. 1987. The suspending river above ground – a serious problem. *Science and Technology Daily (Keji Ribao)*, 20 Oct. 1987, p. 2.

Huanghe Committee Tianshui Station. 1983. Shuitu Baochi Gengzufa De Yanjiu He Pingjia. *Quanguo Shuitu Baochi Gengzuo Xueshu Taolun Hui Ziliao Ji Bian.* Pp. 133–139.

Huanghe Committee Xifeng Station. 1982. Huangtu Gaoyuan Goubao Qu Yong Shao Gong, Shou Xiao Kuai De Shuitu Baochi Cuoshi. *Shuitu Baochi Shiyan Yanjiu Chengguo Huibian 1952–1980.* Pp. 55–64.

Hudson, N.W. 1971. *Soil Conservation*. Batsford, London.

Hudson, N.W. 1981. *Soil Conservation*. 2nd edition, Cornell University Press, Ithaca, NY.

Hudson, N.W. 1982. Soil conservation, research and training requirements in developing tropical countries. *In: Soil Erosion and Conservation in the Tropics*. Am. Soc. Agron. Spec. Publ. No. 43, Madison, WI. Pp. 121–133.

Humphreys, G.S. and Mitchell, P.B. 1983. A preliminary assessment of the role of bioturbation and rainwash on sandstone hillslopes in the Sydney Basin. *In*: Young, R.W. and Nanson, G.C. (Eds), *Aspects of Australian Sandstone Landscapes*. Aust. and NZ Geomorph. Group Special Pub. No. 1, University of Wollongong, Wollongong, Australia.

Hunt, J.S. 1980. Structural stability of mallee soils under cultivation. *J. Soil Cons. NSW* **36**:16–22.

Hurni, H. 1987a. *Erosion – Productivity – Conservation Systems in Ethiopia*. Proceedings of the IV International Conference on Soil Conservation, Venezuela (1985), 20 pp.

Hurni, H. 1987b. Merhabete Integrated Conservation and Development Programme. Programme Document, 'Menschen für Menschen' Foundation and Ministry of Agriculture, Addis Abeba, 51 pp.

Hurni, H. 1988. Degradation and conservation of the soil resources in the Ethiopian highlands. *Mountain Research and Development* **8**(2–3):123–130.

Hurni, H. and Messerli, B. 1981. Mountain research for conservation and development in Simen, Ethiopia. *Mountain Research and Development* **1**(1):49–54.

Hurni, H. and Nuntapong, S. 1983. Agro-forestry improvements for shifting cultivation systems, soil conservation research in northern Thailand. *Mountain Research and Development* **3**:338–345.

IFS (Institute of Forestry Science, Chinese Academy of Forestry Sciences). 1985. *The Prediction of Forest Development and its Environmental Benefits in China in the Year 2000*. Chinese Academy of Forestry Sciences, Beijing. 117 pp.

IGSP (Investigation Group of Shichuan Province). 1987. Establishing the water conservation area in the upper reach of Yangtze river valley as soon as possible. *Bulletin of Soil and Water Conservation (Shuitubaochi Tongbao)* **7**(2):62–64.

IITA. 1973. *1972–73 Report*. International Institute of Tropical Agriculture, Ibadan, Nigeria.

IITA. 1985. *Alley Cropping: A Stable Alternative to Shifting Cultivation*. Ibadan, Nigeria.

INTA (Instituto Nacional de Techologia Agropecuaria). 1957. Use and soil conservation in Argentina. *IDIA* (Buenos Aires) **114**:1–47. (In Spanish.)

INTA. 1958. Soil erosion in the west basin of Carcarana River. *IDIA* **120**:1–28. (In Spanish.)

INTA. 1961. Drought problems in the semi-arid Pampas. *IDIA* **155**:1–69. (In Spanish.)

INTA. 1979. *International Symposium on Soil Conservation in the River Plate Basin*. Academia Nacional de Agronomica y Veterinaria. 109 pp. (In Spanish.)

INTA. 1983. Soil conservation, water conservation and floods in the River Plate Basin (umbrella operation). *IDIA Suppl.* No. 40. 111 pp. (In Spanish.)

IUCN. 1985. Africa: prophecy of plenty. *International Union for Conservation of Nature Bull.* **16**:41.

IUCN. 1986. *The IUCN Sahel Report. A Long-Term Strategy for Environmental Rehabilitation.* International Union for the Conservation of Nature and Natural Resources, Gland, Switzerland.

Jacks, G.V. 1954. *Soil.* T. Nelson and Sons, London.

Jacks, G.V. and Whyte, R.O. 1939. *Vanishing Lands. A World Survey of Soil Erosion.* Doubleday, Doran and Company, New York.

Jackson, S.J. 1987. Soil erosion survey and soil erosion risk in mid-Bedfordshire. *In*: Burnham, C.P. and Pitman, J.I. (Eds), *Soil Erosion*, see *Soil* **3**:95–104.

Jacob, A. 1981. Watershed management: new legislation. *Indian J. Soil Cons.* **9**(2/3):154–163.

Jege, L.K. and Agu, A.N. 1982. Runoff and soil loss from erosion plots in the Ife area of SW Nigeria. *Geo. Eco. Trop.* **6**:161–181.

Jenkinson, D.S. and Ladd, J.N. 1981. Microbial biomass in soil measurement and turnover. *In*: Paul, E.A. and Ladd, J.N. (Eds), *Soil Biochemistry*. Vol. 5, Marcel Dekker, New York and Basel. Pp. 415–471.

Jenny, H. 1941. *Factors of Soil Formation.* McGraw-Hill, New York and London.

Jiang Dechi and Zhu Xianmo. 1962. Water and soil conservation. *In*: Institute of Soil and Fertilizers, Chinese Academy of Agricultural Sciences, *Proc. Agricultural Soils in China.* Science and Technology Press, Shanghai. Pp. 384–440.

Jing Jiemin. 1988. The progress in the experiments of comprehensive controls in the Loess Plateau. *Renmin Ribao*, 29 March 1988, p. 4.

Jing Ke. 1986. The relations between soil erosion and geographical environment in the middle reach of Yellow River. *Geography and Territorial Research (Dilixue He Guetu Yanjou)* **2**(1):26–32.

Jing Ke and Chen Yongzhong. 1983. The preliminary studies on the erosion environment and erosion rates in the Loess Plateau. *Geographical Research (Diliyanjou)* **2**(2):7–10.

Johnson, R.R. 1988. Research. *Journal of Productivity in Agriculture* **1**(1):5–12.

Johnston, A.E. and Mattingly, G.E.G. 1976. Experiments on the continuous growth of arable crops at Rothamstead and Woburn Experimental Stations: effects of treatments on crop yields and soil analyses and recent modifications in purpose and design. *Ann. Agron.* **27**:927–956.

Jordan, W.R. 1983. Whole plant response to water deficits: an overview. *In*: Taylor, H.M., Jordan, W.R. and Sinclair, T.R. (Eds), *Limitations to Efficient Water Use in Crop Production.* American Society of Agronomy, Crop Science Society of America, and Soil Science Society of America, Madison, WI. Pp. 289–317.

Josefaciuk, C.Z. and Josefaciuk, A. 1987. Attempt of soil erosion classification taking into consideration utilitarian purposes. *Roczniki Gleboznawcze* **38**:27–35.

Josefaciuk, C.Z. and Josefaciuk, A. 1988. Strategy of ecosystem and agricultural landscape protection against erosion. Manuscript of report presented for Strategy of Living Resources Conservation in Poland.

Kalm, J.M. 1977. Studies of cultivation techniques at Bouake, Ivory Coast. *In*: Greenland, D.J. and Lal, R. (Eds), *Soil Conservation and Management in the Humid Tropics.* J. Wiley & Sons, UK.

Kanodia, K.C. and Singh, P. 1987. *Prospects of Khas Grass.* Indian Grassland and Fodder Research Institute, Jhansi.

Kattan, Z., Jry Gac and Probst, J.L. 1987. Suspended sediment load and mechanical erosion in the Senegal Basin – estimation of the surface runoff concentration and relative contributions of channel and slope erosion. *J. Hydrology* **92**:59–76.

Kent, R. 1988. Some reflections on the nature of famine emergencies and appropriate preparedness measures. UN–EPPG, Addis Abeba, 40 pp.

Kerin, J. 1987. Land degradation and government. *In*: Chisholm, A.H. and Dumsday, R.G. (Eds), *Land Degradation: Problems and Policies*. Cambridge University Press and CRES, Canberra, Australia.

Ketcheson, J. 1977. Conservation tillage in eastern Canada. *J. Soil Water Cons.* **32**:57–60.

Ketcheson, J.W. and Onderdonk, J.J. 1973. Effect of corn stover on phosphorus in runoff from nontilled soil. *Agron. J.* **65**:69–71.

Khoshoo, T.N. 1982. *Energy from Plants: Problems and Prospects*. Presidential address Botany section. Sixty-ninth Session Indian Science Congress, Calcutta. Pp. 1–112.

Khoshoo, T.N. 1986. *Environmental Priorities in India and Sustainable Development*. Presidential address to the Indian Science Congress Assocs. New Delhi. Pp. i–xxiv and 1–224.

Khoshoo, T.N. 1989. Forestry in India: problems, prospects and the role of tissue culture. *In*: Dhawan V. (Ed.), *Application of Biotechnology in Forestry and Horticulture*. Plenum Press, New York. Pp. 22–24.

Khoshoo, T.N. 1990. Degraded lands for agroecosystems: some policy issues. *In*: Wali, M.K. (Ed.), *Environmental Rehabilitation*. SPB Academic Publishing, The Hague, The Netherlands.

Khosla, P.K. 1983. *Place of Forestry Education in Agricultural Universities*. 12th Convention of Indian Agric. Univ. Ass.

Kinnell, P.I.A. 1983. The effect of kinetic energy of excess rainfall on soil loss from non-vegetated plots. *Australian J. Soil Res.* **21**:445–453.

Klausner, S.D., Zwerman, P.J. and Ellis, D.F. 1974. Surface runoff losses of soluble nitrogen and phosphorus under two systems of soil management. *J. Environ. Qual.* **3**:42–46.

Kovda, V.A. 1983. Loss of productive land due to salinization. *Ambio* **12**:91–93.

Kovda, V.A., Rozanov, B.G. and Onishenko, S.K. 1978. On probability of droughts and secondary salinisation of world soils. *In*: Worthington, E.G. (Ed.), *Arid Land Irrigation in Developing Countries*. Pergamon Press, London. Pp. 237–238.

Kowal, J. 1972. The hydrology of a small catchment basin at Samaru, Nigeria. IV. Assessment of soil erosion under varied land management and vegetation cover. *Nigerian Agric. J.* **7**:143–147.

Kuhnelt, W. 1976. *Soil Biology, With Special Reference to the Animal Kingdom*. Michigan State University Press, East Lansing, MI.

Kunwar, S.S. 1982. *Hugging the Himalayas*. Dasholi Gram Swarajya Mandal, Gopeshwar. 102 pp.

Kussow, W., El-Swaify, S.A. and Mannering, J. 1982. Soil erosion and conservation in the tropics. *Amer. Soc. Agron. Special Publication* **43**. ASA, Madison, Wisconsin.

Lal, R. 1976. *Soil Erosion Problems on an Alfisol in Western Nigeria and their Control*. IITA Monograph No. 1, October.

Lal, R. 1979. Physical characteristics of the soils of the Tropics: Determination and management. *In*: Lal, R. and Greenland, D.J. (Eds), *Soil Physical Properties and Crop Production in the Tropics*. J. Wiley & Sons, UK. Pp. 7–40.

Lal, R. 1980. Soil erosion problems on Alfisols in Western Nigeria. VI. Effects of erosion on experimental plots. *Geoderma* **25**:215–237.

Lal, R. 1982. Effective conservation farming systems for the humid tropics. In: Kussow, W. *et al.* (Eds), Soil erosion and conservation in the tropics. *Amer. Soc. Agron. Special Publication* **43**. ASA, Madison, Wisconsin.

Lal, R. 1984a. Productivity assessment of tropical soils and the effects of erosion. *In*: Rijsberman, F.R. and Wolman, M.G. (Eds), *Quantification of the Effect of Erosion on Soil Productivity in an International Context*. Delft Hydraulics Laboratory, Delft, Netherlands. Pp. 70–94.

Lal, R. 1984b. Soil erosion from tropical arable lands and its control. *Adv. Agron.* **37**:183–248.

Lal, R. 1984c. Mechanized tillage systems effect on soil erosion from an Alfisol in watersheds cropped to maize. *Soil Tillage Res*. **4**:349–360.

Lal, R. 1986. Soil surface management in the tropics for intensive land use and high, sustained production. *Adv. Soil Sci*. **5**:1–105.

Lal, R. 1987a. Managing the soils of sub-Saharan Africa. *Science* **236**:1069–1076.

Lal, R. 1987b. Surface soil degradation and management strategies for sustained productivity in the tropics. *In*: *Management of Acid Tropical Soils for Sustainable Agriculture*. IBSRAM, Bangkok, Thailand.

Lal, R. 1987c. Need for, approaches to, and consequences of land clearing and development in the tropics. *In*: Lal, R., Nelson, M., Scharpenseel, H.W., Sudjadi, M., Garver, C.L. and Niamskul, C. (Eds), *Tropical Land Clearing for Sustainable Agriculture: Proceedings of an IBSRAM Inaugural Workshop*. Bangkok, Thailand. Pp. 15–27.

Lal, R. 1990. *Soil Erosion in the Tropics*. McGraw-Hill, Inc., New York.

Lal, R. 1991. Myths and scientific realities of agroforestry as a strategy for sustainable management for soils in the tropics. *Adv. Soil Sci*. **15**:91–139.

Lal, R. and Kang, B.T. 1982. *Management of Organic Matter in Soils of the Tropics and Subtropics*. XIIth Congress of the International Society of Soil Science, New Delhi, India.

Lang, R.D. 1984. Temporal variations in catchment erosion and implications for estimating erosion rates. *In:* Loughran, R.J. (Ed.), *Drainage Basin Erosion And Sedimentation*. Vol. 1, University of Newcastle, Newcastle, Australia. Pp. 43–50.

Langdale, G.W., Denton, H.P., White, A.W., Gilliam, J.W. and Frye, W.W. 1985. Effects of soil erosion on crop productivity of southern soils. *In*: Follett, R.F. and Stewart, B.A. (Eds), *Soil Erosion and Crop Productivity*. American Society of Agronomy, Crop Science Society of America, and Soil Science Society of America. Madison, WI. Pp. 252–271.

Larson, W.E. 1981. Protecting the soil resource base. *J. Soil Water Cons*. **36**:13–16.

Larson, W.E., Pierce, F.J. and Dowdy, R.H. 1983. The threat of soil erosion to long-term crop production. *Science* **219**:458–465.

Larson, W.E., Pierce, F.J. and Dowdy, R.H. 1985. Loss in long-term productivity from soil erosion in the United States. *In*: El-Swaify, S.A. *et al.*, **24**:262–271.

Latham, M.C. 1984. International nutrition and problems and policies. *In*: *World Food Issues*, Centre for Analysis of World Food Issues, International Agriculture, Cornell University, Ithaca, NY.

Latham, M. 1988. *Soil conservation and soil management: an IBSRAM view. In*: Rimwanich, S. (Ed.).

Lawes, D.A. 1962. The influence of rainfall conservation on the fertility of the loess plain soils of northern Nigeria. *J. Geog. Assoc. of Nigeria* **5**:33–38.

Lee, L.K. 1980. The impact of land ownership factors on soil conservation. *Am. J. Agr. Econ*. **62**:1070–1076.

Lee, M.T., Narayanon, A.S. and Swanson, E.R. 1974. Economic analysis of erosion and sedimentation, Sevenmile Southwest Branch Watershed. University of Illinois, Dept. of Agricultural Economics. *Agr. Econ. Res. Rep. No*. **130**.

Lei Chidi. 1983. *The Forest Resources and the Forestry Development in Kazhou County of Liaoning Province*. The Investigation Team of Natural Resources in Kazhou County of Liaoning Province, Shenyang. 44 pp.

Li Kaiming. 1984. Studies on the patterns of environment, ecology and economy in the Loess Plateau. *Agricultural Developmental Exploration (Nongye Fazhan Tantao)* **8**:15–20.

Li Song. 1987. Longdong Ban Ganhan Diqu Turang Yangfen Liushi De Chubu Yanjiu. *Zhongguo Shuitubaochi* **68**:31–33.

Limpinuntana, V. and Arunin, S. 1986. Salt affected land in Thailand and its agricultural productivity. *In*: Barrett-Lennard, E.G., Malcolm, C.V., Stern, W.R. and Wilkins, S.M.

(Eds), *Forage and Fuel Production from Salt Affected Wasteland*. Proceedings of a seminar held at Cunderdin, Western Australia, 19–27 May, 1984. Elsevier Science Publishers B.V., The Netherlands. Pp. 143–149.

Liu De. 1983. Shandong Sheng Shuitu Baochi Gengzuo Fazhan Lishe, Xianzhuang Ji Qi Shexiang. *Quanguo Shuitu Baochi Gengzuo Xueshu Taolun Hui Ziliao Ji Bian*. Pp. 115–120.

Liu Shuangjin, Shong Qunheng, Mao Yongjiu, Li Yebe, Shen Yongqing and Fu Lishun. 1987. *Developing Trends and Improving Strategies of Ecological Environment in China*. Chinese Committee of Sciences and Technologies, Beijing. 84 pp.

Liu Wanquan. 1986. The development of soil conservation sciences in Yellow river valley. *Soil and Water Conservation in China (Zhongguo Shuitubaochi)* **10**:8–11.

Liu Xiang. 1986. The viewpoint summary in the symposium of developing economy and culture in poverty regions. *Problems of Agricultural Economy (Nongye Jingji Wenti)* **10**:58–61.

Liu Yingqiu. 1988. Exploration on the way of reasonable development and utilization of land. *Geography and Territorial Research (Dilixue He Guetu Yanjou)* **4**(1):26–32.

Lo, A. and El-Swaify, S.A. 1985. Rainfall erosivity in a benchmark soils network. *In*: Silva, J.A. (Ed.), *Soil-based Agrotechnology Transfer*. HITAHR, University of Hawaii, USA. Pp. 160–175.

Logan, T.J. 1977. Establishing soil loss and sediment yield limits for agricultural land. *In*: *Soil Erosion and Sedimentation*. Proc. of the Natl. Symp. on Soil Erosion and Sedimentation by Water. ASAE Publ. 4–77. Pp. 59–68.

Lootens, M. and Lumbu, S. 1986. Suspended sediment production in a suburban tropical basin, Lubumbashi, Zaire. *Hydrological Sciences J.* **31**:39–49.

Łos, M.J. 1985. Erosion of the lower part of the Bzstrzyca river-bed. *Zeszyty Problemowe Postepow Nauk Rolniczych* **311**:23–36.

Lowdermilk, W.C. 1953. Conquest of the land through seven thousand years. *US Agric. Info. Bull. No.* **99**.

Lucas, R.E., Holtman, J.B. and Connor, L.J. 1977. Soil carbon dynamics and cropping practices. *In*: Lockeretz, W. (Ed.), *Agriculture and Energy*. Academic Press, New York. Pp. 333–351.

LUPRD (Land Use Planning and Regulatory Department). Various Reports. Ministry of Agriculture, Addis Abeba.

Lyttleton, T., Lyon, T.L. and Buckman, H.O. 1943. *The Nature and Properties of Soils*. Macmillan, New York (4th edition revised by H.O. Buckman).

Ma Buzhen, Cao Zhirong and Li Denggui. 1986. Guanyu Ziran Xhibei Pohuai Qingkuang De Diaocha. *Shuitu Baochi Shiyan Yanjiu Chengguo Ziliao Xuanbian 1956–1985*. (Dingxi Area Soil Conservation Station). Pp. 369–372.

Ma Zijun. 1987. 'Soil Conservation in the Loess Plateau and Gansu Province,' a speech given on 5 May 1987. Gansu Academia Sinica Biological Research Institute.

Maass, J.M., Jordan, C.F. and Sarukhan, J.F. 1988. Soil erosion and nutrient losses in seasonal tropical agroecosystems under various management techniques. *J. Appl. Ecol.* **25**: 595–607.

Mabbutt, J.A. 1984. A new global assessment of the status and trends of desertification. *Environ. Conserv.* **11**:103–113.

MAFF (Ministry of Agriculture, Fisheries and Food). 1973. *Agricultural Statistics, UK, 1971*. HMSO, London.

MAFF. 1984a. *Soil Erosion by Water*. Leaflet 890. HMSO, London.

MAFF. 1984b. *Agricultural Statistics, UK, 1983*. HMSO, London.

MAFF. 1985. *Soil Erosion by Wind*. Leaflet 891. HMSO, London.

MAFF. 1986. *Agricultural Statistics, UK, 1985*. HMSO, London.

Mannering, J.V. 1981. The use of soil loss tolerances as a strategy for soil conservation. *In*: Morgan, R.P.C. (Ed.), *Soil Conservation: Problems and Prospects*. Wiley, Chichester, UK. Pp. 337–349.

Marques, J.L. 1949. *First South American Congress of Investigators in Agronomic Problems*. Colonia (R. O. del Uruguay).

Marsh, B. 1982. The effect of soil loss on productivity. Western Australian Dept. Agriculture, Perth, Australia. (Unpublished.)

Maruszaczak, H. 1973. Gully erosion in the eastern part of the southern uplands of Poland. *Zeszyty Problemowe Postepow Nauk Rolniczych* **151**:15–30.

Maruszaczak, H. 1985. Zmiany srodowiska przyrodniczego ziem polskich w ostatnim dziesie-cioleciu. *Problemy* **9**:14–19.

Mathur, H.N., Rambabu, Joshie, P. and Singh, B. 1976. Effect of clear felling and reforestation on runoff and peak rate of small watersheds. *Indian Forester* **102**(4):219–226.

Mathur, H.N. *et al.*, 1979. Benefit:cost ratio of fuel cum fodder plantation in Doon Valley. *Indian J. Soil Cons.* **7**(2):53–58.

Mazur, Z. 1985. Differentiation of soils and yields on loess slopes at Jastkow. *Zeszyty Problemowe Postepow Nauk Rolniczych* **292**:65–80.

Mazur, Z. 1988. Variation of soils and crops in a Wavg loess area. *Zeszyty Problemowe Postepow Nauk Rolniczych* **357**:79–94.

Mazur, Z. and Orlik, T. 1985. Crop of some plants under differentiated condition of eroded loess area. *Zeszyty Problemowe Postepow Nauk Rolniczych* **311**:53–71.

Mazur, Z. and Pałys, S. 1985. Water erosion effect on morphology and variability of soil cover in loess areas. *Zeszyty Problemowe Postepow Nauk Rolniczych* **292**:21–37.

Mbagwu, J.S.C. 1982. Studies on soil-loss – productivity relationships of alfisols and ultisols in southern Nigeria. PhD Thesis, Cornell University, USA.

McCalla, T.M. 1942. Influence of biological products on soil structure and infiltration. *Proc. Soil Sci. Soc. Am.* **7**:209–214.

McCormack, D.E., Young, K.K. and Kimberlim, L.W. 1982. Current criteria for determining soil loss tolerance. *ASA Spec. Publ. No.* **45**, American Society of Agronomy, Madison, WI.

McCown, R.L., Jones, R.K. and Peake, D.C.I. 1985. Evaluation of a no-till, tropical legume ley-farming strategy. *In*: Muchow, R.C. (Ed.), *Agro-research for the Semi-arid Tropics: North-west Australia*. University of Queensland Press, Brisbane, Australia. Pp. 450–469.

McFarlane, D.J. undated. Water erosion on potato land during the 1983 growing season Donnybrook. Soil Conservation Branch, Division of Resource Management. *Tech. Report No. 26*. Western Australia Dept. Agriculture, Perth, Australia.

McFarlane, D.J. and Ryder, A.T. 1986. Report on water erosion in the Irishtown–Wongamine area north of Northam during June, 1986. Soil Conservation Branch, Division of Resource Management. *Internal Report 1317/85*, Western Australia Dept. Agriculture, Perth, Australia.

McGarity, J.W. and Storrier, R.R. 1986. Fertilizer. *In*: Russell, J.S. and Isbell, R.F. (Eds), *Australian Soils: The Human Impact*. University of Queensland Press, St Lucia, Australia. Pp. 304–333.

McIntyre, D.S. 1955. Effect of soil structure on wheat germination in a red-brown earth. *Australian J. Agric. Res.* **6**:797–803.

McLaughlin, L. 1988. *Soil Conservation and Rural Development in Dingxi County, Gansu Province, People's Republic of China*. Nanjing University Press, Nanjing, PRC. 193 pp.

McVean, D.N. and Lockie, J.D. 1969. *Ecology and Land Use in Upland Scotland*. Edinburgh University Press, Edinburgh.

Mensah-Bonsu and Obeng, H.B. 1979. A study of runoff and its effects on the water balance in rainfed agriculture in the Upper Volta. *In*: Lal, R. and Greenland, D.J. (Eds), *Soil Properties and Cash Crop Production in the Tropics*. J. Wiley & Sons, Chichester, UK. Pp. 509–520.

Mesfin Wolde-Mariam. 1984. *Rural Vulnerability to Famine in Ethiopia, 1958–1977*. Vikas Publishing House in association with Addis Abeba University. 191 pp.

Messer, J. 1987. The sociology and politics of land degradation in Australia. *In*: Blaikie, P.M. and Brookfield, H.C. (Eds), *Land Degradation and Society*, Methuen and Co., London. Pp. 232–238.

Messerli, B. and Messerli, P. 1978. Wirtschaftliche Entwicklung und ökologische Belastbarkeit im Berggebiet (MAB Schweiz). *Geographica Helvetica* 4:203–210.

Mishra, P.R., Kaushal, R.C., Dayal, S.K.N. and Prabhu Shankar. 1975. *Annual Report*. Central Soil and Water Conservation Research and Training Institute, Dehradun, India.

Miti, T. and Carlson, T. 1972. Vertical and areal distribution of Saharan dust over the western equatorial North Atlantic Ocean. *J. Geophys. Res.* **77**:5255–5265.

Miti, T., Soyer J. and Aloni, K. 1984. Splash erosion in tropical semi-natural ecosystems (Shabe, Zaire). *Zeitschrift für Geomorphologie Supplement Band*. **49**:75–86.

Mo Shiyu. 1981. Mucao Dui Shuitu Baochi Zuoyong De Chubu Fenxi. *Shuitu Baochi Shiyan Yanjiu Chengguo Huibian 1942–1980* **3**:32–38.

MOA. 1968. *Report of the Working Group for Formulation of the Fourth Five Year Plan Proposal for Land and Water Development*. Ministry of Agriculture, New Delhi, India.

MOA. 1984. *Report of the Working Group on Fisheries for Formulation of the Seventh Five Year Plan (1985–90)*. MOA, New Delhi, India.

MOA. 1985. *Indian Agriculture in Brief*. 20th edition, MOA, New Delhi, India.

Moldenhauer, W.C. 1979. Erosion control obtainable under conservation practices. *In*: Peterson, A.E. and Swan, J.B. (Eds), *Universal Soil Loss Equation: Past, Present, and Future*. Soil Science Society of America, Madison, WI. Pp. 33–43.

Moldenhauer, W.C., Lovely, W.G., Swanson, N.P. and Currence, H.D. 1971. Effect of row and tillage systems on soil and water losses. *J. Soil Water Cons.* **26**:193–195.

Molina, J.S. 1963. A new method for the study of *Azotobacter* directly in the soil. *Soil Biology (International News Bulletin). Intl. Soc. Soil Sci.* **3**:19–21.

Molina, J.S. 1968. La decomposition aerobie de la cellulose et la structure active des sols (Resume de 20 ans de recherches). *Ann. Inst. Pasteur* **115**:604–609.

Molina, J.S. 1970. *Papel del* Azotobacter *en la recuperacion biologica de suelos sodicos* (*Utilization of* Azotobacter *in the reclamation of sodic soils*). X International Congress of Microbiology, Mexico.

Molina, J.S. 1973. *Aerobic Decomposition of Cellulose and Production of Polyuronic Colloids*. Presented at GIAM IV (Global Impacts of Applied Microbiology), Sao Paulo, Brazil.

Molina, J.S. 1974. *Reclamation of Sodic Soils by Methods in the Ecology of Soil Microbes*. X International Congress of Soil Science (Moscow).

Molina, J.S. 1976. *Report on Soil Management, Soil Erosion and Soil Conservation in the South of Brazil*. Presented to FAO.

Molina, J.S. 1977. Agua de lluvia para el ganado bovino. *Ciencia e Investigación* **33**:86–97.

Molina, J.S. 1978. Retrieving a world from the Stone Age. *In*: Stone, G.B. (Ed.), *The Spirit of Enterprise* (Rolex International Awards). W.H. Freeman and Co., San Francisco. Pp. 70–77.

Molina, J.S. 1980. *Una nueva conquista del Desierto (A new conquest of the desert)*. Emece Editors, Buenos Aires. (In Spanish.)

Molina, J.S. 1981a. *Hacia una nueva agricultura (Towards a new agriculture)*. Editorial El Ateneo, Buenos Aires. (In Spanish.)

Molina, J.S. 1981b. Domesticating soil microbes to beat the problem of soil erosion. *In*: Stone, G.B. (Ed.), *The Spirit of Enterprise* (Rolex International Awards). Haraps, London.

Molina, J.S. 1986. *Tranqueras Abiertas (Open Gates)*. Editorial El Ateneo, Buenos Aires. (In Spanish.)

Molina, J.S. and Delorenzini, C. 1978. *Alfalfa: fijacion de nitrogeno y decadencia de los alfalfares. (Alfalfa: nitrogen fixation and its decline)*. Edicion Especial Bolsa de Cereales de Buenos Aires. Pp. 181–224. (In Spanish.)

Molina, J.S. and Sauberan, C. 1956. Cellulose bacteria and soil conditioners. *J. Soil Assoc.* (Mother Earth), England **9**:415–417.

Molina, J.S. and Sauberan, C. 1957. Availability of phosphorus in soils. *J. Soil Assoc.* (Mother Earth), England **10**:325–329.

Molina, J.S. and Sauberan, C. 1965. Reclamation of sodic soils by biological methods. *Agrochemistry and Soil Science, Suplementum Symposium on Sodic Soils (UNESCO) Budapest* **14**:411–414.

Molina, J.S. and Spaini, L.S. 1948. Influencia de un coloide organico sobre la estabilidad en agua de diversos suelos argentinos. (Influence of an organic colloid upon the water stability of different Argentine soils.) *Rev. Arg. Agron.* **15**:113-116. (In Spanish.)

Molina, J.S. and Spaini, L.S. 1949. Coloides producidos en la descomposicion aerobia de la celulosa y su influencia sobre la estructura del suelo. (Influence of organic colloids produced in the aerobic decomposition of cellulose and their influence upon soil structure.) *Rev. Arg. Agron.* **16**:33–49. (In Spanish.)

Molina, J.S. and Spaini, L.S. 1950. Colloid production during aerobic decomposition of cellulose and their action upon the structure of different soil types. V International Congress Microbiology. *Arquivos* **2**:594–601.

Molina, J.S. *et al*. 1974. El plan florentino ameghino. Resultados de un plan intgegral de lucha contra las inundaciones, la sequia y la erosion de los suelos. *Ciencia e Investigación* (Buenos Aires) **30**:313–326.

Molnar, R.I. 1964. Soil conservation – economic and social considerations. *J. Australian Instit. Agric. Sci.* **30**:247–257.

Morgan, R.P.C. 1974. Nature provides, man erodes. *Geog. Mag.* **46**:528–535.

Morgan, R.P.C. 1979. *Soil Erosion*. Longman, London.

Morgan, R.P.C. 1980. Soil erosion and conservation in Britain. *Prog. in Phys. Geog.* **4**:24–47.

Morgan, R.P.C. 1985. Assessment of soil erosion risk in England and Wales. *Soil Use and Management* **1**:127–131.

Morgan, R.P.C. 1986. Soil erosion in Britain: the loss of a resource. *The Ecologist* **16**(1):40–41.

Morgan, R.P.C. 1987. Sensitivity of European soils to ultimate physical degradation. *In*: Barth, H. and L'Hermite, P. (Eds), *Scientific Basis for Soil Protection in the European Community*. Proc. of Commission for the European Communities Symp. Oct. 1986, Berlin. Elsevier, London.

Mosley, G. 1984. Agriculture and conservation – a conservation viewpoint. *In*: Gilmour, D., Hamer, I. and Bouchier, J. (Eds), *Agriculture and Conservation: Achieving a Balance*. Proceedings of a conference at Clyde Cameron College, Wodonga, Australia. Pp. 78–81.

Murray–Darling Basin Ministerial Council. 1987. *Murray–Darling Basin Environmental Resources Study*, Canberra, Australia.

Murthy, B.N. 1980. Sedimentation studies in reservoirs. *Tech. Report* **20**, Vol I. Central Board of Irrigation and Power, New Delhi, India.

Musgrave, G.W. and Neal, O.R. 1937. Rainfall and relative losses in various forms. *Trans. Am. Geophysical Union* **18**:349–355.

Musto, J.C. 1979. La erosion hidrica en el ambito argentino de la Cuenca del Plata. Relatos Simposio Internacional 'La erosion del Suelo en la Cuenca del Plata' (INTA), *IDIA*, Nov. 1979. (In Spanish.)

Myers, B.A. 1950. Flatland conservation. *Agric. Eng.* **31**:78–79.

NADC (National Agricultural Division Committee). 1981. *Agricultural Division of China.* Agricultural Press, Beijing. 334 pp.

Nair, P.K.R. 1987. Agroforestry in the context of land clearing and development in the tropics. *In*: Lal, R., Nelson, M., Scharpenseel, H.W., Sudjadi, M., Garver, C.L. and Niamskul, C. (Eds), *Tropical Land Clearing for Sustainable Agriculture: Proceedings of an IBSRAM inaugural Workshop*. Bangkok, Thailand. Pp. 29–44.

Naklicki J. 1971. The effect of gully vicinity on soil moisture and plant yields on the example of Slawin. *Zeszyty Problemowe Postepow Nauk Rolniczych* **119**:135–141.

Narayana, V.V.D. and Rambabu. 1983. Estimation of soil erosion in India. *J. Irrigation and Drainage Engineering* **109**(4):419–434.

NAS. 1974. *Productive Agriculture and a Quality Environment.* National Academy of Sciences, Washington, DC.

NAS. 1982. *Impact of Emerging Agricultural Trends on Fish and Wildlife Habitat.* NAS, National Academy Press, Washington, DC.

National Soil Erosion Research Laboratory. 1987. *User Requirements, USDA Water Erosion Prediction Project (WEPP)*, draft 6.3. Agric. Research Service, USDA, Indiana, USA.

NCA. 1976. *Report of National Commission on Agriculture.* Ministry of Agriculture, New Delhi, India.

Neil, D.T. and Galloway, R.W. 1989. Estimation of sediment yields from farm dam catchments. *Australian Journal of Soil and Water Conservation* **2**(1):46–51.

Newson, M.D. 1980. The erosion of drainage ditches and its effect on bedload yields in Mid-Wales. *Earth Surf. Proc. Landforms* **5**:275–290.

Newson, M.D. 1981. Mountain streams. *In*: Lewin, J. (Ed.), *British Rivers*. Allen and Unwin, London. Pp. 59–89.

Ngatunga, E.L.N., Lal, R. and Uriyo, A.P. 1984. Effects of surface management on runoff and soil erosion from some plots at Mlingano, Tanzania. *Geoderma* **33**:1–12.

Nicou, R. and Charreau, C. 1980. *Mechanical Impedance to Land Preparation as a Constraint to Food Production in the Tropics (With Special Reference to Fine Sandy Soils in West Africa). Priorities for Alleviating Soil-Related Constraints to Food Production in the Tropics.* International Rice Research Institute, New York State College of Agriculture and Life Sciences. Cornell University and University Consortium of Soils for the Tropics (Joint Publication). Pp. 371–388.

Niewiadomski, W. 1968. *Studies on Soil Erosion in Northern Poland.* Panstwowe Wydawnictwa Rolnicze i Lezne, Warszawa, Pp. 29–49.

Niewiadomski, W. and Baranska, L. 1977. Effectiveness of mineral fertilizing on eroded slope. *Zeszyty Problemowe Postepow Nauk Rolniczych* **193**:123–134.

Niewiadomski, W. and Grabarczyk, S. 1977. The structure of soil usage as a factor of soil conservation against water erosion. *Zeszyty Problemowe Postepow Nauk Rolniczych* **193**:135–155.

Niewiadomski, W. and Poradowski, J. 1959. Obserwacje na erozja wietrzna okresu luty-marzec 1956 na polu gospodarstwa doswiadczalnego Posorty. *Roczniki Nauk Rolniczych* **73**:213–222.

Nix, H.A. 1976. Climate and crop productivity in Australia. *In*: *Proceedings of the Symposium on Climate and Rice*. The International Rice Research Institute, Los Banos, Philippines. Pp. 495–507.

Nix, J. 1985. *Farm Management Pocket Book*. 16th edition, Wye College, London.

NSBC (National Statistics Bureau of China). 1987. *The Publication of Sampling Investigation of the Population in China*. Renmin Ribao, 12 Nov.

NSESPRPC (National Soil Erosion–Soil Productivity Research Planning Committee). 1981. Soil erosion effects on soil productivity: a research perspective. *J. Soil Water Cons.* **32**:82–90.

OECD (Organisation for Economic Co-operation and Development). 1985. *State of the Environment*. Paris, France.

Oldeman, L.R., Hakkeling, R.T.A. and Sombroek, W.G. 1990. *World Map on the Status of Human-Induced Soil Degradation: an Explanatory Note*. International Soil Reference and Information Centre, Wageningen, and United Nations Environment Programme, Nairobi.

Olive, L.J. and Walker, P.H. 1982. Processes in overland flow – erosion and production of suspended material. *In*: O'Loughlin, E.M. and Cullen, P. (Eds), *Prediction in Water Quality*. Australian Academy of Science, Canberra, Australia. Pp. 87–119.

Orlik, T. 1971. Some problems of agricultural economy on the eroded loess soils. *Zeszyty Problemowe Postepow Nauk Rolniczych* **119**:103–121.

Orlik, T. and Czerwinski, S. 1985. Root mass of some plants cultivated on different parts of the earth's surface sculpture. *Zeszyty Problemowe Postepow Nauk Rolniczych* **311**:73–81.

Oschwald, W.R. and Siemens, J.C. 1976. Soil erosion after soybeans. *In*: Hill, L.D. (Ed.), *World Soybean Research*. Interstate Printers and Publishers, Danville, IL. Pp. 74–81.

O'Sullivan, T.E. 1985. Farming systems and soil management: the Philippines/Australian Development Assistance Program experience. *In*: Craswell, E.T., Remenyi, J.V. and Nallana, L.G. (Eds), *Soil Erosion Management: Proceedings of a Workshop held at PCARRD, Los Banos, Philippines*, 3–5 December 1984. ACIAR, Proceedings No. 6.

OTA. 1982. *Impacts of Technology on US Cropland and Rangeland Productivity*. Office of Technology Assessment, Government Printing Office, Washington, DC.

Outhet, D. and Morse, R. 1984. Management of sediment in New South Wales, Australia. *Water International* **9**:169–171.

Overrein, L.N. 1978. Changes in soil productivity through acidification. *In*: *Symposia Papers*, Vol. 3 of the 11th Congress International Soil Science Society, Edmonton, Canada, June 1978. Pp. 260–277.

Pałys, S. 1971. The erosion of the upper and middle parts of the Wieprz river against the background of the general watershed characteristics. *Zeszyty Problemowe Postepow Nauk Rolniczych*, **119**:69–89.

Pałys, S. 1985. Wplyw erozji rzecznej na rolnictwo i go spodarke wodna. *Wiadomosci melioracyjne i lakarskie* **28**:245–248.

Pankhurst, C.E. and Sprent, J.R. 1975. Effects of water stress on the respiratory and nitrogen-fixing activity of soybean root nodules. *J. Exp. Bot.* **26**:287–304.

Pearsall, W.H. and Pennington, W. 1973. *The Lake District*. New Naturalist Series, 53, Collins, London.

Peng Fude. 1986a. The outline of 11 poor regions. *Problems of Agricultural Economy (Nongye Jingji Wenti)* **8**:23–24.

Peng Fude. 1986b. Some problems of agricultural development in poor mountain areas of China. *Agricultural Technological Economy (Nongye Jishu Jingji)* **4**:7–9.

Peng Fude. 1987. Problems and strategies about land resources in China. *Regional Study and Exploitation (Chuyu Yanju He Kaifa)* **6**(3):1–5.

Pierce, F.J., Dowdy, R.H., Larson, W.C. and Graham, W.A.P. 1984. Soil productivity in the corn belt: an assessment of erosion's long-term effects. *J. Soil Water Cons.* **39**:131–136.

Pieri, C. 1987. Management of acid soils in Africa. *In*: Sanchez, P.A., Stoner, E.R. and Pushparajah, E. (Eds), *Management of Acid Tropical Soils For Sustainable Agriculture: Proceedings of an IBSRAM Inaugural Workshop*. Bangkok, Thailand. Pp. 13–39.

Piggin, C.M. and Pareira, V. 1985. The use of *Leucaena* in Nusa Tenggara Timur. *In*: Craswell, E.T. and Tangendjaja, B. (Eds), *Shrub Legume Research in Indonesia and Australia: Proceedings of an International Workshop held at Balai Penelitian Ternak, Ciawi-Bogor, Indonesia, 2/2/84*. ACIAR Proceedings Series No. 3.

Pimentel, D. 1976. Land degradation: effects on food and energy resources. *Science* **194**:149–155.

Pimentel, D. 1984. Energy flow in the food system. *In*: Pimentel, D. and Hall, C.W. (Eds), *Food and Energy Resources*. Academic Press, New York. Pp. 1–23.

Pimentel, D. and Hall, C.W. (Eds) 1989. *Food and Natural Resources*. Academic Press, San Diego.

Pimentel, D., Terhune, E.C., Dyson-Hudson, R., Rochereau, S., Samis, R., Smith, E., Denman, D., Reifschneider, D. and Shepard, M. 1976. Land degradation: effects on food and energy resources. *Science* **194**:149–155.

Pimentel, D., Moran, M.A., Fast, S., Weber, G., Bukantis, R., Balliett, L., Boveng, P., Cleveland, C., Hindman, S. and Young, M. 1981. Biomass energy from crop and forest residues. *Science* **212**:1110–1115.

Pimentel, D., Dazhong, W., Eigenbrode, S., Lang, H., Emerson, D. and Karasik, M. 1986a. Deforestation: interdependency of fuelwood and agriculture. *Oikos* **46**:404–412.

Pimentel, D., Allen, J., Beers, A., Guinand, L., Linder, R., McLaughlin, P., Meer, B., Musonda, D., Perdue, D., Poisson, S., Siebert, S., Stoner, K., Salazar, R. and Hawkins, A. 1986b. World food economy and the soil erosion crisis. *Environmental Biology Rept. 86–2*. Cornell University, Ithaca, New York.

Pimentel, D., Allen, J., Beers, A., Guinand, L., Linder, R., McLaughlin, P., Meer, B., Musonda, D., Perdue, D., Poisson, S., Siebert, S., Stoner, K., Salazar, R. and Hawkins, A. 1987. World agriculture and soil erosion. *Bioscience* **37**(4):277–283.

Planning Commission. 1981. *Report on Development of Backward Areas*. New Delhi, India.

Planning Commission. 1982. *Report of the Task Force for the Study of Eco-development in the Himalayan Regions*. New Delhi, India.

Podwinska, Z. 1970. Rozmieszczenie wodnych mtynow zbozowych w malopolsce w XV wieku. *Kwartalnik Historii Kultury Materialnej* **18**:3.

Pollard, R.W., Sharp, B.M.H. and Madison, F.W. 1979. Farmers' experience with conservation tillage: a Wisconsin survey. *J. Soil Water Cons.* **34**:215–219.

Pope, A.P. III, Bhide, S. and Heady, E.O. 1983. Economics of conservation tillage in Iowa. *J. Soil Water Cons.* **38**:370–373.

Posner, J.L. 1981. *Cropping Systems and Soil Conservation in Tropical America*. Report to Rockefeller Foundation, New York.

Prochal, P. 1973. Factors influencing the intensity of washout and the movement of rubble in the watershed of the stream Wierchomlin Wiellea. *Zeszyty Problemowe Postepow Nauk Rolniczych* **151**:31–48.

Prochal, P. 1984. *Melioracje Przeciw Erozyjne*. Akademia Rolnicza w Krakowie, Krakow. 191 pp.

Prospero, T. and Carlson, T. 1972. Vertical and areal distribution of Saharan dust over the western equatorial North Atlantic Ocean. *J. Geophys. Res.* **77**:5255–5265.

Prove, B.G. 1984. Soil erosion and conservation in the sugar-canelands of the wet tropical coast. *Erosion Research Newsletter* **10**:6–7.

Puckridge, D.W. and French, R.J. 1983. The annual legume pasture in cereal-ley farming systems of southern Australia: a review. *Agric. Ecosystems and Environment* **9**:229–267.

Quansah, C. 1981. The effect of soil type, slope, rain intensity and their interactions on splash detachment and transport. *J. Soil Sci.* **32**:215–224.

Rambabu, Agarwal, M.C., Vishwanathan, M.K. and Joshie, P. 1980. *Economics and Benefit–Cost Ratio of* Eucalyptus *Plantation for Fuel Purposes in Denuded Lands of Doon Valley.* National Symposium on Soil Conservation and Water Management in 1980s. Indian Assoc. of Soil and Water Conservationists. Dehradun, India.

Rao, C.S., Sivamohan, M.V.K., Chandy, S. and Rao, R.S. 1988, and 1990. *Pilot Project for Watershed Development in Rainfed Areas, Maheshwaram (Vol. 1–1988; and Special Studies 1, 2, 3–1990)*, Administrative Staff College of India and Government of Andhra Pradesh, Hyderabad.

Rao, K.L. 1975. *India's Water Wealth: Its Assessment, Uses and Projections*. Orient Longman Ltd, New Delhi, India.

RCEP. 1979. *Agriculture and Pollution*. Seventh Report, Royal Commission on Environmental Pollution. HMSO, London.

Reed, A.H. 1979. Accelerated erosion of arable soils in the U.K. by rainfall and run-off. *Outlook on Agriculture* **10**:41–48.

Reed, A.H. 1983. The erosion risk of compaction. *Soil and Water* **11**(3):29–33.

Reed, A.H. 1986. Soil loss from tractor wheelings. *Soil and Water* **14**(4):12–14.

Reganold, J.P., Elliot, L.F. and Unger, Y.L. 1987. Long-term effects of organic and conventional farming on soil erosion. *Nature* **330**:370–372.

Reniger, A. 1950. Proba oceny nasilenia i zasiegow potencjalnej erozji gleb w Polsce. *Zeszyty Problemowe Postepow Nauk Rolniczych* **54**:125–142.

Repelewska-Pakalowa, J. and Pekala K. 1988. The eolian soil erosion on the Lublin upland in 1981–1985. *Zeszyty Problemowe Postepow Nauk Rolniczych* **357**:7–16.

Rickson, R.E. and Stabler, P.J. 1985. Community responses to non point pollution from agriculture. *Journal of Environmental Management* **20**:281–293.

Rickson, R., Saffigna, P., Vanclay, F. and McTainsh, G. 1987. Social bases of farmers' responses to land degradation. *In*: Chisholm, A.H. and Dumsday, R.G. (Eds), *Land Degradation: Problems and Policies*. Cambridge University Press and CRES, Canberra, Australia. Pp. 187–200.

Rimwanich, S. (Ed.) 1988. *Land Conservation for Future Generations*. Proceedings of the 5th International Soil Conservation Conference 18–29 January 1988, Bangkok, Thailand. Dept. of Land Devp., Ministry of Agr. and Cooperatives. 1310 pp.

Ring, P.J. 1982. Soil erosion, rehabilitation and conservation – a comparative agricultural engineering/economic case study. *In*: *Conference on Agricultural Engineering*, The Institution of Engineers, Australia, Barton, Australia.

Riquier, J. 1982. A world assessment of soil degradation. *Nature and Resources* **18**(2):18–21.

Robertson, G. 1987. Contributions from the physical and biological sciences. *In*: Chisholm, A.H. and Dumsday, R.G. (Eds), *Land Degradation: Problems and Policies*. Cambridge University Press and CRES, Canberra, Australia. Pp. 305–314.

Robertson, G.P., Herrera, D. and Roswall, T. (Eds) 1982. *Nitrogen Cycling in Ecosystems of Latin America and the Caribbean*. Martinus Nijoff/Dr W. Junk Publishers, The Hague.

Robinson, A.R. 1971. Sediment: our greatest pollutant? *Agr. Eng.* **52**:406–408.

Robinson, D.A. and Boardman, J. 1988. Cultivation practice, sowing season and soil erosion on the South Downs, England: a preliminary study. *J. Agric. Sci. Camb.* **110**:169–177.

Rodda, J.C. 1970. Rainfall excesses in the UK. *Trans. Inst. Br. Geogs.* **49**:49–60.

Romkens, M.J.M., Nelson, D.W. and Mannering, J.V. 1973. Nitrogen and phosphorus composition of surface runoff as affected by tillage method. *J. Environ. Qual.* **2**:292–295.

Roose, E.J. 1977a. Adaption of soil conservation techniques to the ecological and socio-economic conditions of West Africa. *Agron. Trop.* **32**:132–140.

Roose, E.J. 1977b. Application of the USLE of Wischmeier and Smith in West Africa. *In*: Greenland, D.J. and Lal, R. (Eds), *Soil Conservation and Management in the Humid Tropics*. J. Wiley & Sons, Chichester, UK. Pp. 177–188.

Roose, E. 1987. Reseau erosion. *ORSTOM Bulletin* **7**.

Rose, C.W. 1987. Controlling erosion for cropping system experiments. *In*: Latham, M. (Ed.), *Soil Management Under Humid Conditions in Asia, ASIALAND: Proceedings of the First IBSRAM Regional Seminar*. Khon Kaen, Phitsanulok, Thailand, 13–20 October 1986. Pp. 297–315.

Rose, C.W., Saffigna, P.G., Hairsine, P.B., Palis, R.G., Okwach, G., Proffitt, A.P.B. and Lovell, C.J. 1988. Erosion processes and nutrient loss. *In*: Rimwanich, S. (Ed.).

Rosewell, C.J. 1986. *Evaluation of the Universal Soil Loss Equation – Bare Fallow Plot Experiments*. Report SCS, Sydney, Australia.

Rosewell, C.J. and Edwards, K. 1988. SOIL-LOSS – a program to assist in the selection of management practices to reduce erosion. *Soil Conservation Service of NSW Technical Handbook No.* **11**, SCS, Sydney, Australia.

Sajise, P.E. 1982. Our critical uplands ... and what we can do about them. *Upland World* **1**(2):4–10.

Sanchez, P.A. 1984. Nutrient dynamics following rainforest clearing and cultivation. *In*: *Proceedings of an International Workshop on Soils, Townsville, Australia 12–16 September 1983*. ACIAR Proceedings Series No. 2. Pp. 52–56.

Sanchez, P.A. 1987. Management of acid soils in the humid tropics of Latin America. *In*: Sanchez, P.A., Stoner, E.R. and Pushparajah, E. (Eds), *Management of Acid Tropical Soils for Sustainable Agriculture*. Proceedings of an IBSRAM inaugural workshop. Bangkok, Thailand. Pp. 63–107.

Sanchez, P.A. and Benites, J.R. 1987. Low-input cropping for acid soils of the humid tropics: a transition technology between shifting and continuous cultivation. *In*: Latham, M., Ahn, P. and Elliott, C.R. (Eds), *Land Development and Management of Acid Soils in Africa*, Proceedings of the Second Regional Workshop on Land Development and Management of Acid Soils in Africa. Lusaka and Kasama, Zambia, 9–16 April 1987. IBSRAM Proceedings No. 7. Pp. 85–106.

Sancholuz, L.A. 1984. Land degradation in Mexican maize fields. PhD Thesis, University of British Columbia, Vancouver.

Sanders, D.W. 1988. Environmental degradation and socio-economic impacts: past, present, and future approaches to soil conservation. *In*: Rimwanich, S. (Ed.).

Sattaur, O. 1989. Grass grows into a hedge against erosion. *New Scientist* 13 May.

Sauberan, C. 1955. La chala de maiz como mejoradora de los suelos. (Corn stubble as an improver of soils). *Rev. Arg. Agron.* **22**:154–157.

Sauberan, C. and Molina, J.S. 1955. La Chala de maiz como mejoradora de los Suelos. (Com *stulolote* as an improver of soils). *Rev. Arg. Argon.* **22**:154–157.

Sauberan, C. and Molina, J.S. 1956. Deben quemarse los rastrojos? (Must stubble be burnt?). *Ciencia e Investigación* (Buenos Aires) **11**:483–489. (In Spanish.)

Sauberan, C. and Molina, J.S. 1957. *Agotamiento, erosion y recuperazion de suelos en la Republica Argentina (Exhaustion, erosion and reclamation of soils in Argentina)*. Editorial Hombre y Suelo.

Sauberan, C. and Molina, J.S. 1958. *Agotamiento, Erosion y Recuperacion de Suelos en la Republica Argentina. (Exhaustion, Erosion and Reclamation of Soils in Argentina)*. Editorial hombre y suelo (Buenos Aires). Tercer Premio Nacional (Third National Prize). (In Spanish.)

Sauberan, C. and Molina, J.S. 1962. *Soluciones Para los Problemas del Campo. (Solution for the Country's Problems).* Primer Premio Nacional (First National Prize). (In Spanish.)

Sauberan, C. and Molina, J.S. 1964. Suelos salitrosos (Alkali-soils). *J. Soil Assoc.* (Mother Earth), England **13**:311–315.

Sauberan, C. and Molina, J.S. 1965. *Hacia la Formacion de Grupos CREA Chacareros. (Formation of Farmers' CREA Groups).* Numero estadistico de la Bolsa de Cereales de Buenos Aires (In Spanish.)

Sauer, E.L. and Case, H.C.M. 1954. Soil conservation pays off. Results of ten years of conservation farming in Illinois. *Univ. Ill. Agr. Expt. Sta. Bull. No.* **575**.

Saunders, I. and Young, A. 1983. Rates of surface processes on slopes, slope retreat and denudation. *Earth Surf. Processes and Landforms* **8**:473–501.

Schertz, D.L., Moldenhauer, D.W., Franzmeier, D.P. and Sinclair, H.R. 1985. Field evaluation of the effect of soil erosion on crop productivity. *In: Erosion and Soil Productivity*. Proc. Nat. Symp. American Society of Agricultural Engineers, St Joseph, MI. Pp. 9–17.

Schmidt, B.L., Allmaras, R.R., Mannering, J.V. and Papendick, R.I. 1982. *Determinants of Soil Loss Tolerance*. ASA Special Publication No. 45. American Soc. Agron. Wisconsin, USA.

Schmuck, A. 1969. *Meteorologia i klimatologia dla WSR*. Panstwowe Wydawnictwo Naukowe Warszawa. 316 pp.

Schultz, T.W. 1987. The long view in economic policy: the case of agriculture and food. *International Center for Economic Growth, Occasional Papers No.* **1**:16.

Schuman, G.C., Spomer, R.G. and Piest, R.F. 1973. Phosphorus losses from four agricultural watersheds on Missouri valley loess. *Soil Sci. Soc. Am. Proc.* **37**(3):424–427.

Seckler, D.W. 1981. *Economic Evaluation of the Fakot Project Report on Operational Research Project, Watershed Management*. Central Soil and Water Conservation Research and Training Institute, Dehradun, India.

Shah, S.L. 1981. *Agricultural Planning and Development in the North-Western Himalayas, India. Nepal's Experience in Hill Agricultural Development*. Ministry of Food and Agriculture, HMG, Kathmandu, Nepal. Pp. 160–168.

Shah, M.M., Fischer, G., Higgins, G.M., Kassam, A.H. and Naiken, L.A. 1985. *People, Land, and Food Production-Potentials in the Developing World*. International Institute for Applied Systems Analysis. CP-85-11. Laxenburg, Austria.

Shalhevet, J., Mantell, J., Bielorai, H. and Shimshi, D. 1979. *Irrigation of Field and Orchard Crops under Semi-arid Conditions*. International Irrigation Information Center, Volcani Center, Bet Dagan, Israel. Distributed by Pergamon Press, Elmsford, NY.

Shanan, L. 1987. The impact of irrigation. *In*: Wolman, M.G. and Fournier, F.G.A. (Eds), *Land Transformation in Agriculture*, SCOPE 32. John Wiley & Sons, Chichester, England. Pp. 115–131.

Sharma, A.K., Pradhan, I.P., Nema, J.P. and Tejawni, K.G. 1980. *25 Years Research on Soil and Water Conservation in Ravine Lands of Gujarat*. Monograph No. 2. Central Soil and Water Conservation Research and Training Institute, Vasad, India.

Shaw, A.G.L. 1967. History and development of Australian agriculture. *In*: D.B. Williams (Ed.), *Agriculture in the Australian Economy*. Sydney University Press, Sydney, Australia.

Shi Deming. 1985a. Soil erosion and conservation in the red soil area. *In*: Li Chingkui, (Ed.), *Red Soils in China*. Science Press, Beijing, Pp. 237–253.

Shi Deming. 1985b. Preservation of soils through conservation practices. *Soil and Water Conservation in China (Zhongguo Shuitubaochi)* 4:2–5.

Shi Deming. 1987. Degeneration and prevention of red soils in China. *Soil and Water Conservation in China (Zhongguo Shuitubaochi)* **12**:2–5.

Shrader, W.D., Johnson, H.P. and Timmons, J.F. 1963. Applying erosion control principles. *J. Soil Water Cons.* **18**:195–200.

Siemens, J.C. and Oschwald, W.R. 1978. Corn–soybean tillage systems: erosion control, effects on crop production, costs. *Trans. Am. Soc. Agr. Eng.* **21**:293–302.

Silvey, S. 1981. The contribution of new wheat, barley and oat varieties to increasing yield in England and Wales, 1947–78. *J. National Inst. Agric. Botany* **15**:399–412.

Simmons, I.G. 1987. Transformation of the land in pre-industrial time. *In*: Wolman, M.G. and Fournier, F.G.A. (Eds), *Land Transformation in Agriculture*, SCOPE 32. John Wiley & Sons, Chichester, England. Pp. 45–77.

Sinden, J.A. 1987. *Influence of Soil Conservation on Land Values in the Murray–Darling Basin*. Proceedings Soil Conservation Service NSW Annual Conference, Sydney, Australia.

Sinha, B. 1984a. Need of integrated approach to development of catchment and command of irrigation projects. *Water International* **9**(4):158–160.

Sinha, B. 1984b. *Role of Watershed Management in Water Resources Development Planning*. Paper presented at the workshop on the management of river and reservoir sedimentation in Asian countries. East–West Centre, Honolulu, Hawaii.

Sivamohan, M.V.K., Someshwariah, K. and Venkataramana, M. 1988. *Pilot Project for Watershed Development in Rainfed Areas (Benchmark Socio-economic Survey–1986)*, Administrative Staff College of India, Hyderabad.

Skrodzki, M. 1978. Rolnictwo na glebach ulegajacych erozji. *In*: Zieminicki (Ed.), *Erozja wodna. Panstwowe Wydawnictwo Rolnicze i Leane Warszawa*. Pp. 46–59.

Smith, D.D. 1946. The effect of contour planting on crop yield and erosion losses in Missouri. *J. Am. Soc. Agron.* **38**:810–819.

Smith, M. 1986. *Agriculture and Nature Conservation in Conflict: The Less Favoured Areas of France and the UK*. Arkleton Trust.

Soil Conservation Service (USDA). 1948. *Guide for Soil Conservation Surveys*. Govt. Printing Office, Washington, DC.

Soil Conservation Service of NSW. 1982. *Soil Data Card Handbook*. Technical Handbook No. 4. SCSNSW, Sydney, Australia.

Spears, J. 1988. *Forestry Research Priorities in the 1990s*. Background discussion paper for proposed Rockefeller Foundation Meeting on Forestry Research, Bellagio, November 1988.

Speirs, R.B. and Frost, C.A. 1985. Soil erosion in the east of Scotland. *Research and Development in Agriculture* **2**:161–167.

Speirs, R.B. and Frost, C.A. 1987. Soil water erosion on arable land in the UK. *Research and Development in Agric.* **4**:1–11.

Spomer, R.G., Piest, R.F. and Heinemann, H.G. 1976. Soil and water conservation with western Iowa tillage systems. *Trans. Am. Soc. Agr. Eng.* **19**:108–112.

SSEW (Soil Survey of England and Wales). 1983. Soil erosion. *In*: *Rothamsted Experimental Station Report for 1982*, part 1. RES, Harpenden, Herts, UK. Pp. 242–243.

SSEW. 1984. Soil erosion. *In*: *Rothamsted Experimental Station Report for 1983*. RES, Harpenden, Herts, UK. Pp. 205–206.

SSEW. 1985. Soil erosion. *In*: *Rothamsted Experimental Station Report for 1984*. RES, Harpenden, Herts, UK. Pp. 208–209.

Stammers, R. and Boardman, J. 1984. Soil erosion and flooding on Downland areas. *The Surveyor* **164**:8–11.

Stamp, L.D. 1948. *The Land of Britain, Its Use and Misuse*. Longman, Green and Co, London.

Stanley, P.I. and Hardy, A.R. 1984. The environmental implications of current pesticides usage in cereals. *In*: Jenkins, D. (Ed.), *Agriculture and the Environment*. ITE Symp. Feb. 1984. NERC, Cambridge, England. Pp. 66–72.

Stocking, M. 1984. *Erosion and Soil Productivity: A Review*. Consultants' Working Paper No. 1. Food and Agriculture Organization of the United Nations, Rome, Italy.

Stocking, M. 1987. Measuring land degradation. *In*: Blaikie, P. and Brookfield, H. (Eds), *Land Degradation and Society*. Methuen, London. Pp. 49–63.

Stocking, M. 1988. Quantifying the on-site impact of soil erosion. *In*: Rimwanich, S. (Ed.).

Stocking, M. and Peake, L. 1987. Erosion-induced loss in soil productivity: trends in research and international cooperation. *In*: Pla Sentis, I. (Ed.), *Soil Conservation and Productivity* (Proc. IV Int. Conf. on Soil Conservation, Maracay, Venezuela, 1985), Sociedad Venezolana de la Cienca del Suelo, Maracay, Venezuela. Pp. 399–438.

Stoneman, T.C. 1962. Loss of structure in wheatbelt soils. *J. Dept Agric. West. Aust.* **3**:493–497.

Stoneman, T.C. 1973. Soil structure changes under wheatbelt farming systems. *J. Dept Agric. West. Aust.* **14**:209–214.

Strzemski, M. 1964. Uwagi ogolne o przemianach srodowiska geograficznego Polski jako tla przyrodniczego rozwoju rolnictwa od polowy trzeciego tysiaclecia p.n.e. do naszych czasow. *In*: Hensel, W. (Ed.), *Zarys historii gospodarstwa wiejskiego w Polsce*. i H. Lowmianski/PWRIL, Warszawa.

Subramanya, S. and Sastry, K.N.R. undated. *Indigenous Knowledge About the Use of 'Vetiveria Zizanioides' for Conserving Soil and Moisture*. Karnataka State Watershed Development Commission, Bangalore.

Sukmana, S. 1987. Alang-alang land in Indonesia: problems and prospects. *In*: Latham, M. (Ed.), *Soil Management Under Humid Conditions in Asia, ASIALAND: Proceedings of the First IBSRAM Regional Seminar*. Khon Kaen, Phitsanulok, Thailand, 13–20 October 1986. Pp. 397–411.

Sun Jianxuan. 1985. *Preliminary Descriptions of the Terms of Soil and Water Conservation*. Water Conservancy and Electrical Power Press, Beijing. 272 pp.

Swaminathan, M.S. 1983. Our greatest challenge – feeding a hungry world. *In*: Bixler, G. and Shemilt, L.W. (Eds), *Perspectives and Recommendations*. Chemistry and World Food Supplies: the New Frontiers. Chemrawn II. International Rice Research Institute, Philippines. Pp. 25–46.

Swanson, E.R. and Harshbarger, C.E. 1964. An economic analysis of effects of soil loss on crop yields. *J. Soil Water Cons.* **19**:183–186.

Szabolcs, I. 1964. Salt-affected soils in Hungary. *Proc. Symp. on Sodic Soils, Budapest-Agrochemistry and Soil Science. Suppl.* **14**:275–290.

Szpakowska, B., Pemkowiak, J. and Zyczynska-Baloniak, I. 1986. Comparison of some physicochemical parameters of humic substances isolated from three different aquatic ecosystems. *Archiv Hydrobiol.* **108**:259–267.

Szumanski, A. 1977. Changes in the course of the lower San channel river in XIX and XX centuries and their influence on the morphogenesis of its floodplain. *Studia Geomorphologica Carpatho-Blacanica* **11**:139–153.

TAC (Technical Advisory Committee). 1988. *Sustainable Agricultural Production: Implications for International Agricultural Research*. Paper for TAC Workshop held in Rome, 12–14 January 1988.

Taff, S.J. and Runge, C.F. 1988. Wanted: A Leaner and Meaner CRP. *Choices First Quarter 1988*. Pp. 16–18.

Tang Defu. 1985. *Discussion on Soil Erosion and Conservation in Northeastern China*. Northeastern Agricultural College, Harbin. 8 pp.

Tang Keli. 1985. The perspectives of soil erosion and conservation in the Loess Plateau. *In*: *The Situations, Strategies and Perspectives of Soil Sciences in China*. Chinese Association of Soil Science, Chinese Association of Science and Technology, Beijing. Pp. 45–48.

Tang Keli, Shi, D. and Zhu. X. 1988. Soil erosion and soil conservation in China. *In*: Rimwanich, S. (Ed.).

Tang Keli, Zhang Zhongzi, Kong Xiangling, Chao Xuan and Shi Rueiyan. 1987. A study of soil loss and soil degradation in the Loess Plateau. *Bulletin of Soil and Water Conservation (Shuitubaochi Tongbao)* **7**(6):12–18.

Tejwani, K.G. 1979a. Soil and water conservation: promise and performance in 1970s. *Indian J. Soil Cons.* **7**(2):80–86.

Tejwani, K.G. 1979b. Malady-remedy analysis of soil and water conservation in India. *Indian J. Soil Cons.* **7**(1):29–45.

Tejwani, K.G. 1980a. Soil and water conservation. *In*: *Handbook of Agriculture*. 5th edition, Indian Council of Agric. Research, New Delhi, India. Pp. 120–157.

Tejwani, K.G. 1980b. A quarter century (1955–1979) of soil and water conservation research in India. *In*: Gupta, R.K. *et al.*, (Eds), *Proceedings of the National Symposium on Soil Conservation and Water Management in the 1980s*. Indian Association of Soil and Water Conservationists. Dehradun, India. Annex. Pp. 1–48.

Tejwani, K.G. 1981. *Manpower Needs for Watershed Management in India*. Proceedings of the National Seminar on Watershed Management. Forest Research Institute and Colleges, Dehradun, India. Pp. 193–202.

Tejwani, K.G. 1982. Soils policy in India. Needs and direction. Transactions of the 12th International Congress of Soil Science. New Delhi, India. *Symposia Papers* **III**:98–105.

Tejwani, K.G. 1984a. Reservoir sedimentation in India – its causes, control and future course of action. *Water International*, **9**(4):150–154.

Tejwani, K.G. 1984b. *Biophysical and Socio-economic Causes of Land Degradation and Strategy to Foster Watershed Degradation in the Himalayas*. IUFRO Symposium of effects of forest land use on erosion and slope stability. East–West Center, Honolulu, Hawaii. Pp. 55–60.

Tejwani, K.G. 1984c. Watershed Management. *In*: Soil research in retrospect and prospect. *Bull. Indian Soc. Sci.* **14**:169–175.

Tejwani, K.G. 1986. Training, research, demonstration in watershed management. *FAO Conservation Guide* **14**:201–219.

Tejwani, K.G. 1987. Watershed management in the Indian Himalaya. *In*: Khoshoo, T.N. (Ed.), *Perspectives in Environmental Management*. Oxford and IBH Publishing Co. Pvt. Ltd, New Dehli, India.

Tejwani, K.G. and Rambabu. 1981. Unpublished data. Central Soil and Water Conservation Research and Training Institute, Dehradun, India.

Tejwani, K.G. and Rambabu. 1982. Evaluation of the environmental benefits of soil and water conservation programmes. *Indian J. Soil Cons.* **10**(2/3):80–90.

Tejwani, K.G., Gupta, S.K. and Mathur, H.N. 1975. *Soil and Water Conservation Research, 1956–70*. Indian Council of Agric. Research, New Delhi, India.

Thomas, G.W. 1988. Elephant grass for soil erosion control and livestock feed. *In*: Moldenhauer, W.C. and Hudson, N.W. (Eds), *Conservation Farming on Steep Lands*, Soil and Water Conservation Society, Ankeny, Iowa.

Thomson, L.A.J. 1987. Australian acacias for saline, alkaline soils in the hot, dry subtropics and tropics. *In*: *Australian Acacias in Developing Countries: Proceedings of an International Workshop held at the Forestry Training Centre, Gympie, Qld, Australia, 4–7 August 1986*. ACIAR Proceedings No. 16. Pp. 66–69.

Tian Houmo. 1985. The sharp increase of population and soil erosion in the Loess Plateau. *Bulletin of Soil and Water Conservation (Shuitubaochi Tongbao)* **3**:7–11.

Tian Houmo. 1987. On the ecological conservation in the Loess Plateau. *Acta Ecologica Sinica (Shengtai Xuebao)* **7**(4):376–378.

Tian Xinyuan. 1988. The comprehensive control of small watersheds and its benefits in Shichuan province. *Soil and Water Conservation in China (Zhongguo Shuitubaochi)* **3**:19–20.

Tisdall, J.M. and Oades, J.M. 1982. Organic matter and water-stable aggregates in soils. *J. Soil Sci.* **33**:141–163.

TMB [Territorial Management Bureau]. 1984. *The Outline of the Territorial Resources in China*. National Planning Committee, Beijing. 222 pp.

Tothill, J.C. 1987. Application of agroforestry to African crop–livestock farming systems. *ILCA Bulletin No.* **29** – December 1987, Pp. 20–23.

Tran-Vinh-An and Nguba, H. 1971. Contribution à l'étude utile de quelques sols du Zaire. *African Soils* **16**:91–103.

Troeh, F.R., Hobbs, J.A. and Donahue, R.L. 1980. *Soil and Water Conservation for Productivity and Environmental Protection*. Prentice-Hall, Inc., Englewood Cliffs, NJ.

Turski, R. and Dobrzanski, B. 1968. Properties of humic acids of soils from eroded areas. *Polish Journal of Soil Science* **1**:11–17.

Turski, R. and Wincenciak, C. 1969. Studies in infrared on humic acids of soil on eroded areas. *Polish Journal of Soil Science* **2**:35–41.

UNEP (United Nations Environment Programme). 1980. *Annual Review*. UNEP, Nairobi, Kenya.

UNEP. 1982a. *World Soils Policy*. Nairobi 1982, Na.82-5947-1553C. Printed in Kenya (UNEP).

UNEP. 1982b. *Development and Environment in the Wider Caribbean Region: A Synthesis*. UNEP Regional Reports and Studies No. 14.

USDA (United States Department of Agriculture). 1965. *Losses in Agriculture*. Agricultural Handbook No. 291, Agr. Res. Serv., Government Printing Office, Washington, DC.

USDA. 1980. *Report and Recommendations on Organic Farming*. USDA Study Team, Government Printing Office, Washington, DC.

USDA. 1981. *A Time to Choose*. Government Printing Office, Washington, DC.

USDA. 1985. *Agricultural Statistics 1985*. Government Printing Office, Washington, DC.

USDA-ARS and EPA-ORD. 1976. *Control of Water Pollution from Cropland*. EPA Report No. EPA-600/2-75-0266, ARS Report No. ARS-H-5-2. 2 vols. Government Printing Office, Washington, DC.

USDC. 1983. *Climate Impact Assessment*. United States. Annual Summary 1982. US Department of Commerce, Washington, DC.

USDI. 1982. *Manual of Stream Channelization Impacts on Fish and Wildlife*. US Dept. of Interior, Fish and Wildlife Service. Biol. Serv. Program. July 1982.

Uttar Pradesh Government. 1985. *Hill Development at a Glance*. State Planning Institute, Lucknow, India.

Valentin, C. 1986. Surface crusting of arid sandy soils. *In*: Callebaut, F., Gabriels, D. and DeBoodt, M. (Eds), *Assessment of Surface Sealing and Crusting*. University Ghent, Belgium.

Venkataraman, C., Tejwani, K.G. and Deshmukh, G.R. 1980. *An Annotated Bibliography of Scientific Contributions of Central Soil and Water Conservation Research and Training Institute, Dehradun, India*.

Verma, B., Chinnamani, S., Bhola, S.N., Rao, D.H., Prasad, S.N. and Prakash, C. 1986. *Twenty-five Years of Research on Soil and Water Conservation in Ravine Lands of*

Rajasthan. Central Soil and Water Conservation Research and Training Institute Research Centre, Kota, India.

Vimalkishore, Singh, G. and Sastry, G. 1980. Cited from *Annual Report*. Central Soil and Water Conservation Research and Training Institute. Dehradun, India.

Vohra, B.B. 1981. *A Policy for Land and Water*. Sardar Patel Memorial Lectures, 1980. Dept. of Envir., New Delhi, India.

Volk, B.G. and Loeppert, R.H. 1982. Soil organic matter. *In*: Kilmer, V.J. (Ed.), *Handbook of Soils and Climate in Agriculture*. CRC Press, Boca Raton, FL. Pp. 211–268.

von Uexkull, H.R. 1987. Acid soils in the humid tropics: managing the soil surface. *In*: Latham, M. (Ed.), *Soil Management Under Humid Conditions in Asia, ASIALAND: proceedings of the First IBSRAM Regional Seminar*. Khon Kaen, Phitsanulok, Thailand, 13–20 October 1986. Pp. 333–349.

Walker, D.J. and Young, D.L. 1986. Assessing soil erosion productivity damage. *In*: *Soil Conservation. Assessing the National Resources Inventory*. Vol. 2., National Academy Press, Washington, DC. Pp. 21–62.

Walker, P.H. 1980. Soil morphology, genesis and classification in Australia. *In*: Abbott, T.S., Hawkins, C.A. and Searle, P.G.E. (Eds), *National Soils Conference 1980 Review Papers*, Aust. Soc. Soil Sci. Inc., Glen Osmond, Australia.

Walling, D.E. 1984. The sediment yield of African rivers. *IAHS* **144**:265–283.

Walling, D.E. and Webb, B.W. 1981. Water quality. *In*: Lewin, J. (Ed.), *British Rivers*. Allen and Unwin, London. Pp. 126–169.

Wang Changgui. 1985. The outline of energy resources in China. *Energy Information (Nongyuan Xinxi)* **11**:1–4.

Wang Guangren, Zhang Hongting, Qi Zhongzheng and Liu Xiao. 1987. To speed up the comprehensive controls of new erosion along with production conservation in Shanxi province. *Bulletin of Soil and Water Conservation (Shuitubaochi Tongbao)* **7**(5):1–7.

Wang Huayun. 1984. The development of soil conservation in the Loess Plateau. *Soil and Water Conservation in China (Zhongguo Shuitubaochi)* **9**:4–7.

Wang Huayun. 1986. Commemorating the 40th anniversary of harnessing Yellow river by the people. *Soil and Water Conservation in China (Zhongguo Shuitubaochi)* **10**:2–7.

Wang Siangqun. 1986. Reducing 200 million tons of inlet sediment in Yellow river annually. *Renmin Ribao*, 16 Nov. 1986, p. 1.

Wang Yongan. 1987a. A plan for conservation protective forest system in Yangtze river valley. *Bulletin of Soil and Water Conservation (Shuitubaochi Tongbao)* **7**(1):33–37.

Wang Yongan. 1987b. Establishing forest vegetation and its ecological and economical benefits in Yangtze river valley. *Ecological Economy (Shengtai Jingji)* **6**:28–30.

Wang Zhan and Chen Chuanguo. 1982. Yangtze river is following Yellow river's steps, ecological conditions are getting worse. *Journal of Ecology (Shengtaixue Zhazhi)* **1**(2):30–32.

Wasson, R.J. and Galloway, R.W. 1984. Erosion rates near Broken Hill before and after European settlement. *In*: Loughran, R.J. (Ed.), *Drainage Basin Erosion and Sedimentation Volume 1*. University of Newcastle, Newcastle, Australia. Pp. 213–220.

Watson, A. 1987. Water-induced soil erosion on arable land in NE Scotland. *Grampian Farming, Forestry and Wildlife Advisory Group Magazine* **7**:14–16.

Wen Dazhong. 1986. *Some Strategies of Agricultural Development in Western Liaoning*. Institute of Forest and Soil Sciences, Chinese Academy of Science, Shenyang. 47 pp.

Wen Dazhong. 1987. Studies on energy flows through an agroecosystem in a semi-arid mountain-hilly land area of western Liaoning. *In*: *Proc. Intl. Symp. on Dryland Farming*. Yangling, China.

Wen Dazhong. 1989. Food and fuel resources in a poor rural area in China. *In*: Pimentel, D. and Hall, C. (Eds), *Food and Natural Resources*. Academic Press, San Diego.

Wen Dazhong, Jiao Zhenjia and Yu Pairu. 1978. The investigation report of forestry in agricultural area of Hailun county in Heilongjiang province. *In*: *Proc. Comprehensive Investigations of Natural Resources of Hailun County in Heilongjiang Province*. The Comprehensive Investigations Team of Hailun, Chinese Academy of Sciences, Harbin.

Wijewardene, R. 1984. Agro-forestry in Tropical Asia. *Pesquisa Agropecuaria Brasileira 19* s/n, June 1984, Pp. 315–324.

Wijewardene, R. and Waidyanatha, P. 1984. *Systems, Techniques and Tools. Conservation Farming for Small Farmers in the Humid Tropics*. Department of Agriculture, Sri-Lanka and the Commonwealth Consultative Group on Agriculture for the Asia–Pacific Region.

Wilkinson, B., Broughton, W. and Parker-Strutton, J. 1969. Survey of wind erosion on sandy soils in the E Midlands. *Experimental Husb*. **18**:53–59.

Williams, J.R. 1985. The physical components of the EPIC model. *In*: El-Swaify *et al.*, **25**:272–284.

Williams, M.A.J. 1973. The efficacy of creep and slopewash in tropical and temperate Australia. *Australian Geographical Studies* **11**:62–68.

Wilson, S.J. and Cooke, R.U. 1980. Wind erosion. *In*: Kirkby, M.J. and Morgan, R.P.C. (Eds), *Soil Erosion*. John Wiley, Chichester, UK. Pp. 217–249.

Winogradsky, S. 1949. *Microbiology du Sol*. Masson et Cie, Paris.

Wischmeier, W.H. and Mannering, J.V. 1965. Effect of organic matter content of the soil on infiltration. *J. Soil Water Cons*. **20**:150–152.

Wischmeier, W.H. and Smith, D.D. 1978. Predicting rainfall erosion losses – a guide to conservation planning. US Department of Agriculture, Agriculture Handbook No. 537, USDA, Washington, USA.

Wit, C.T. de 1958. Transpiration and crop yields. *Versl. Landbouwk. Onderz. No*. 64.6.

World Bank. 1991. Extracts from 'Role of vetiver grass in soil and moisture conservation' by G.M. Bharad and B.C. Bathkal, in *Vetiver Newsletter* No. 6, June 1991, The World Bank, Washington.

World Bank. 1990. *Vetiver Grass: The Hedge Against Erosion*. 3rd edition, The World Bank, Washington.

WCED (World Commission on Environment and Development). 1987a. *Our Common Future*. Oxford University Press, Oxford and New York.

WCED. 1987b. *Food 2000. Global Policies for Sustainable Agriculture*. A Report to the WCED. Zed Books Ltd, London and New Jersey.

WRI (World Resources Institute). 1985. *Tropical Forests: A Call for Action, Part 1*. Report of an International Task Force convened by the World Resources Institute, The World Bank, and the United Nations Development Programme, October 1985.

WRI. 1987. *World Resources*. Washington, DC.

Wright, P. and Bonkoungou, E.G. 1985/86. Soil and water conservation as a starting point for rural forestry: the OXFAM project in Ouahigouye, Burkina Faso. *Rural Africana* **23/24**:79–86.

Wu Youzheng and Pan Jinlin. 1986. Improving erosion controls and creating a better ecological environment. *Journal of Ecology (Shengtaixue Zhazhi)* **5**(3):47–50.

Wu Zhenfeng. 1981. *The Evolution of Shanxi Geography*. Shanxi People Press, Xian. 250 pp.

Wyman, C., Cross, B., Boyle, J.R., Preston, S.B. and Golson, E. 1978. Ecological analysis for development purposes of three broadly characterized fragile environments. *In*: *Science and Technology for Managing Fragile Environments in Developing Nations*. The Office of

International Studies, School of Natural Resources, University of Michigan, Ann Arbor, MI. Pp. 25–147.

Xie Lianhui. 1988. The reviving loess land. *Renmin Ribao*, 4 Feb.

Xu Zhemin, Qi Zhankui, Miao Wenli, Zhang Mingyi and Chui Mingtao. 1985. *A Study of Mulching Cultivation Methods Employed on the Loess Plateau*. Wheat Research Institute of Shanxi, Taiyuan. 31 pp.

Yao Huandou. 1985. The ecological and economic benefits of *Hippophae rhamnoides* in Youyu country. *Soil and Water Conservation in China (Zhonggue Shuitubaochi)* **1**:23.

Yang Sheng and Lian Sheng. 1987. Shanwang Gou Xiao Liuyu Zonghe Zhili De Xiaoyi. *Zhongguo Shuitubaochi* **69**:47–52.

Yang Songwang. 1983. Nongue Gengzuo Jishu Shi Nongtian Zhili Jingi Youxiao Cuoshi. *Quanguo Shuitubaochi Gengzuo Xueshu Taolunhui Ziliao Xuanbian*.

Yang Yansheng, Shi Deming and Lu Xixi. 1987. Serious soil and water loss in Three Gorge Region. *Bulletin of Soil and Water Conservation (Shuitubaochi Tongbao)* **7**(4):41–47.

Yang Zhenhuai. 1986. Solidarity urged for opening up new prospects in conservation practices. *Soil and Water Conservation in China (Zhongguo Shuitubao chi)* **8**:7–12.

Ye Zhenou, Ma Zhongxiao, Liu Jiusheng and Wan Yi. 1986. Hantitian Peifei Baoshang, Gaochan Wenchan Shiyan Yanjiu Baogao. *Shuitu Baochi Shiyan Yanjiu Chengguo Ziliao Xuanbian 1956–1980*, vol. 1. Pp. 157–175.

Yost, R.S., El-Swaify, S.A., Dangler, E.W. and Lo, A. 1985. The influence of simulated soil erosion and restorative fertilization on maize production on an Oxisol. *In*: El-Swaify *et al.*, **24**:248–261.

Young, G.J. 1983. Soil erosion and its impact upon water resources. *In*: Holmes, J.W. (Ed.), *The Effects of Changes in Land Use Upon Water Resources*. Adelaide, Australia. Pp. 103–110.

Yu Chunzhu. 1983. The trace element losses and controls in the soil of the middle reach of Yellow river. *Soil and Water Conservation in China (Zhonggue Shuitubaochi)* **4**:12–15.

Yu Chunzhu, Liu Yaohong, Dai Minggun and Peng Lin. 1985. The influences of erosion and conservation on agro-ecological environment in the Loess Plateau. *Journal of Ecology (Shengtaixue Zhazhi)* **4**(5):6–9.

Yudelman, M., Greenfield, J.C. and Magrath, W.B. 1990. *New Vegetative Approaches to Soil and Moisture Conservation*, World Wildlife Fund, The Conservation Foundation, Washington.

Zak, J. 1978. Ziemie Polskie w swarozytnosci. *In*: Topolski, J. (Ed.), *Dzieje Polski*. PWN, Warszawa, Poland.

Zakaria, M.N., Yew, F.K., Pushparajah, E. and Karim, B.A. 1987. Current programs, problems and strategies for land clearing and development in Malaysia. *In*: Lal, R., Nelson, M., Scharpenseel, H.W., Sudjadi, M., Garver, C.L. and Niamskul, C. (Eds), *Tropical Land Clearing for Sustainable Agriculture: Proceedings of an IBSRAM Inaugural Workshop. Bangkok, Thailand*. Pp. 141–152.

Zhang Gueliang. 1987. Extending 30 km^2 of new land in Yellow river mouth annually. *Science and Technology Daily (Keji Ribao)*, March 1.

Zhang Jinhui. 1987. The cost–benefit analysis of the comprehensive controls in Wangmaogou small watershed. *Soil and Water Conservation in China (Zhonggue Shuitubaochi)* **7**:40–43.

Zhang Maosheng. 1983. Shitu Baochi Gengzuo Jishu Zai Fazhan Ganhan Shan Qu Nongye Shengchan Lu De Zuoyong. *Quanguo Shuitu Baochi Gengzuo Xueshu Taolun Hui Ziliao Ji Bian*. Pp. 67–69.

Zhang Maosheng. 1987. Technical measures for controlling slope fields in the semi-arid region. *In: Proc. Intl. Symposium on Dryland Farming, Yangling*. Pp. 347–350.

Zhang Minghuan. 1984. The experiences and the lessons in managing land and controlling water and soils. *Bulletin of Soil and Water Conservation (Shuitubaochi Tongbao)* **3**:61–64.

Zhang Qinghai. 1985. Saving and developing rural energy. *In*: He Kang (Ed.), *Agricultural Almanac of China in 1984*. Agricultural Press, Beijing. Pp. 374–375.

Zhang Zhongzhi. 1988. The soil erosion in the Loess Plateau is aggravating. *Kexue Bao*, 15 April.

Zheng Shaoxiang. 1988. The urgency of controlling Ganjian river valley. *Science and Technology Daily (Keji Ribao)*, March 24.

Zhou Guangyu. 1986. *The Forests in Shandong Province*. Chinese Forestry Press, Beijing. 386 pp.

Zhu Junfeng. 1985. *Natural Resources and Agricultural Division of Shanbei Protective Plantation Region*. Chinese Forestry Press, Beijing. 1174 pp.

Zhu Shiguang. 1987. Soil loss and environmental evolution in the historical period. *Geography and Terrestrial Research (Dilixue He Guetu Yanjou)* **3**(1):46–51.

Zhu Xianmo. 1984a. Some problems on territorial management in the Loess Plateau. *Bulletin of Soil and Water Conservation (Shuitubaochi Tongbao)* **5**:1–6.

Zhu Xianmo. 1984b. Proper development and careful protection of land resources in the Loess Plateau. *Scientia Geographica Sinica (Dili Xuebao)* **4**(8):91–105.

Zhu Xianmo. 1986. *Land Resources in the Loess Plateau of China*. Shannxi Science and Technology Press, Xian. 250 pp.

Zhuo Dakang. 1987. The urgency of studying to control Yellow river. *Kexue Bao*, 10 Nov.

Ziemnicki, S. 1968. *Melioracje przeciwerozyjne*. Panstwowe Wydawnictwa Ronicze i Lesne Warszawa. 360 pp.

Ziemnicki, S. 1972. An example of colmatation application for land drainage. *Zeszyty Problemowe Postepow Nauk Rolniczych* **130**:7–18.

Ziemnicki, S. 1978. *Ochrona gleb przed erozja*. Panstwowe Mydawnictwo Rolnicze i Lesne. Warsawa. 183 pp.

Ziemnicki, S. and Naklicki, J. 1971. The present state and development of three gullies on the Lublin plateau. *Zeszyty Problemowe Postepow Nauk Rolniczych* **119**:23–45.

Ziemnicki, S. and Orlik, T. 1971. Characteristics of period runoff from a rolling watershed. *Zeszyty Problemowe Postepow Nauk Rolniczych* **119**:5–22.

Zyczynska-Baloniak, I. 1980. Dissolved organic matter in water of lake and channel in the agricultural landscape. *Polish Ecological Studies* **6**:155–166.

INDEX

acidity of soil 177–8, 235, 258, 269, 271–2
  see also soil pH
acid rain 258, 269
afforestation 203, 208, 221, 272
  China 81–2, 90
  Ethiopia 35, 38, 39, 43, 54
  India 126, 133, 134
  see also forests and forestry
agroforestry 3, 81, 238, 271
  West Africa 20–3, 25
Alfisols 7, 17–18, 20, 299
  humid tropics 235, 236, 242
alkalinity of soil 111–12, 131, 187–8, 190, 263
alley cropping 271
  see also intercropping
aluminium 183, 235, 267, 269
Aridisols 7, 235, 236, 248

bacteria 96, 175
  cellulose decomposition 186–7, 189–90
  nitrogen fixation 175, 178, 182, 186–7, 189
bunding 294–5, 299
  India 123, 133, 138
bushfires 11, 153

calcium (Ca)
  lost through erosion 230, 241, 280
  soil content 2, 16–17, 175, 177, 187, 281
cation exchange capacity 2, 13, 16, 18, 241, 281
cellulose 177, 181, 189
  decomposition 175–6, 178, 182–3, 186–7, 189–90
check-dams 79, 80, 82, 123
climate classification 233–4
colloids 13, 16, 175, 176, 182–3, 186–7, 189
compaction of soil 228, 258, 264, 280
  Argentina 181, 185
  by machinery in UK 199, 201, 202–3, 205
  West Africa 7, 8, 13, 21
contour planting 3–4, 22–4, 80, 82, 148, 205, 271
  China 82, 96, 97
  hedges 294, 299, 301, 305
  soil loss rate 177, 178, 284–7
  yield 290

Coordinating Boards 255
copper (Cu) 13, 69, 241
crop residues 2, 158, 205, 212, 285
  animal feed 43–4
  burning 162, 172–5, 181
  as fuel 4, 35, 37, 76, 87, 105, 280
  see also stubble
crop rotation 3–4, 23, 148, 203, 269
  alfalfa 174–8, 183, 185–7
  Argentina 173–8, 183, 185–7
  China 80, 91–3, 97
  grass 91, 92, 97, 166, 175–6, 187, 232
  legumes 20, 22, 25, 93, 97, 273
  soil erosion decrease 92–3, 231–2, 284–5, 287, 289
  UK 203, 213
crusting of soil 8, 13, 199, 258

dams 136, 224
  siltation damage 159–61, 237, 246–7, 268, 283
deforestation 22, 88, 217, 252, 257–8
  Ethiopia 27, 29, 34–5
  India 120, 121, 134, 144
  salination 268
desertification 68–9, 121, 261, 264, 265, 273
deserts 42, 63, 110, 154
direct drilling 212
disease 31–3, 52, 294, 298
dredging 173, 252, 283
drought 189
  Australia 148–9, 155, 156
  China 88–90, 95, 98, 99
  famine in Ethiopia 27–30, 33
  India 144
  vetiver grass 294, 296
  West Africa 11, 13
  yields 262, 267, 284, 288
dung *see* manure and dung

engineering techniques 21, 82, 85, 271
  China 82, 85, 94, 98, 103

Lightning Source UK Ltd.
Milton Keynes UK
UKOW06f2014281114

242336UK00001B/3/P

9 780521 104715